T0305531

Structural Equation Modeling Using R/SAS

There has been considerable attention to making the methodologies of structural equation modeling available to researchers, practitioners, and students along with commonly used software. Structural Equation Modeling Using R/SAS aims to bring it all together to provide a concise point-of-reference for the most commonly used structural equation modeling from the fundamental level to the advanced level. This book is intended to contribute to the rapid development in structural equation modeling and its applications to real-world data. Straightforward explanations of the statistical theory and models related to structural equation models are provided, using a compilation of a variety of publicly available data, to provide an illustration of data analytics in a step-by-step fashion using commonly used statistical software of R and SAS. This book is appropriate for anyone who is interested in learning and practicing structural equation modeling, especially in using R and SAS. It is useful for applied statisticians, data scientists and practitioners, applied statistical analysts and scientists in public health, and academic researchers and graduate students in statistics, while also being of use to R&D professionals/practitioners in industry and governmental agencies.

Key features:

- Extensive compilation of commonly used structural equation models and methods from fundamental to advanced levels
- Straightforward explanations of the theory related to the structural equation models
- Compilation of a variety of publicly available data
- Step-by-step illustrations of data analysis using commonly used statistical software R and SAS
- Data and computer programs are available for readers to replicate and implement the new methods to better understand the book contents and for future applications
- Handbook for applied statisticians and practitioners

Ding-Geng Chen, Ph.D. Professor and Executive Director in Biostatistics College of Health Solutions Arizona State University, USA.

Yiu-Fai Yung, Ph.D. Senior Manager, Advanced Analytics R & D, SAS Institute Inc.

Structural Equation Modeling Using R/SAS

A Step-by-Step Approach with Real Data Analysis

Ding-Geng Chen and Yiu-Fai Yung

CRC Press
Taylor & Francis Group
Boca Raton London New York

CRC Press is an imprint of the
Taylor & Francis Group, an **informa** business
A CHAPMAN & HALL BOOK

Designed cover image: © Ding-Geng Chen and Yiu-Fai Yung

First edition published 2024
by CRC Press
6000 Broken Sound Parkway NW, Suite 300, Boca Raton, FL 33487-2742

and by CRC Press
4 Park Square, Milton Park, Abingdon, Oxon, OX14 4RN

CRC Press is an imprint of Taylor & Francis Group, LLC

© 2024, Ding-Geng Chen and Yiu-Fai Yung

ISBN: 9781032431239 (hbk)
ISBN: 9781032431451 (pbk)
ISBN: 9781003365860 (ebk)

DOI: 10.1201/9781003365860

Typeset in Palatino
by codeMantra

Publisher's note: This book has been prepared from camera-ready copy provided by the authors.

Contents

List of Figures

Preface

There are many books on structural equation modeling (SEM). In general, we can classify them into five related categories. Category 1 is on theoretical development that requires advanced mathematical knowledge from the readers to understand the book contents. Category 2 is to develop and illustrate SEM using *Mplus* which requires the readers to have *Mplus* software, and it is expensive for most of the readers. Category 3 is to use *AMOS* software on SEM, which is also expensive for most of the readers. Category 4 is to use *SAS* for SEM since *SAS* is mostly commonly available in academic and industry environments, and Category 5 is to use *R* for SEM since *R* is freely available as an open-source software, especially with the developed *R* packages *lavaan* (i.e., *latent variable analysis*, publicly available at `https://lavaan.ugent.be/`). This book is the combination of Categories 4 and 5 to discuss the development of SEM using *R* package *lavaan* and the *SAS* procedure *CALIS* with a step-by-step implementation in real data analysis.

This book originated from our many years of research and teaching as well as many training workshops in structural equation modeling, and it is now created as a research reference book for future teaching and learning. In this book, we intend to make the methodologies of SEM clear and available to researchers, practitioners, and students, along with commonly used software. We aim to bring it all together to provide a concise point-of-reference for most commonly used structural equation modeling from the fundamental level to the advanced level.

We provide straightforward explanations of the statistical theory and models related to structural equation modeling, a compilation of a variety of publicly available data, and an illustration of data analytics in a step-by-step fashion using commonly used statistical software of *R/SAS*. We will have the data and computer programs available for readers to replicate and implement the new methods. We envisage that this book will contribute to the rapid development of structural equation modeling and its applications to real-world data.

Structure of the Book

This book is structured into 12 chapters. Chapters 1–7 focus on the fundamental structural equation modeling and Chapters 8–12 for advanced topics in SEM with real-world applications.

Specifically, we start this book in _Chapter 1_ with an overview of the classical _linear regression_ model and then go along to the _path analysis_, which is commonly seen as a special modeling in general structural equation model. This is to bring all readers to the same level of understanding of regression model before we introduce general SEM. More importantly, the use of _path analysis_ opens up the possibilities of analyzing more sophisticated regression models that can handle _latent variables, parameter constraints, mediation effects, cross-group comparisons_, and so on. _Chapter 2_ is then to extend the _path analysis_ in _Chapter_ 1 from analyzing only _Observed_ variables (or _manifest_ variables) to include the analysis of _latent_ variables (i.e., _unobserved_ variables). In this chapter, we focus on a popular type of latent variable modeling—_confirmatory factor analysis (CFA)_. With the understanding of latent variables, _Chapter 3_ discusses mediation analysis, which is an important statistical method frequently used by social and behavioral scientists to understand the complex relationships between predictors and outcome variables by adding mediators. We illustrate mediation analysis using the famous _Industrialization and Political Democracy_ data set from Bollen (1989) on structural equation modeling.

All the illustrations from Chapters 1 to 3 are based on the data that are from a multivariate normal distribution implicitly with the maximum likelihood estimation to estimate the parameters in structural equation modeling. However, this normality assumption might not be practically possible. _Chapter 4_ is then to explore the structural equation modeling for non-normal data with the robust estimation when the data are not normally distributed. In _Chapter 5_, we further explore the robust estimation in SEM for categorical data (such as binary, ordinal and nominal from questionnaires using Likert-type scales) where there is an ordinality in this type of data.

Extending the SEM for single-group on _Chapters 1–5_, _Chapters 6_ and 7 discuss the multi-group data analysis. In multi-group SEM, we focus on whether or not the SEM components, such as the measurement model and/or the structural model, are invariant (i.e., equivalent) across the multi-group of interest, such as gender, age, culture, and intervention treatment groups.

Chapters 8–12 are structured for advanced topics in SEM with real world applications. _Chapter 8_ is to illustrate a full structural equation model to analyze the relationship between features associated with pain-related disability (PRD) in people and painful temporomandibular disorder (TMD). The main purpose of this chapter is to demonstrate the two-stage process of structural equation modeling with the first stage to establish the measurement model and the second stage to fit a full structural equation model that builds upon the measurement model. In _Chapter 9_, we illustrate latent growth-curve modeling with a longitudinal data collected from 405 Hong Kong women on breast cancer post-surgery assessment presented in Byrne (2012). As one of the most powerful models in structural equation modeling to analyze longitudinal data, latent growth-curve modeling is used to model longitudinal growth trajectories over time and tests differences between groups (such as, in different intervention groups, age groups, etc.). As a general model for longitudinal growth, this model can incorporate both the within-individual and between-individual variations along with the longitudinal trajectories. *Chapter 10* is to introduce a full longitudinal mediation modeling. In this chapter, we extend the classical mediation analysis, which is often based on cross-sectional data collected at a single time point to infer causal mediation effects. However, causal interpretations of effects presume some temporal order of the variables, and therefore the full longitudinal mediation modeling can be used to analyze the longitudinal studies, where data are collected in different time points with the temporal order of some variables. In this chapter, a full longitudinal mediation model is presented to analyze several potential longitudinal relationships and examine the auto-regressive mediation effects on self-enhancement effect over time, multi-path mediation process with cross-lagged effects and feedback effects simultaneously.

In *Chapter 11*, we discuss multi-level structural equation modeling for multi-level data produced from most of the research in social sciences and public health, which are clustered and the data within cluster arecorrelated. In this chapter, we introduce the implementation in *lavaan* for analyzing multi-level *GALO* data used in Hox et al. (2018), which was collected from 1377 pupils within 58 schools from an educational study. Therefore, the 1377 pupils were nested within the 58 schools, where the 58 schools acted as clusters. Due to clustering, pupils' outcomes within the schools are more dependent on each other due to school-level factors. The last *Chapter 12* is to illustrate the sample size determination and power analysis in structural equation modeling. Studies should be well designed with adequate sample size and statistical power to address the research question or objective. We start with a simple case with two group comparison to illustrate the principles of sample size determination and then proceed to the general structural equation models. We make use of the latent growth-curve modeling in *Chapter* 9 to determine the sample size needed to detect the surgical effect between *Mastectomy* and *Lumpectomy* on *Mood* for women on breast cancer post-surgery assessment. In this chapter, we illustrate the steps to determine the required sample size to detect a significant effect on *Mood* with statistical power at a pre-specified level (such as at power 80%).

Acknowledgments

We owe a great deal of gratitude to many who helped in the completion of the book. We gratefully acknowledge the professional support of Mr. David Grubbs and his team from Chapman & Hall/CRC, who made the publication of this book a reality.

This work is partially supported by the National Research Foundation of South Africa (Grant Number 127727) and South African DST-NRF-SAMRC SARChI Research Chair in Biostatistics (Grant Number 114613).

This book is written using *bookdown* package (Xie, 2019), and we thank the package creator, Dr. Yihui Xie, to have created such a wonderful package for public use. We also thank the *SAS* Institute for supporting the project and the *R* creators and community for creating wonderful open-source *R* computing environment so that this book can be made possible. **Thank You All!**

Ding-Geng Chen, Ph.D.
Arizona State University, Phoenix, AZ, USA
University of Pretoria, Pretoria, South Africa
Yiu-Fai Yung, Ph.D.
SAS Institute Inc, Cary, NC, USA

Authors

Dr. Ding-Geng Chen is an elected fellow of American Statistical Association, an elected member of the International Statistical Institute, and an elected fellow of the Society of Social Work and Research. Currently, he is the Executive Director and Professor in Biostatistics at the College of Health Solutions, Arizona State University. Dr. Chen is also an extraordinary Professor and the SARChI research chair in Biostatistics at the Department of Statistics, University of Pretoria, South Africa, and he is also an Honorary Professor at the School of Mathematics, Statistics and Computer Science, University of KwaZulu-Natal, South Africa. He was the Wallace H. Kuralt distinguished Professor in Biostatistics in the University of North Carolina-Chapel Hill, a Professor in Biostatistics at the University of Rochester, and the Karl E. Peace endowed eminent scholar chair in Biostatistics at Georgia Southern University. He is also a senior statistics consultant for biopharmaceuticals and government agencies with extensive expertise in Monte-Carlo simulations, clinical trial biostatistics, and public health statistics. Dr. Chen has more than 200 referred professional publications, co-authored and co-edited 36 books on clinical trial methodology and analysis, meta-analysis, data sciences, causal inferences, and public health applications. He has been invited nationally and internationally to give speeches on his research.

Dr. Yiu-Fai Yung is a senior manager in Advanced Analytics R&D at SAS Institute Inc. He develops software for causal analysis, factor analysis, and structural equation modeling. He has led several statistical modeling workshops at conferences such as SAS user group meetings, Joint Statistical Meetings, and International Meeting of Psychometric Society. Prior to joining SAS, he taught statistics in the Department of Psychology at University of North Carolina-Chapel Hill. He has published research articles in *Psyhometrika, Multivariate Behavioral Research, British Journal of Mathematical and Statistical Psychology, and Structural Equation Modeling*. He has also served as reviewers and associated editors for professional research journals.

1

Linear Regression to Path Analysis

We start this book with an overview of the classical linear regression model and then go along to the path analysis, which is commonly seen as a special modeling approach in the general structural equation model (SEM). This is to bring all readers to the same level of understanding of the regression model before we introduce general SEM.

We illustrate these two analyses using a data set from Byrne (2012). The original study (Byrne, 1994) was to test for the validity and invariance of factorial structure within and across gender for elementary and secondary teachers on *Maslach Burnout Inventory (MBI)* developed in Maslach and Jackson (1981) and Maslach and Jackson (1986). The data set consists of a calibration sample of elementary male teachers ($n = 372$).

As discussed in Maslach and Jackson (1981), the *MBI* measures three dimensions of burnout, which include *Emotional Exhaustion, Depersonalization,* and *Reduced Personal Accomplishment.* Typically, we use *burnout* to indicate the inability to function effectively in one's job as a consequence of prolonged and extensive job-related stress. Specifically, *emotional exhaustion* represents the feelings of fatigue that develop as one's energies become drained; *depersonalization* represents the development of negative and uncaring attitudes toward others; and *reduced personal accomplishment* represents a deterioration of self-confidence and dissatisfaction in one's achievements.

As a measurement tool, the *MBI* is a 22-item instrument structured on a 7-point Likert-type scale that ranges from 1 (*feeling has never been experienced*) to 7 (*feeling experienced daily*). The 22 items are measures for three subscales, each representing one facet of *burnout*: the *Emotional Exhaustion* subscale (*EE*) comprises nine items, the *Depersonalization* (*DP*) subscale five, and the *Personal Accomplishment* (*PA*) subscale eight. Note that the *PA* subscale is a negative measure of *Reduced Personal Accomplishment.* For illustration of classical regression and path analysis in this chapter, we created and analyzed the three sum scores of items that correspond to the three subscales of *MBI*. The data set is saved in *mbi.csv* under the subdirectory *data* where we stored all the data sets used in the book.

Note: We will use *R* and *SAS* packages for analyzing data in this book. Typically, *R* code is presented and explained first and then supplemented by SAS code for the same model fitting. Specifically, for structural equation modeling, we use R package *lavaan* (i.e., latent variable analysis) and the *CALIS* procedure of SAS throughout this book. In addition, all data sets used in the book are stored in the subfolder *data*.

DOI: 10.1201/9781003365860-1

Remember to install this *R* package to your computer using *install. packages("lavaan")* and load this package into *R* session using *library(lavaan)* before running all the *R* programs in this chapter as follows.

```
# Load the lavaan package into R session
library(lavaan)
```

We will explain the *R* coding and functions along with the example using *lavaan*. For comprehensive understanding of *lavaan*, readers can refer to the webpage: `https://www.lavaan.ugent.be/`. For general inquiries about SAS analytic software products, readers can go to the webpage: `http://support.sas.com/` and for the documentation of the *CALIS* procedure: `https://support.sas.com/rnd/app/stat/procedures/calis.html`.

1.1 Descriptive Data Analysis

1.1.1 Data and Summary Statistics

We first need to read the data into *R*. Since this data set is saved in the *.csv* file format, we can read the data set into *R* using *R* command *read.csv* as follows:

```
# Read the data  into R session using 'read.csv'
dMBI = read.csv("data/mbi.csv", header=T)
# Check the data dimension
dim(dMBI)
```

```
## [1] 372    3
```

Note again the data set *mbi.csv* is stored under the subfolder *data*, which is why we read the data into *R* session with the *R* code *read.csv("data/mbi.csv", header=T)*. The argument *header=T* is to read in the data into *R* when the first row (i.e., the *header*) in the data set is not data values, but variable *label*.

As seen from the *R* output, this data set contains 372 observations and 3 variables.

We can print the first six observations using *R* function *head* to see the data along with the variable names. As seen from the following output, there are three variables named *sPA*, *sDP* and *sEE*, representing the summed scores of *PA*, *DP* and *EE*.

```
# Print the first 6 observations
head(dMBI)
```

```
##    sPA sDP sEE
## 1  49  11  27
## 2  48   6  14
## 3  43  17  45
## 4  48   5  39
## 5  52  23  38
## 6  52  11  18
```

We can summarize the data using *R* function *summary*. From the summary output, we can see the data distribution for all variables in the data. As seen from the summary output, the variables are summarized as the *min.* (i.e., the minimum value), *1st Qu.* (i.e., the first quantile), *Median* (i.e., the data median), *Mean* (i.e., the mean value), *3rd Qu.* (i.e., the third quantile), and *Max.* (i.e., the maximum value).

```
# Summary of the data
summary(dMBI)
```

```
##       sPA             sDP            sEE
##   Min.   :29.0   Min.   : 5   Min.   : 9.0
##   1st Qu.:44.8   1st Qu.: 7   1st Qu.:22.0
##   Median :50.0   Median :10   Median :30.0
##   Mean   :48.2   Mean   :11   Mean   :30.8
##   3rd Qu.:53.0   3rd Qu.:13   3rd Qu.:38.0
##   Max.   :56.0   Max.   :31   Max.   :60.0
```

Now we demonstrate how to use SAS code to read the csv file into a SAS-data set *dMBI* and produce similar statistical summary output to that of the R code.

```
/******* Read the CSV file into an external SAs
         data set *******/
proc import datafile='mbi.csv'
   out  = dMBI
   dbms = csv
   replace;
run;


libname U "C:\";


data u.dMBI;
  set work.dMBI;
run;
```

```
/******* Print 6 obs of the data set and some sumary
         statistics ******/
proc print data='c:\dMBI'(obs=6);
run;

proc means data='c:\dMBI' N min q1 median mean q3 max std;
run;
```

SAS output for printing of the first six observations and summary statistics are shown in the following tables:

Obs	sPA	sDP	sEE
1	49	11	27
2	48	6	14
3	43	17	45
4	48	5	39
5	52	23	38
6	52	11	18

Variable	N	Minimum	Lower Quartile	Median	Mean	Upper Quartile	Maximum	Std Dev
sPA	372	29.0000000	44.5000000	50.0000000	48.2500000	53.0000000	56.0000000	5.7278226
sDP	372	5.0000000	7.0000000	10.0000000	10.9919355	13.0000000	31.0000000	5.0706533
sEE	372	9.0000000	22.0000000	30.0000000	30.8413978	38.0000000	60.0000000	10.9305135

1.1.2   Preliminary Graphical Analysis

Typical preliminary analysis consists of plotting the data to identify whether or not a potential linear or nonlinear relationship exists between the outcome and explanatory variables. Let's consider a simple hypothetical scenario with the MBI data in which we examine whether the Emotional Exhaustion (EE) and Depersonalization (DP) are factors that contribute to the Personal Accomplishment (PA).

To illustrate this, we can make use of R function *plot* to scatterplot the data and then overlay a linear model fit on the top of the scatterplot. For the relationship between EE and PA, we can make use of the following R code:

```
# Scatterplot the data
plot(sPA~sEE,xlab="Emotional Exhaustion",
     ylab="Personal Accomplishment", dMBI)
# Add the linear model fit
abline(lm(sPA~sEE,dMBI), lwd=3, col="red")
```

Because there is no systematic nonlinear trend in the scatter plot from Figure 1.1, a simple linear model seems to be most appropriate even though there exists a large variation. As indicated by the declining trend, higher

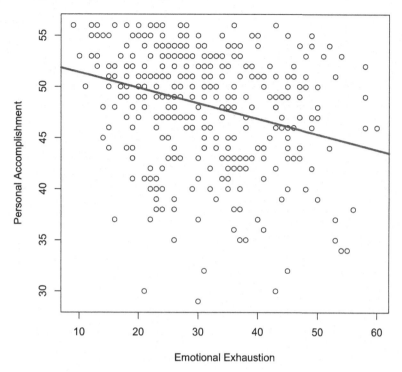

FIGURE 1.1
Illustration between "Emotional Exhaustion" and "Personal Accomplishment".

degree of *Emotional Exhaustion (EE)* is associated with reduced *Personal Accomplishment (PA)*.

Similarly, we can plot the *Depersonalization (DP)* to the *Personal Accomplishment (PA)* with the following *R* code chunk:

```
# Scatterplot the data
plot(sPA~sDP,xlab="Depersonalization",
    ylab="Personal Accomplishment", dMBI)
# Add the linear model fit
abline(lm(sPA~sDP,dMBI), lwd=3, col="red")
```

Again, a linear line seems to be most appropriate even though there exists a large variation in the scatter plot from Figure 1.2. As indicated by the declining trend, higher degree of *Depersonalization (DP)* is also accompanied with reduced *Personal Accomplishment (PA)*.

Keep in mind that data graphics are enormously helpful for understanding the regression models behind the mathematical equations and we recommend the exploratory and graphical data analysis as the first step in data analysis.

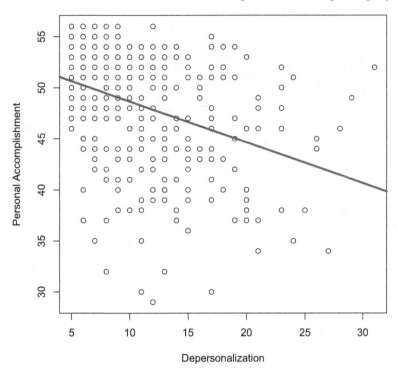

FIGURE 1.2
Illustration among "Depersonalization" with "Personal Accomplishment".

The following SAS code produces similar graphical outputs:

```
/******* Plot sPA against sEE *********/
proc sgplot data='c:\dMBI' noautolegend;
    scatter x=sEE y=sPA;
    reg x=sEE y=sPA /lineattrs=(color=red);
    yaxis label = 'Personal Accomplishment';
    xaxis label = 'Emotional Exhaustion';
run;

/******* Plot sPA against sDP *********/
proc sgplot data='c:\dMBI' noautolegend;
    scatter x=sDP y=sPA;
    reg x=sDP y=sPA /lineattrs=(color=red);
    yaxis label = 'Personal Accomplishment';
    xaxis label = 'Depersonalization';
run;
```

Because the SAS graphical outputs are similar to Figures 1.1 and 1.2, they are not shown here.

1.2 Review of Multiple Linear Regression Model

Before we analyze any real data using the linear regression model, let's briefly review the theoretical background of the regression models without theoretical proof. There are many books with more comprehensive discussions on regression model, interested readers can be referred to Gelman and Hill (2007) and Fox and Weisberg (2016 and 2019), to list a few.

1.2.1 Multiple Linear Regression Model

Suppose that we have a data set with N observations (e.g., $N = 372$ in the *dMBI* data set) from participants in a study with outcome variable y, for example, the variable *sPA* in Figures 1.1 and 1.2, which are related and predicted from p explanatory variables of $x_1, \ldots, x_p$ (i.e., two explanatory variables in the data *dMBI*, which are $x_1 = sEE$ and $x_2 = sDP$). The general multiple linear regression (MLR) model can be written in the following equation:

$$y_i = \beta_0 + \beta_1 x_{1i} + \cdots + \beta_p x_{pi} + \epsilon_i \tag{1.1}$$

where $i = 1, \ldots, N$ for the ith observation, β_0 is the intercept parameter, $\beta_1, \ldots, \beta_p$ are regression coefficients, ϵ_i is an error term for response i. In addition, the error term is independent of all explanatory variables (or regressors) Xs.

The MLR model in Equation (1.1) is often represented in the following matrix form,

$$y = X\beta + \epsilon \tag{1.2}$$

where y is a $N \times 1$ vector of the observed response variable, X is a $N \times (p+1)$ design matrix including all the p explanatory variables with the first column as 1 representing the intercept (i.e., β_0), $\beta = (\beta_0, \beta_1, \ldots, \beta_p)$ is a $(p+1)$-vector of regression parameters, and ϵ is the error term. Specifically,

$$y = \begin{pmatrix} y_1 \\ y_2 \\ \vdots \\ y_N \end{pmatrix}, \quad \epsilon = \begin{pmatrix} \epsilon_1 \\ \epsilon_2 \\ \vdots \\ \epsilon_N \end{pmatrix},$$

$$X = \begin{pmatrix} 1 & x_{11} & \cdots & x_{p1} \\ 1 & x_{12} & \cdots & x_{p2} \\ \vdots & \vdots & \ddots & \vdots \\ 1 & x_{1N} & \cdots & x_{pN} \end{pmatrix} \tag{1.3}$$

1.2.2   The Method of Least Squares Estimation (LSE)

Fundamentally, by estimating the parameter β, we try to find values (i.e., estimates) for the elements in β such that the systematic component (i.e., $X\beta$) explains as much of the variation in the response (i.e., y) as possible. In a sense, this is to find parameter values that make the error as small as possible. In statistics, this is referred to as the *least squares estimation*, i.e., to find β so that the sum of the squared errors is as small as possible. Mathematically, the least squares estimate (LSE) of β, denoted by $\hat{\beta}$, is obtained by minimizing the sum of squared errors (SSE):

$$
\begin{aligned}
\text{SSE} &= \sum_i \epsilon_i^2 = \epsilon'\epsilon \\
&= (y - X\beta)'(y - X\beta) \\
&= y'y - 2\beta X'y + \beta'X'X\beta
\end{aligned}
\tag{1.4}
$$

Taking the derivative of the sum of squared errors (SSE) with respect to β and setting to zero leads to

$$
X'X\hat{\beta} = X'y \tag{1.5}
$$

If $X'X$ is *invertible*, we have

$$
\hat{\beta} = (X'X)^{-1}X'y. \tag{1.6}
$$

The analytic expression of $\hat{\beta}$ in Equation (1.6) thus guarantees the LSE of β when $X'X$ is invertible.

1.2.3   The Properties of LSE

Without delving into mathematical details, we list some well-known statistical properties and terminology of the least squares estimation in the following sections. The statistical properties assume the following: (1) The error term (ϵ) is independent of the explanatory variables (xs) and (2) the observations are sampled independently. More assumptions will be stated when necessary.

1.2.3.1   Unbiasedness

The LSE $\hat{\beta}$ is unbiased—that is, $E(\hat{\beta}) = \beta$. The sampling variance of $\hat{\beta}$ is expressed as: $\text{var}(\hat{\beta}) = (X'X)^{-1}\sigma^2$ if $\text{var}(\epsilon) = \sigma^2 I$ (homoscedasticity assumption).

1.2.3.2 The Predicted Values

The predicted values are calculated as

$$\hat{y} = X\hat{\beta} = X(X'X)^{-1}X'y = Hy,$$

where $H = X(X'X)^{-1}X'$ is called the *hat-matrix*, which turns observed y into "y-hat" $\hat{y}$ (i.e., the predicted values).

1.2.3.3 The Residuals for Diagnostics

The regression residuals are defined as

$$\hat{\epsilon} = y - \hat{y} = y - X\hat{\beta} = (I - H)y,$$

which are the key components for model diagnostics.

1.2.3.4 Residual Sum of Squares (RSS)

The RSS is defined as

$$\text{RSS} = \hat{\epsilon}'\hat{\epsilon} = y'(I - H)'(I - H)y = y'(I - H)y,$$

which is used to estimate the residual variance and to evaluate the goodness of model fitting.

1.2.3.5 Variance Estimate

It can be shown that

$$E(\hat{\epsilon}'\hat{\epsilon}) = \sigma^2(N - (p + 1))$$

where $p+1$ is the number of columns of the design matrix X, i.e., the number of parameters in the linear model. This yields an unbiased estimate of σ^2 by:

$$\hat{\sigma}^2 = \frac{\hat{\epsilon}'\hat{\epsilon}}{N - (p + 1)} = \frac{\text{RSS}}{N - (p + 1)} \qquad (1.7)$$

where $N - (p + 1)$ is the *degrees of freedom* of the model.

1.2.3.6 R^2

R^2 is called the *coefficient of determination*. It represents the percentage of response variation that is explained by the systematic component X. Mathematically, it is defined as:

$$R^2 = \frac{\sum(\hat{y}_i - \bar{y})^2}{\sum(y_i - \bar{y})^2} = 1 - \frac{\sum(y_i - \hat{y}_i)^2}{\sum(y_i - \bar{y})^2}. \tag{1.8}$$

R^2 ranges from 0 to 1, with values closer to 1 indicating a better fit of the model. Therefore, R^2 is frequently used as a goodness-of-fit measure of the regression model.

1.2.4   Assumptions in Regression Models

The basic assumptions for classical linear regression are independence of observations, independence between the regressors and errors, homogeneity variance at each level of regressors (homoscedasticity), and normally distributed residuals. Except for the normality assumption, all other assumptions are required to derive the statistical properties of LSE described in the previous section. With the normality assumption, statistical inferences such as hypothesis testing and confidence interval construction are made possible. When any of these assumptions are violated, the statistical inferences might be wrong in parameter estimate, standard errors, t-values, and p-values.

Besides these basic assumptions, a relatively inconspicuous assumption in classical regression models in Equation (1.1) is that all explanatory variables or regressors (i.e., all x's) are fixed variables, in the sense that they are *controlled* in the design of the study rather than being sampled from observational studies. In the classical regression model, the only random variables are the response y and the error ϵ. The question is about whether the regression modeling is still valid in practical research when both xs and y are sampled from a population.

Fortunately, when the regressors are random, the statistical theory described in previous sections can still be maintained with conditional interpretations of the regression modelings and statistical assumptions (e.g., homoscedasticity in conditional variance $\text{var}(y|X)$), and with a multivariate normality assumption among the regressors and y. See Section 2.5 of Jennrich (1995) for a more detailed explanation.

1.3   Path Analysis Model

In contrast, the tradition of structural equation modeling, of which *path analysis* is considered as a special case, assumes all variables in the analysis are sampled and hence are random. This section introduces *path analysis* and demonstrates how it can be used to estimate classical regression models. More importantly, the use of *path analysis* opens up the possibilities of analyzing more sophisticated regression models that can handle latent variables, parameter constraints, mediation effects, cross-group comparisons, and so on. These and other topics and applications will be discussed in subsequent chapters. The next section demonstrates the traditional structural equation modeling approach to the classical regression problems by using a simple path model.

1.3.1 Implied Covariance Structures of a Simple Path Model

A simple linear regression model can be viewed as a single path model, as represented by the following simple path diagram:

$$x \longrightarrow y \leftarrow \epsilon \tag{1.9}$$

Often, the error term for an outcome (or endogenous) variable is omitted for clarity so that the following representation is equivalent:

$$x \longrightarrow y \tag{1.10}$$

Treating y and x as random variables, we can write the simple linear regression model as:

$$y = \beta_0 + \beta_1 x + \epsilon \tag{1.11}$$

where ϵ is a random error term, which is independent of x, and β_0 and β_1 are the intercept and slope parameters, respectively.

Traditionally, structural equation modeling approaches the modeling problem as a covariance structure analysis. Let us first denote the population covariance matrix for y and x as Σ, which is a 2×2 symmetric matrix defined by the following equivalent expressions:

$$\Sigma \equiv \begin{pmatrix} \text{Var}(y) & \text{Cov}(y, x) \\ \text{Cov}(x, y) & \text{Var}(x) \end{pmatrix} \equiv \begin{pmatrix} \sigma_y^2 & \sigma_{yx} \\ \sigma_{xy} & \sigma_x^2 \end{pmatrix} \tag{1.12}$$

where $\text{Var}(y)$ and $\text{Var}(x)$ denote the population variances of y and x, respectively, and $\text{Cov}(y,x)$ denotes the population covariance between y and x. We call Σ an *unstructured* covariance matrix because it contains elements that are *not* structured by the parameters that characterize the functional relationship between y and x. That is, Σ is not structured in terms of β_0 and β_1 in Equation (1.11).

In contrast, when the regression model (1.11) is hypothesized for relating y and x, a *structured* covariance matrix is implied from the model. To

derive the covariance structures, we utilize the following well-known results of
the Var() and Cov() operators:

$$\text{Var}(a + bx) = b^2\text{Var}(x)$$
$$\text{Cov}(a + bx, c + dy) = bd\text{Cov}(x, y)$$
$$\text{Cov}(x, x) \equiv \text{Var}(x)$$

where a, b, c, and d are fixed numbers and x and y are random variables.

Applying these results to the simple linear regression model (1.11), the
following equations are derived:

$$\text{Var}(y) = \text{Var}(\beta_0 + \beta_1 x + \epsilon) = \beta_1^2\sigma_x^2 + \sigma^2$$
$$\text{Cov}(y, x) = \text{Cov}(\beta_0 + \beta_1 x + \epsilon, x) = \beta_1\text{Cov}(x, x) + \text{Cov}(x, \epsilon) = \beta_1\sigma_x^2$$

where $\sigma^2 \equiv \text{Var}(\epsilon)$. With these analytic results and definitions under the linear
regression model, an implied *structured* covariance matrix is derived as:

$$\Sigma(\theta) = \begin{pmatrix} \beta_1^2\sigma_x^2 + \sigma^2 & \beta_1\sigma_x^2 \\ \beta_1\sigma_x^2 & \sigma_x^2 \end{pmatrix} \tag{1.13}$$

where $\theta = (\beta_1, \sigma^2, \sigma_x^2)'$ is the parameter vector for the structured covariance
matrix $\Sigma(\theta)$.

1.3.2 What Does It Mean by Modeling a Covariance Structure?

In covariance structure analysis or structural equation modeling, the null
hypothesis is usually of the following form:

$$H_0 : \Sigma = \Sigma(\theta)$$

In common words, the equality hypothesis of the *unstructured* and *structured*
population covariance matrices states that the variances and covariances of
the analysis variables are exactly prescribed as functions of θ. Because the
structured covariance matrix is usually derived or implied from a model (e.g.,
the simple linear regression model in the previous section), in practice this null
hypothesis is equivalent to asserting that the hypothesized model is exactly
true in the population.

Let us use a numerical example to illustrate how the hypothesized covariance structure is applied to derive the population covariance matrix. Assuming
that the linear regression model (Equation 1.11) is true (or, equivalently, H_0
is true) and $\theta = (\beta_1, \sigma^2, \sigma_x^2)' = (2, 5, 10)'$ is known, the population covariance

matrix is evaluated by applying the covariance structures, as shown in the following equations:

$$\Sigma \equiv \begin{pmatrix} \sigma_y^2 & \sigma_{yx} \\ \sigma_{xy} & \sigma_x^2 \end{pmatrix} = \Sigma(\theta)$$

$$= \begin{pmatrix} \beta_1^2 \sigma_x^2 + \sigma^2 & \beta_1 \sigma_x^2 \\ \beta_1 \sigma_x^2 & \sigma_x^2 \end{pmatrix}$$

$$= \begin{pmatrix} 2^2 \times 10 + 5 & 2 \times 10 \\ 2 \times 10 & 10 \end{pmatrix} = \begin{pmatrix} 45 & 20 \\ 20 & 10 \end{pmatrix}$$

The demonstration so far follows the following hypothetical-deductive sequence:

- Step (1): There is a hypothesized *true* path model that accounts for the relationships of variables in question.

- Step (2): The hypothesized model implies or derives analytically a covariance structure $\Sigma(\theta)$ that can produce the population *unstructured* covariance matrix exactly—that is, an assertion of the null hypothesis: $H_0 : \Sigma = \Sigma(\theta)$.

- Step (3): By plugging in the known θ value in the implied covariance structure matrix $\Sigma(\theta)$, the *unstructured* covariance Σ is evaluated numerically.

In an empirical study where we start with a random sample, we attempt to make statistical inferences by steps that somewhat "reverts" the hypothetical-deductive sequence:

- Step (3'): Compute the sample covariance matrix S and use this quantity to "represent" the unstructured covariance matrix (Σ) in the population.

- Step (2'): Given the hypothesized covariance structure $\Sigma(\theta)$ in Step (2), find the *best* numerical value, say $\hat{\theta}$, for θ in Step (2) so that $\Sigma(\hat{\theta})$ can reproduce the sample *unstructured* covariance matrix S as much as possible.

- Step (1'): Compare $\Sigma(\hat{\theta})$ with S to make an inferential statement about the hypothesized model in Step (1).

Steps (2') and (1') represent, respectively, the *estimation* and *model assessment* processes of structural equation modeling. We now use a numerical example to demonstrate the inferential steps (3'), (2'), and (1').

The hypothesized model is still the linear regression model (Equation 1.11) for the y and x variables. For didactic purposes, suppose that you are so lucky that your data yield a sample covariance matrix that is exactly the same as true population covariance matrix. That is,

$$S = \begin{pmatrix} 45 & 20 \\ 20 & 10 \end{pmatrix} \tag{1.14}$$

Now, according to Step (2'), you want to find the best $\hat{\theta}$ so that $\Sigma(\hat{\theta})$ can reproduce S as much as possible. Indeed, for this special case of a *saturated* model, we can find a $\hat{\theta}$ value that reproduces the sample covariance matrix *exactly*. The reason is that there are three distinct parameters in θ, and they can be solved analytically by three equations that correspond to the three distinct elements in the sample covariance matrix. That is, we are trying to solve the following equation for the unknown parameter values:

$$S = \begin{pmatrix} 45 & \text{-symmetric-} \\ 20 & 10 \end{pmatrix} = \begin{pmatrix} \beta_1^2 \sigma_x^2 + \sigma^2 & \text{-symmetric-} \\ \beta_1 \sigma_x^2 & \sigma_x^2 \end{pmatrix} \tag{1.15}$$

Essentially, you can start with the (2,2)-element and find directly that $\hat{\sigma}_x^2 = 10$. Then with the (2,1)-element, $\hat{\beta}_1 = 20/\hat{\sigma}_x^2 = 20/10 = 2$; and finally, with the (1,1)-element, $\hat{\sigma}^2 = 45 - \hat{\beta}_1^2 \hat{\sigma}_x^2 = 45 - 40 = 5$. We now complete the estimation step (2') with $\hat{\theta} = (\hat{\beta}_1, \hat{\sigma}^2, \hat{\sigma}_x^2)' = (2, 5, 10)'$—not surprisingly, with the sample covariance matrix coincides with the population covariance matrix, the estimation *recovers* the true parameter values.

To evaluate the model fit and hence making an inference about the null hypothesis, according to (1'), we would need to compare $\Sigma(\hat{\theta})$ with S and see if their discrepancy (if any) warrants any doubts about the hypothesized linear regression model. In this very special case of a saturated model, $\Sigma(\hat{\theta})$ reproduces the sample covariance matrix S perfectly and so we fail to reject the hypothesized model.

Some important notes about this demonstration are now in order:

- A *saturated* covariance structure model contains as many parameters in θ as the distinct elements in the covariance matrix. In contrast, a *restricted* covariance structure model contains fewer parameters in θ than the number of distinct elements in the covariance matrix. The difference between these two numbers is the model degrees of freedom. A saturated model has zero degrees of freedom—as is in the current demonstration (i.e., the number of parameters in θ and the number of distinct elements in the covariance matrix are both 3).

- A saturated model always has a perfect fit for random samples (i.e., $\Sigma(\hat{\theta}) = S$) when the equations are consistent for solving for θ. This statement is true even when your sample covariance matrix S is very different from the true population covariance matrix Σ. For this reason of trivial perfect model fit, evaluation of the fit of saturated models is usually not of practical importance in structural equation modeling. However, it might still be meaningful to interpret the parameter estimates of a saturated model.

- In general, estimation cannot be done by solving simple mathematical equations, as is demonstrated here for our simple example with a saturated model. Instead, one has to define a fitting or discrepancy function of $\Sigma(\theta)$

and S and then to find the optimal value $\hat{\theta}$ of the function given S by some optimization techniques. Section 1.3.3 discusses more various discrepancy functions for estimating θ.

- Evaluation of model fit can be done by various fit statistics, which measure the *distance* between $\Sigma(\theta)$ and S in various ways. Section 1.3.4 covers this topic in more detail.

1.3.3 Discrepancy Functions for Parameter Estimation and Model Fitting

The previous section explains that in structural equation modeling (SEM), we are interested in making an inference about whether the hypothesized model holds in the population. The null hypothesis is phrased as a statement about the model-implied covariance structures, that is:

$$H_0 : \Sigma = \Sigma(\theta) \qquad (1.16)$$

In a random sample, we start with a sample covariance matrix S. With S and the hypothesized covariance structures $\Sigma(\theta)$ as input information, we estimate the model parameters θ with an optimal value $\hat{\theta}$. To support the hypothesized model, we hope not to reject this H_0 and look for a large p-value to accept the H_0—this would happen if $\Sigma(\hat{\theta})$ can reproduce S closely.

The previous section demonstrates a simple saturated model that we can write out the implied covariance structures explicitly so that we can estimate θ analytically. In general, however, analytic solutions are not possible because the mathematical expression for the implied covariance structures often involves complex mathematical operations based on matrix algebra, such as vectors, inverse matrix, weighted matrix, determinants, traces, and other operators. Interested readers can refer to Bollen (1989) and later chapters for more comprehensive understanding of the construction of general implied covariance structures.

The remaining part of this section aims to describe estimation in a more general setting that applies to restricted or saturated models, with simple or complex covariance structures, alike. In general, the estimation in SEM is through the minimization of an objective (discrepancy) function, $F\left(S, \Sigma(\theta)\right)$. The corresponding estimated parameters are obtained as the $\hat{\theta}$ value that minimizes $F\left(S, \Sigma(\theta)\right)$ at $F_{\min} = F\left(S, \Sigma(\hat{\theta})\right)$. Hence, $F_{\min}$ measures the extent to which S differs from $\Sigma(\theta)$ at an optimal choice of $\theta = \hat{\theta}$.

From an inferential point of view, $F_{\min}$ reflects the lack of fit of the hypothesized model to the population. The difference, $S - \Sigma(\hat{\theta})$, is called the residual covariance matrix, which reflects the elementwise discrepancies between the unstructured and structured covariance matrices. Certainly, both $F_{\min}$ and $S - \Sigma(\hat{\theta})$ are random quantities in the sense that they are subject to sampling fluctuations. Hence, statistical inferences about the hypothesized model must be based on additional statistical assumptions and tests, which will be covered in the next section.

Several commonly used discrepancy functions are now introduced. The maximum likelihood (ML) discrepancy function is derived from the multivariate normal likelihood of the data and is formulated as follows:

$$F_{ML}\left(S, \Sigma(\theta)\right) = \log|\Sigma(\theta)| - \log|S| + \text{trace}\left(S\Sigma(\theta)^{-1}\right) - p \qquad (1.17)$$

where $|S|$ and $|\Sigma(\theta)|$ are the determinants of S and $\Sigma(\theta)$, respectively, and p is the dimension of S or $\Sigma(\theta)$.

Another estimator that is also based on the multivariate normality of the data is the *Normal-Theory Generalized Least Squares (NTGLS)*, or simply *GLS*, estimator, whose discrepancy function is formulated as follows:

$$F_{\text{GLS}}\left(S, \Sigma(\theta)\right) = \frac{1}{2} \times \text{trace}\left[S^{-1}(S - \Sigma(\theta))\right]^2 \qquad (1.18)$$

An estimator that is suitable for any arbitrary distribution is the Asymptotically Distribution-Free (ADF) estimator (Browne, 1984) or commonly called the *Weighted Least Squares (WLS)* estimator, whose discrepancy function is formulated as follows:

$$F_{\text{ADF}}\left(S, \Sigma(\theta)\right) = \text{vecs}(S - \Sigma(\theta))'W\,\text{vecs}(S - \Sigma(\theta)) \qquad (1.19)$$

where vecs(.) extracts the lower triangular elements of the symmetric matrix in the argument and puts the elements into a vector, and W is a symmetric weight matrix. Because S is of dimension p, vecs(S) is a vector of order $p^* = p(p+1)/2$ and W is a $p^* \times p^*$ symmetric matrix. For optimality reasons, W is usually computed as the inverse of a consistent estimate of the asymptotic covariance matrix of vecs($\sqrt{n}S$) in applications.

Many variants of discrepancy functions have been proposed, but the three described here are the most basic and popular ones. Although the ADF estimation is suitable for any arbitrary distributions, simulation results (Hu et al., 1992) show that it needs to have an extremely large sample size to achieve desirable statistical properties. Thus the maximum likelihood estimator with the discrepancy function F_{ML} is still the most popular one in applications.

1.3.4   Evaluation of the Structural Model and the Goodness-of-Fit Statistics

After the estimation of model parameters, we want to make some inferential statements about the hypothesized model. There are several goodness-of-fit statistics, which are all computed after the model estimation.

1.3.4.1 χ^2 Test of Model Fit

The chi-square (χ^2) statistic is computed as the product of the sample size and the minimized discrepancy function—that is $\chi^2 = N \times F_{\min}$. Some software might use $(N - 1)$ in the product when only covariance structures are fitted. Recall that the null hypothesis being tested is $H_0 : \Sigma = \Sigma(\theta)$. The chi-square statistic is being used to determine whether the sample presents evidence against the null hypothesis. Under the null hypothesis of a true null model, the chi-square statistic is distributed as a χ^2 variate with specific model degrees of freedom. When the observed p-value of the chi-square statistic is very small (e.g., less than a conventional α-level of 0.05), the data present an extreme event under H_0. Such an extreme event (associated with a small p value) suggests that the null hypothesis H_0 might not be tenable and should be rejected. Conversely, the higher the p-value associated with the observed χ^2 value, the closer the fit between the hypothesized model (under H_0) and the perfect fit (Bollen, 1989).

However, the chi-square test statistic is sensitive to sample size. Because the χ^2 statistic equals $N \times F_{\min}$, this value tends to be substantial when the model does not hold (even minimally) and the sample size is large (Jöreskog and Sörbom, 1993). The conundrum here, however, is that the analysis of covariance structures is grounded in large sample theory. As such, large samples are critical to obtaining precise parameter estimates, as well as to the tenability of asymptotic distributional approximations (MacCallum et al., 1996).

Findings of well-fitting hypothesized models, where the χ^2 value approximates the degrees of freedom, have proven to be unrealistic in most SEM empirical research. That is, despite relatively negligible difference between a population covariance matrix (Σ) and a hypothesized model with population structured covariance matrix $\Sigma(\theta)$, it is not unusual that the hypothesized model would be rejected by the χ^2 test statistic with a practically large enough sample size.

Hence, the χ^2-testing scheme of the sharp null hypothesis $\Sigma = \Sigma(\theta)$ in practical structural equation modeling has been deemphasized in favor of the use of various fit indices, which measures how good (or bad) the structural model approximates the *truth* as reflected by the unstructured model. The next few sections describe some of these popular fit indices and their conventional uses.

1.3.4.2 Loglikelihood

Two loglikelihood values are usually reported in SEM software, one for the hypothesized model (under H_0) and the other for the saturated (or unrestricted) model (under H_1). Some software packages might also display the loglikelihood value of the so-called baseline model—usually the uncorrelatedness model, which assumes covariances among all observed variables are zeros. Although the magnitudes of these loglikelihood values are not indicative of model fit, they are often computed because they are the basis of other fit indices such as the *Information Criteria* that are described in Section 1.3.4.6. Moreover, under multivariate normality, the χ^2 statistic for model fit test is -2 times of the difference between the loglikelihood values under H_0 and H_1.

1.3.4.3   Root Mean Square Error of Approximation (RMSEA) (Steiger and Lind, 1980) and the Standardized Root Mean Square Residual (SRMR)

They belong to the category of absolute indices of fit. However, Browne et al. (2002) have termed them, more specifically, as "absolute misfit indices" (p. 405).

By convention, RMSEA ≤ 0.05 indicates a good fit, and RMSEA values as high as 0.08 represents reasonable errors of approximation in the population (Browne and Cudeck, 1993). In elaborating on these cutpoints, RMSEA values ranging from 0.08 to 0.10 indicate mediocre fit, and those >0.10 indicate poor fit.

The Root Mean Square Residual (RMR) represents the average residual value derived from the fitting of the variance–covariance matrix for the hypothesized model $\Sigma(\theta)$ to the variance–covariance matrix of the sample data (S). However, because these residuals are relative to the sizes of the observed variances and covariances, they are difficult to interpret. Thus, they are best interpreted in the metric of the correlation matrix (Hu and Bentler; Jöreskog and Sörbom, 1989), which is represented by its standardized version, the Standardized Root Mean Square Residual (SRMR).

The SRMR represents the average value across all standardized residuals and ranges from 0.00 to 1.00. In a well-fitting model, this value should be small (say, .05 or less).

1.3.4.4   Comparative Fit Index (CFI) (Bentler, 1995)

As an alternative index in model fitting, CFI is normed in the sense that it ranges from 0.00 to 1.00, with values close to 1.00 being indicative of a well-fitting model. A cutoff value close to 0.95 has more recently been advised (Hu and Bentler, 1999). Computation of the CFI is as follows:

$$\text{CFI} = 1 - \frac{\chi_H^2 - df_H}{\chi_B^2 - df_B} \qquad (1.20)$$

where H = the hypothesized model, and B = the baseline model. Usually, the baseline model refers to the uncorrelatedness model, in which the structured covariance matrix is a diagonal matrix in the population—that is, covariances among all observed variables are zeros.

1.3.4.5   Tucker-Lewis Fit Index (TLI) (Tucker and Lewis, 1973)

In contrast to the CFI, the TLI is a nonnormed index, which means that its values can extend outside the range of 0.00–1.00. This index is also known as

Bentler's non-normed fit index (Bentler and Bonett, 1980) in some software packages. Its values are interpreted in the same way as for the CFI, with values close to 1.00 being indicative of a well-fitting model. Computation of the TLI is as follows:

$$\text{TLI} = \frac{\left(\frac{\chi_B^2}{df_B} - \frac{\chi_H^2}{df_H}\right)}{\left(\frac{\chi_B^2}{df_B} - 1\right)} \qquad (1.21)$$

1.3.4.6 The Akaike's Information Criterion (AIC) (Akaike, 1987) and the Bayesian Information Criterion (BIC) (Raftery, 1993; Schwarz, 1978)

The AIC and BIC are computed by the following formulas:

$$\text{AIC} = -2\,(\text{log-likelihood}) + 2k$$
$$\text{BIC} = -2\,(\text{log-likelihood}) + \ln(N)k$$

where log-likelihood is computed under the hypothesized model H_0 and k is the number of independent parameters in the model. Unlike various fit indices described previously (such as RMSEA, SRMR, CFI, and TLI), AIC and BIC are not used directly for assessing the absolute fit of a hypothesized model. Instead, AIC and BIC are mainly used for model selection.

Suppose that there are several competing models that are considered to be reasonable for explaining the data. These models can have different numbers of parameters, and they can be nested or nonnested within each other. The AIC and BIC values are computed for each of these computing models. You would then select the *best* model, which has the *smallest* AIC or BIC value.

As can be seen from either formula for AIC and BIC, the first term (−2 times of the loglikelihood) is a measure of model misfit and the second term (a multiple of the number of parameters in a model) is a measure of model complexity. Because you can always reduce, or at least never increase, model misfit (that is—minimize the first term) by increasing the model complexity with additional parameters, we can interpret AIC or BIC as a criterion that selects the best model based on the optimal balance between model fit (or misfit) and model complexity. The two information criteria differ in that the BIC assigns a greater penalty of model complexity than the AIC (say, whenever $N \geq 8$), and is more apt to select parsimonious models (Arbuckle, 2007).

1.3.4.7 Which Goodness-of-Fit Indices to Use?

Different goodness-of-fit indices reflect different aspects of model fit. Therefore, in practice, they are usually used together to examine the model fit. The χ^2 statistic provides the strictest statistical criterion for testing the perfect fit of the hypothesized model. In practice, it is usually supplemented by the RMSEA, SRMR. CFI, and TLI. The RMSEA (and its 90% confidence interval) and SRMR are used to gauge the absolute fit of the model, with RMSEA having a built-in adjustment of model complexity by incorporating the model degrees of freedom into its formula. The CFI and TLI measure incremental fit as compared with a baseline model, which is usually that of the uncorrelatedness model. The RMSEA, SRMR, CFI, and TLI all have some established conventional levels for claiming a good model fit. We will adopt these conventions in interpreting the SEM results. In contrast, the AIC and BIC do not use any conventional levels for measuring model fit, as their values (i.e., the smallest the better) are used to select the *best* model among a set of competing models.

1.4 Data Analysis Using R

1.4.1 Multiple Linear Regression

To examine the relationship among *EE*, *DP* with *PA*, we can make use of linear regression. The implementation of linear regression in *R* is very straightforward. The *R* function *lm* (i.e., *linear model*) will allow you to perform this linear regression.

```
# multiple linear regression
mlm <- lm(sPA~sEE+sDP, dMBI)
# summary of the model fitting
summary(mlm)
```

```
##
## Call:
## lm(formula = sPA ~ sEE + sDP, data = dMBI)
##
## Residuals:
##     Min      1Q  Median      3Q     Max
## -18.993  -2.945   0.759   3.925  11.115
##
## Coefficients:
##               Estimate Std. Error t value  Pr(>|t|)
## (Intercept)   53.8942     0.8525   63.22  < 0.0001 ***
## sEE           -0.0710     0.0305   -2.33    0.02 *
## sDP           -0.3143     0.0657   -4.79    0.0000025 ***
## ---
## Signif. codes:
## 0 '***' 0.001 '**' 0.01 '*' 0.05 '.' 0.1 ' ' 1
##
## Residual standard error: 5.33 on 369 degrees of freedom
## Multiple R-squared:  0.138,   Adjusted R-squared:  0.133
## F-statistic: 29.5 on 2 and 369 DF,  p-value: 0.00000000000136
```

As seen from this output, we obtained a statistically significant regression with the associated p-value $= 1.358\mathrm{e}{-12}$ from the *F-statistic* $= 29.45$ on 2 and 369 DFs. Even though it is statistically significant, the R^2 is 0.138, which is small and this is consistent with large variations in Figures 1.1 and 1.2. The estimated residual variance is $\hat{\sigma}^2 = 5.333\hat{\,}2 = 28.4409$.

With this regression, the *sEE* (i.e., $\hat{\beta}_1 = -0.07098$, p-value $= 0.0204$) and *sDP* (i.e., $\hat{\beta}_2 = -0.31432$, p-value $= 2.47\mathrm{e}{-06}$) are both significant predictors to the *sPA*. Specifically, for every unit of *sEE* increases, the *sPA* decreases by 0.07098 units and for every unit of *sPA* increases, the *sPA* reduces by 0.31432 units.

1.4.2 Path Analysis

A path model representation of the multiple linear regression is shown as follows:

$$sEE \longrightarrow sPA \longleftarrow sDP \qquad (1.22)$$

This path model can be fitted using *R* library *lavaan*. The implementation in *lavaan* is to first to write a *model* component and then this *model* component will be called by function *sem* for model fitting the observed data.

For *path analysis*, the *model* component is very similar to the linear regression (i.e., *lm*) to specify the response variable (i.e. *sPA*) on the left of the *R* operator ˜ and the explanatory variables (i.e., *sEE* and *sDP*) on the right of the operator ˜ as follows:

```
# Path model specification
PathMod <- 'sPA ~ 1+ sEE + sDP'
# Call "sem" to fit the path model
fit.PathMod <- sem(PathMod, data = dMBI)
# Print the model summary
summary(fit.PathMod, fit.measures=TRUE)
```

```
## lavaan 0.6-12 ended normally after 11 iterations
##
##   Estimator                                         ML
##   Optimization method                           NLMINB
##   Number of model parameters                         4
##
##   Number of observations                           372
##
## Model Test User Model:
##
##   Test statistic                                 0.000
##   Degrees of freedom                                 0
```

```
##
## Model Test Baseline Model:
##
##   Test statistic                                      55.095
##   Degrees of freedom                                       2
##   P-value                                              0.000
##
## User Model versus Baseline Model:
##
##   Comparative Fit Index (CFI)                          1.000
##   Tucker-Lewis Index (TLI)                             1.000
##
## Loglikelihood and Information Criteria:
##
##   Loglikelihood user model (H0)                    -1149.062
##   Loglikelihood unrestricted model (H1)            -1149.062
##
##   Akaike (AIC)                                      2306.124
##   Bayesian (BIC)                                    2321.799
##   Sample-size adjusted Bayesian (BIC)               2309.108
##
## Root Mean Square Error of Approximation:
##
##   RMSEA                                                0.000
##   90 Percent confidence interval - lower               0.000
##   90 Percent confidence interval - upper               0.000
##   P-value RMSEA <= 0.05                                   NA
##
## Standardized Root Mean Square Residual:
##
##   SRMR                                                 0.000
##
## Parameter Estimates:
##
##   Standard errors                                   Standard
##   Information                                       Expected
##   Information saturated (h1) model                Structured
##
## Regressions:
##                    Estimate  Std.Err  z-value  P(>|z|)
##   sPA ~
##     sEE             -0.071    0.030   -2.339    0.019
##     sDP             -0.314    0.065   -4.805    0.000
##
```

```
## Intercepts:
##                      Estimate  Std.Err  z-value  P(>|z|)
##     .sPA               53.894    0.849   63.473    0.000
##
## Variances:
##                      Estimate  Std.Err  z-value  P(>|z|)
##     .sPA               28.216    2.069   13.638    0.000
```

As seen from this output, we obtained a seemingly excellent model fit with 1 for both *CFI* and *TLI* and 0 for both *RMSEA* and *SRMR*. But this is not a surprise because this path model is a *saturated* model from the point of view of structural equation modeling. That is, the model degrees of freedom is zero— which means that you explain the distinct variances and covariances of the three observed variables by the same number of parameters in the path model.

To see why the model degrees of freedom is zero, we notice that there are six path model parameters: variance of *sEE*, variance of *sDP*, covariance between *sEE* and *sDP*, effect of *sEE* on *sPA* (β_1), effect of *sDP* on *sPA* (β_2), and the error variance of *sPA* (σ^2). To fit the path model as a structural equation model, these *six* path model parameters are used to explain the *six* observed distinct variance and covariance elements in the sample covariance matrix. Hence, the model does not summarize the sample statistics in more precise terms and the fit should be perfect. Therefore, for saturated models, it is often not of interest to assess the model fit. Instead, the main focus should be on the interpretations of the parameter estimates.

With this *path analysis*, the *sEE* (i.e., $\hat{\beta}_1 = -0.071$, *p*-value = 0.019) and *sDP* (i.e., $\hat{\beta}_2 = -0.314$, *p*-value = 0.000) are both significant predictors of *sPA*. The estimated residual variance is $\hat{\sigma}^2 = 28.216$. You can verify that all these estimates are very similar to that of the multiple regression analysis in Section 1.4.1.

In this example, we fit a saturated path model by structural equation modeling and the results are very similar to that of the multiple regression analysis. You may then wonder what you can gain additionally by using SEM. As mentioned earlier, the answer is that SEM is more flexible and general to deal with many different modeling situations, including latent variable model, multiple-group analysis, mediation analysis, constrained parameters, and so on. These topics will be discussed in later chapters. We now use another example to illustrate this argument for SEM.

Suppose now you hypothesize that the effects of *sEE* and *sDP* on *sPA* are equal in the multiple regression model. In the scatter plots that are shown earlier, this means that the slopes of the linear regressions are the same in these plots. With this equality imposed in the path model, the model is no longer saturated. Instead, this restricted model uses fewer parameters to explain the observed variances and covariances and hence is a more precise model. You can use SEM to fit such a model easily.

With R, you can set the specific equality constraint by inserting the same parameter name (beta1) for the two slopes (or regression coefficients), as shown in the following *lavaan* code:

```
# Path model specification
PathMod1 <- 'sPA ~ 1+ beta1*sEE + beta1*sDP'
# Call "sem" to fit the path model
fit.PathMod1 <- sem(PathMod1, data = dMBI)
# Print the model summary
summary(fit.PathMod1, fit.measures=TRUE)
```

```
## lavaan 0.6-12 ended normally after 11 iterations
##
##   Estimator                                         ML
##   Optimization method                           NLMINB
##   Number of model parameters                         4
##   Number of equality constraints                     1
##
##   Number of observations                           372
##
## Model Test User Model:
##
##   Test statistic                                 7.912
##   Degrees of freedom                                 1
##   P-value (Chi-square)                           0.005
##
## Model Test Baseline Model:
##
##   Test statistic                                55.095
##   Degrees of freedom                                 2
##   P-value                                        0.000
##
## User Model versus Baseline Model:
##
##   Comparative Fit Index (CFI)                    0.870
##   Tucker-Lewis Index (TLI)                       0.740
##
## Loglikelihood and Information Criteria:
##
##   Loglikelihood user model (H0)              -1153.018
##   Loglikelihood unrestricted model (H1)      -1149.062
##
##   Akaike (AIC)                                2312.036
##   Bayesian (BIC)                              2323.792
##   Sample-size adjusted Bayesian (BIC)         2314.274
```

```
##
## Root Mean Square Error of Approximation:
##
##    RMSEA                                          0.136
##    90 Percent confidence interval - lower         0.061
##    90 Percent confidence interval - upper         0.231
##    P-value RMSEA <= 0.05                           0.032
##
## Standardized Root Mean Square Residual:
##
##    SRMR                                           0.032
##
## Parameter Estimates:
##
##    Standard errors                             Standard
##    Information                                 Expected
##    Information saturated (h1) model          Structured
##
## Regressions:
##                     Estimate  Std.Err  z-value  P(>|z|)
##    sPA ~
##       sEE    (bet1)   -0.137    0.019   -7.093    0.000
##       sDP    (bet1)   -0.137    0.019   -7.093    0.000
##
## Intercepts:
##                     Estimate  Std.Err  z-value  P(>|z|)
##    .sPA              54.001    0.857   62.989    0.000
##
## Variances:
##                     Estimate  Std.Err  z-value  P(>|z|)
##    .sPA              28.822    2.113   13.638    0.000
```

First, we can look at the estimation result for the effects of *sEE* and *sDP*. As required by the model specification, the effect estimates for both are the same: $\hat{\beta}_1 = -0.137$ (SE = 0.019, $p = 0.000$). But is this a good model? To answer this, we can turn to the model fit statistics.

The χ^2 statistic is 7.912 (df = 1, $p = 0.005$). Therefore, from a purely statistical point of view, this restricted model should be rejected in favor of the saturated (unrestricted) model. Notice that, the model degrees of freedom is 1 which reflects the single equality constraint over a saturated model (i.e, the original path model). The results of fit indices: SRMR = 0.032, RMSEA = 0.136, CFI = 0.870, and TLI = 0.740, do not support this restricted model consistently. Therefore, it may be safer to stay with the original saturated path model to interpret these effects.

1.5 Data Analysis Using SAS

Analyzing the current multiple regression model can be done by using the *REG* procedure of SAS, as shown in the following:

```
proc reg data='c:\dMBI';
   model sPA = sEE sDP;
run;
```

SAS produces the following results for the regression analysis:

The REG Procedure
Model: MODEL1
Dependent Variable: sPA

Number of Observations Read	372
Number of Observations Used	372

Analysis of Variance					
Source	DF	Sum of Squares	Mean Square	F Value	Pr > F
Model	2	1675.55227	837.77613	29.45	<.0001
Error	369	10496	28.44498		
Corrected Total	371	12172			

Root MSE	5.33338	R-Square	0.1377
Dependent Mean	48.25000	Adj R-Sq	0.1330
Coeff Var	11.05364		

Parameter Estimates					
Variable	DF	Parameter Estimate	Standard Error	t Value	Pr > |t|
Intercept	1	53.89416	0.85253	63.22	<.0001
sEE	1	-0.07098	0.03047	-2.33	0.0204
sDP	1	-0.31432	0.06568	-4.79	<.0001

To analyze the corresponding path model by structural equation modeling, the *CALIS* procedure of SAS can be used, as shown in the following:

```
proc calis data='c:\dMBI';
   path  sPA <== sEE sDP;
run;
```

In the *PATH* statement, you can also write out each path separately:

```
proc calis data='c:\dMBI';
   path  sPA <== sEE,
         sPA <== sDP;
run;
```

To constrain the equality effects on *sPA* from *sEE* and *sDP*, you can use the following code:

```
proc calis data='c:\dMBI';
   path  sPA <== sEE = beta1,
         sPA <== sDP = beta1;
run;
```

For the constrained model, *PROC CALIS* produces the following output for parameter estimates:

The CALIS Procedure
Covariance Structure Analysis: Maximum Likelihood Estimation

PATH List							
Path			Parameter	Estimate	Standard Error	t Value	Pr > |t|
sPA	<===	sEE	beta1	-0.13748	0.01941	-7.0831	<.0001
sPA	<===	sDP	beta1	-0.13748	0.01941	-7.0831	<.0001

Variance Parameters						
Variance Type	Variable	Parameter	Estimate	Standard Error	t Value	Pr > |t|
Exogenous	sDP	_Add1	25.71153	1.88780	13.6198	<.0001
	sEE	_Add2	119.47613	8.77221	13.6198	<.0001
Error	sPA	_Add3	28.89981	2.12189	13.6198	<.0001

Covariances Among Exogenous Variables						
Var1	Var2	Parameter	Estimate	Standard Error	t Value	Pr > |t|
sEE	sDP	_Add4	30.79387	3.29182	9.3547	<.0001

Squared Multiple Correlations			
Variable	Error Variance	Total Variance	R-Square
sPA	28.89981	32.80795	0.1191

Essentially, these are the same estimates as that of *lavaan*. For example, the estimate of the constrained path coefficient beta1 is -0.137 and the error variance estimate of *sPA* is 28.90.

PROC CALIS produces quite a long list of fit statistics in the following fit summary table:

Fit Summary		
Modeling Info	Number of Observations	372
	Number of Variables	3
	Number of Moments	6
	Number of Parameters	5
	Number of Active Constraints	0
	Baseline Model Function Value	0.5173
	Baseline Model Chi-Square	191.9066
	Baseline Model Chi-Square DF	3
	Pr > Baseline Model Chi-Square	<.0001
Absolute Index	Fit Function	0.0213
	Chi-Square	7.8907
	Chi-Square DF	1
	Pr > Chi-Square	0.0050
	Z-Test of Wilson & Hilferty	2.5733
	Hoelter Critical N	181
	Root Mean Square Residual (RMR)	1.4429
	Standardized RMR (SRMR)	0.0387
	Goodness of Fit Index (GFI)	0.9862
Parsimony Index	Adjusted GFI (AGFI)	0.9170
	Parsimonious GFI	0.3287
	RMSEA Estimate	0.1363
	RMSEA Lower 90% Confidence Limit	0.0604
	RMSEA Upper 90% Confidence Limit	0.2312
	Probability of Close Fit	0.0325
	ECVI Estimate	0.0485
	ECVI Lower 90% Confidence Limit	0.0336
	ECVI Upper 90% Confidence Limit	0.0837
	Akaike Information Criterion	17.8907
	Bozdogan CAIC	42.4852
	Schwarz Bayesian Criterion	37.4852
	McDonald Centrality	0.9908
Incremental Index	Bentler Comparative Fit Index	0.9635
	Bentler-Bonett NFI	0.9589
	Bentler-Bonett Non-normed Index	0.8906
	Bollen Normed Index Rho1	0.8766
	Bollen Non-normed Index Delta2	0.9639
	James et al. Parsimonious NFI	0.3196

This fit summary table can be trimmed to show only the interested statistics by using some options in the FITINDEX statement. See Chapter 2 for an example.

The *CALIS* output includes all fit statistics that are shown in the *lavaan* output. In most cases, fit indices and other modeling information are labeled similarly in both software output. Two exceptions are noted here. First, the

TLI in *lavaan* is labeled as "Bentler-Bonett Non-normed Index" in *CALIS*. Second, the BIC in *lavaan* is labeled as "Schwarz Bayesian Criterion" in *CALIS*.

There are also some numerical discrepancies in fit indices between *CALIS* and *lavaan*. First, the CFI and TLI (Bentler-Bonett Non-normed Index) in *CALIS* shows much higher values (and thus more favorable model fit) than the corresponding values in *lavaan*. The reason is that the baseline model of *CALIS* fits an uncorrelatedness model for **all** observed variables, while the baseline model of *lavaan* retains the correlation between the two explanatory variables (*sDP* and *sEE*). As a result, the CFI and TLI in *CALIS* are computed by comparing the fit of the hypothesized model with a much worse baseline model than that of *lavaan*, resulting in a more favorable incremental fit as reflected by the CFI or TLI. However, fit indices such as RMSEA and SRMR, whose computations does not involve the fit of a baseline model, are essentially the same in *CALIS* and *lavaan*.

Second, the AIC and BIC values in *CALIS* and *lavaan* are quite different. But these differences are more apparent than real. The AIC and BIC formulas used by *lavaan* were shown in Section 1.3.4.6. However, *CALIS* uses formulas that differ only by an additional constant term for the log-likelihood of the saturated model. Hence, when you order the fit of competing models by their AIC (or BIC) values, the same order would result and the same *best* model would be selected either by *CALIS* or by *lavaan*.

1.6 Discussions and Further Readings

In this chapter, we started with the commonly used classical multiple regression. In the regression models, we typically assume that the explanatory variables (i.e., the *x*s variables) are fixed without measurement errors. This non-random assumption is too restrictive and not true in practice. Therefore, the *path analysis* is more appropriate in real data analysis.

In this chapter, we then introduced the *path analysis* as a special analysis in the general SEM. We discussed the model fitting with MLE as well as the implementations in *R* and *SAS*. We illustrated the data analysis using real data, so interested readers can follow these analyses for their own research data.

1.7 Exercises

The built-in data set *PoliticalDemocracy* in *R* package *lavaan* is the 'famous' *Industrialization and Political Democracy* data set used in Bollen (1989). The data set contains various measures of political democracy and industrialization in developing countries. It has 75 observations from 11 variables, and the format is as follows:

- y1: Expert ratings of the freedom of the press in 1960

- y2: The freedom of political opposition in 1960

- y3: The fairness of elections in 1960

- y4: The effectiveness of the elected legislature in 1960

- y5: Expert ratings of the freedom of the press in 1965

- y6: The freedom of political opposition in 1965

- y7: The fairness of elections in 1965

- y8: The effectiveness of the elected legislature in 1965

- x1: The gross national product (GNP) per capita in 1960

- x2: The inanimate energy consumption per capita in 1960

- x3: The percentage of the labor force in industry in 1960

1. Create summary factors from the subscores: Create summary factors as follows:

   - *ind60* factor measured by three variables: x1, x2 and x3
   - *dem60* factor measured by four variables: y1, y2, y3 and y4
   - *dem65* factor measured by four variables: y5, y6, y7 and y8

2. Regression model: Perform a regression model using *dem65* as the response variable and *ind60* and *dem60* as explanatory variables.

3. Path analysis: Perform a *path analysis* using *dem65* as the response variable and *ind60* and *dem60* as explanatory variables.

4. Conclusions: Compare the results from the regression analysis and path analysis.

2

Latent Variables—Confirmatory Factor Analysis

Extending the *Path Analysis* in Chapter 1 from analyzing only *Observed* variables (or *manifest* variables), we will include the analysis of *latent* variables (i.e., *unobserved* variables) in this chapter, which focuses on a popular type of latent variable modeling—*confirmatory factor analysis (CFA)*. For example, the three scale variables (i.e., *DP*, *EE*, and *PA*) in the burnout *MBI* example of Chapter 1 can be considered as latent variables that are *unobserved*, but are reflected by their corresponding subscale items (9, 5, and 8 items, respectively). These items are called measurement items (or variables) in a confirmatory factor analysis. The latent variables that are measured or reflected by these items are also known as *factors*.

To illustrate the *classical regression* and *path analysis* in Chapter 1, we create subscale scores *sDP*, *sEE*, and *sPA* by summing up their corresponding subscale items. The sum-score approach uses *sDP*, *sEE*, and *sPA* directly as observed variables in a regression type of analysis. In contrast, the *CFA* in this chapter uses a different approach that has at least two distinctions:

- Unlike the sum-score approach that creates sum scores (i.e., *sDP*, *sEE*, and *sPA*) explicitly as proxies for latent factors, the *DP*, *EE*, and *PA* factors in *CFA* are not explicitly computed in the analysis.

- While the sum scores approach implicitly assumes that subscale items contribute equally to compute the corresponding construct, subscale items might reflect the corresponding underlying factor differently in a *CFA*.

To illustrate the CFA, we consider two data examples in this chapter. First, we will re-use the original data (named in *elemmbi.csv*), which contains the 22 item-level data for the 372 elementary male teachers. We will still use this data set to illustrate the CFA among these *latent* variables from *EE* and *DP* to *PA* with *measurement* model. Furthermore, we will use the built-in data set called *HolzingerSwineford1939* in *lavaan* in this chapter to illustrate the CFA analysis among the *latent* measures among *visual*, *textual* and *speed* measured from 9 sub-measures.

Note to readers again: We will use R package *lavaan* (i.e., *latent variable analysis*) in this chapter. Remember to install this R package to your computer

DOI: 10.1201/9781003365860-2   31 using *install.packages("lavaan")* and load this package into *R* session using *library(lavaan)* before running all the *R* programs in this chapter as follows.

```
# Load the lavaan package into R session
library(lavaan)
```

2.1 Data Descriptions

2.1.1 Maslach Burnout Inventory

As discussed in Chapter 1, the *MBI* measures three dimensions of burnout which include the *Emotional Exhaustion (EE)*, the *Depersonalization (DP)*, and the *Reduced Personal Accomplishment (PA)*. As a measurement tool, the *MBI* is a 22-item instrument structured on a 7-point Likert-type scale that ranges from 1 (*feeling has never been experienced*) to 7 (*feeling experienced daily*). The data we use for illustration in this chapter is in data set named *elemmbi.csv*, which contains the 22 item-level data for the 372 elementary male teachers.

We first read the data into *R*. Since this data set is saved in the *.csv* file format, we can read the data set into *R* using *read.csv* as follows:

```
# Read the data into R session using 'read.csv'
dMBI = read.csv("data/ELEMmbi.csv", header=F)
colnames(dMBI) = paste("I",1:22,sep="")
# Check the data dimension
dim(dMBI)
```

```
## [1] 372   22
```

```
# Print the data summary
summary(dMBI)
```

```
##        I1              I2              I3
##  Min.   :1.00    Min.   :1.00    Min.   :1.00
##  1st Qu.:3.00    1st Qu.:4.00    1st Qu.:2.00
##  Median :4.00    Median :5.00    Median :3.00
##  Mean   :4.37    Mean   :4.87    Mean   :3.53
##  3rd Qu.:6.00    3rd Qu.:6.00    3rd Qu.:5.00
##  Max.   :7.00    Max.   :7.00    Max.   :7.00
##        I4              I5              I6
##  Min.   :2.0     Min.   :1.0     Min.   :1.00
##  1st Qu.:6.0     1st Qu.:1.0     1st Qu.:2.00
```

```
## Median :7.0    Median :2.0    Median :2.00
## Mean    :6.3   Mean    :2.2   Mean    :2.71
## 3rd Qu.:7.0    3rd Qu.:3.0    3rd Qu.:4.00
## Max.    :7.0   Max.    :7.0   Max.    :7.00
##        I7            I8              I9
## Min.    :2.00   Min.    :1.00   Min.    :1.00
## 1st Qu.:6.00    1st Qu.:2.00    1st Qu.:6.00
## Median :6.00    Median :2.00    Median :7.00
## Mean    :6.31   Mean    :3.04   Mean    :6.03
## 3rd Qu.:7.00    3rd Qu.:4.00    3rd Qu.:7.00
## Max.    :7.00   Max.    :7.00   Max.    :7.00
##        I10           I11             I12
## Min.    :1.0    Min.    :1.00   Min.    :1.0
## 1st Qu.:1.0     1st Qu.:1.00    1st Qu.:5.0
## Median :2.0     Median :2.00    Median :6.0
## Mean    :2.2    Mean    :2.24   Mean    :5.7
## 3rd Qu.:3.0     3rd Qu.:3.00    3rd Qu.:6.0
## Max.    :7.0    Max.    :7.00   Max.    :7.0
##        I13           I14             I15
## Min.    :1.00   Min.    :1.00   Min.    :1.00
## 1st Qu.:2.00    1st Qu.:3.00    1st Qu.:1.00
## Median :3.50    Median :4.00    Median :1.00
## Mean    :3.59   Mean    :4.03   Mean    :1.77
## 3rd Qu.:5.00    3rd Qu.:5.00    3rd Qu.:2.00
## Max.    :7.00   Max.    :7.00   Max.    :7.00
##        I16           I17             I18
## Min.    :1.00   Min.    :2.00   Min.    :1.0
## 1st Qu.:1.00    1st Qu.:6.00    1st Qu.:5.0
## Median :2.00    Median :7.00    Median :6.0
## Mean    :2.47   Mean    :6.41   Mean    :5.7
## 3rd Qu.:3.00    3rd Qu.:7.00    3rd Qu.:7.0
## Max.    :7.00   Max.    :7.00   Max.    :7.0
##        I19           I20             I21          I22
## Min.    :1.00   Min.    :1.00   Min.    :2.00   Min.    :1.00
## 1st Qu.:6.00    1st Qu.:1.00    1st Qu.:5.00    1st Qu.:1.00
## Median :6.00    Median :2.00    Median :6.00    Median :2.00
## Mean    :5.95   Mean    :2.24   Mean    :5.85   Mean    :2.58
## 3rd Qu.:7.00    3rd Qu.:3.00    3rd Qu.:7.00    3rd Qu.:3.00
## Max.    :7.00   Max.    :7.00   Max.    :7.00   Max.    :7.00
```

As seen from the R output, this data set contains 372 observations and 22 columns. These 22 columns correspond to the 22 items in the *MBI*. A confirmatory factor model with three factors for these 22 items is hypothesized and is graphically described in Figure 2.1.

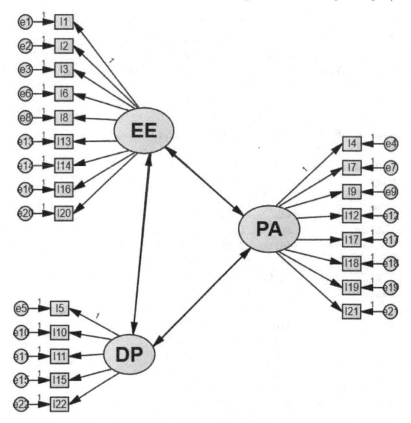

FIGURE 2.1
Relationship among Latent Variables EE, DP and PA.

As seen from Figure 2.1,

1. *EE* is measured (reflected) by the nine items of 1, 2, 3, 6, 8, 13, 14, 16, and 20, as seen by the single-headed arrows from *EE* to these nine items. These nine items are correlated, which can be seen in Figure 2.2.

2. *DP* is measured (reflected) by the five items of 5, 10, 11, 15, and 22, as seen from the single-headed arrows from *DP* to these five items. These five items are correlated, which can be seen in Figure 2.3.

3. *PA* is measured (reflected) by eight items of 4, 7, 9, 12, 17, 18, 19, and 21, as seen from the single-headed arrows from *PA* to these eight items. These eight items are correlated, which can be seen in Figure 2.4.

4. In addition, the three latent factors *EE*, *DP* and *PA* are correlated as denoted by the double-headed arrows in Figure 2.1.

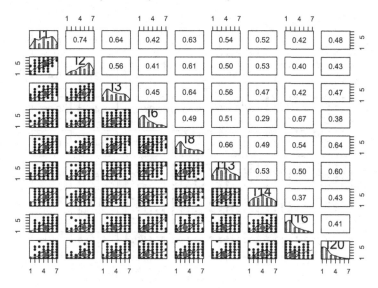

FIGURE 2.2
Correlations among Items for Latent Variable EE.

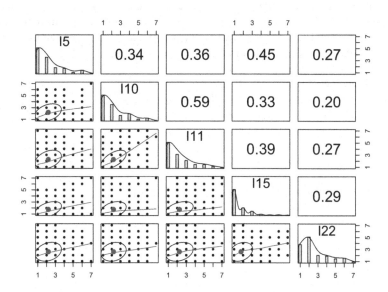

FIGURE 2.3
Correlations among Items for Latent Variable DP.

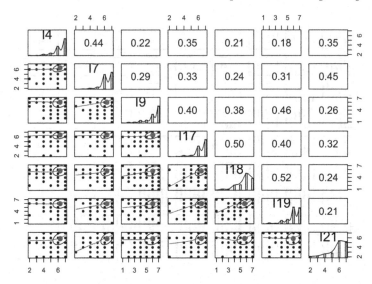

FIGURE 2.4
Correlations among Items for Latent Variable PA.

2.1.2   HolzingerSwineford1939

The built-in data set *HolzingerSwineford1939* in *R* package *lavaan* consists of mental ability test scores of seventh- and eighth-grade children from two different schools (Pasteur and Grant-White). It has 301 observations from 15 variables, and the format is as follows:

- id: Child identifier

- sex: Gender

- ageyr: Age, year part

- agemo: Age, month part

- school: School (Pasteur or Grant-White)

- grade: Grade

- x1: Visual perception

- x2: Cubes

- x3: Lozenges

- x4: Paragraph comprehension

- x5: Sentence completion

- x6: Word meaning

- x7: Speeded addition

- x8: Speeded counting of dots

- x9: Speeded discrimination of straight and curved capitals.

The data can be loaded into R as follows:

```
# load the data from R package "lavaan" and name it as "dHS"
dHS = HolzingerSwineford1939
# data dimension
dim(dHS)
```

```
## [1] 301  15
```

```
# print the data summary
summary(dHS)
```

```
##       id               sex            ageyr
## Min.   :  1    Min.   :1.00    Min.   :11
## 1st Qu.: 82    1st Qu.:1.00    1st Qu.:12
## Median :163    Median :2.00    Median :13
## Mean   :177    Mean   :1.51    Mean   :13
## 3rd Qu.:272    3rd Qu.:2.00    3rd Qu.:14
## Max.   :351    Max.   :2.00    Max.   :16
##
##      agemo              school           grade
## Min.   : 0.00    Grant-White:145    Min.   :7.00
## 1st Qu.: 2.00    Pasteur    :156    1st Qu.:7.00
## Median : 5.00                       Median :7.00
## Mean   : 5.38                       Mean   :7.48
## 3rd Qu.: 8.00                       3rd Qu.:8.00
## Max.   :11.00                       Max.   :8.00
##                                     NA's   :1
##       x1               x2              x3
## Min.   :0.67    Min.   :2.25    Min.   :0.25
## 1st Qu.:4.17    1st Qu.:5.25    1st Qu.:1.38
## Median :5.00    Median :6.00    Median :2.12
## Mean   :4.94    Mean   :6.09    Mean   :2.25
## 3rd Qu.:5.67    3rd Qu.:6.75    3rd Qu.:3.12
## Max.   :8.50    Max.   :9.25    Max.   :4.50
##
```

```
##          x4                 x5                 x6
##   Min.    :0.00    Min.    :1.00    Min.    :0.14
##   1st Qu.:2.33    1st Qu.:3.50    1st Qu.:1.43
##   Median :3.00    Median :4.50    Median :2.00
##   Mean    :3.06    Mean    :4.34    Mean    :2.19
##   3rd Qu.:3.67    3rd Qu.:5.25    3rd Qu.:2.71
##   Max.    :6.33    Max.    :7.00    Max.    :6.14
##
##          x7                 x8                 x9
##   Min.    :1.30    Min.    : 3.05    Min.    :2.78
##   1st Qu.:3.48    1st Qu.: 4.85    1st Qu.:4.75
##   Median :4.09    Median : 5.50    Median :5.42
##   Mean    :4.19    Mean    : 5.53    Mean    :5.37
##   3rd Qu.:4.91    3rd Qu.: 6.10    3rd Qu.:6.08
##   Max.    :7.43    Max.    :10.00    Max.    :9.25
##
```

The CFA model commonly used for this data is the three correlated latent variables (or factors), each with three indicators as graphically described in Figure 2.5.

As seen from Figure 2.5, these three latent factors are as follows:

- *Visual* factor measured by three items: x1, x2 and x3.

- *Textual* factor measured by three items: x4, x5 and x6.

- *Speed* factor measured by three items: x7, x8 and x9.

2.2   CFA Model and Estimation

This section explains the construction of a CFA model with more formal details. For simplicity in presenting these details, we will use the CFA model depicted in Figure 2.5 for the *HolzingerSwineford1939* data as a running example. This data set has only nine items: x1 to x9. After presenting the formal details of the CFA model in Sections 2.2.1–2.2.5, we will discuss the CFA model for the *MBI* data in Section 2.2.6.

2.2.1   The CFA Model

In the CFA model depicted in Figure 2.5 for the *HolzingerSwineford1939* data, there are nine observed or manifest variables that reflect three underlying factors. Suppose for the moment that all the manifest variables are

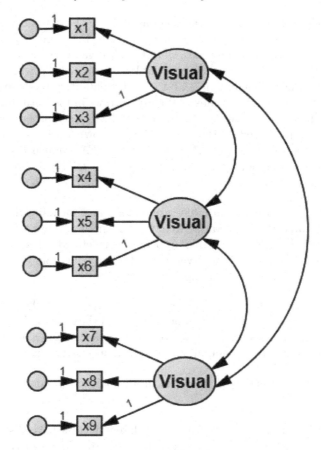

FIGURE 2.5
Diagram of the Three-Factor CFA for the Holzinger & Swineford Data.

centered at zero, the functional relationships between the variables and the factors are prescribed by the following equations in the CFA model:

$$x1 = b_1 \times \text{Visual} + e_1$$
$$x2 = b_2 \times \text{Visual} + e_2$$
$$x3 = b_3 \times \text{Visual} + e_3$$
$$x4 = b_4 \times \text{Textual} + e_4$$
$$x5 = b_5 \times \text{Textual} + e_5$$
$$x6 = b_6 \times \text{Textual} + e_6$$
$$x7 = b_7 \times \text{Speed} + e_7$$
$$x8 = b_8 \times \text{Speed} + e_8$$
$$x9 = b_9 \times \text{Speed} + e_9 \qquad (2.1)$$

The functional relationships among variables in these equations are represented by the 18 directed arrows in Figure 2.5. For example, in the first equation there are two directed arrows to the observed variable x1—one from $Visual$ and the other from e_1. The effect of $Visual$ on x1 is b_1, which is a slope parameter or a factor loading in a CFA model. In addition to these equations for specifying functional relations or directed arrows in the path diagram, the three latent factors are assumed to be correlated with each other. Covariances of factors are represented by the bi-directional arrows in Figure 2.5.

As presented in Equation (2.1), specifying a CFA is much like specifying a set of linear regression equations. However, the following two important distinctions are noted:

- In a CFA, all observed variables are *endogenous*, meaning that they are being explained by other variables in the system of equations. In the model equations, all observed variables appear on the left side of the equal signs. In the path diagram, all observed variables are being pointed to by latent factors and their corresponding error terms.

- In a CFA, all latent factors are *exogenous*, meaning that they serve as explanatory variables of the outcome variables in the system of equations. In the model equations, all latent factors appear on the right side of the equal signs. In the path diagram, all latent factors are never being pointed to.

Followed from these two distinctions are two important conceptions about a CFA:

- In a CFA, the observed variables and latent factors have distinct roles. The latent factors *explain* the variance and covariances of the observed variables while the observed variables *reflect* or *measure* the latent factors. This is why a CFA is also referred to as a measurement model (of latent factors).

- In a CFA, the explanatory variables are latent (unobserved) in the set of linear equations so that the usual least-squares regression formulas (as described in Chapter 1) cannot be applied to estimate the factor loadings (or slope parameters). Instead, a CFA model is usually fitted by using covariance structure analysis techniques.

In the next few sections, we will show how the covariance structure of a CFA is derived and uses mathematical expressions to understand how latent factors *explain* the variances and covariances among the observed variables.

2.2.2   Covariance Structure Parameters of the CFA Model

To understand the covariance structure in a CFA model, let us first recognize three types of parameters in Equation (2.1):

- Nine b's for the factor loadings or slope parameters
- Six variance and covariance parameters for the three latent factors
- Nine variance parameters for the error terms e's

These add up to 24 $(9 + 6 + 9)$ parameters, which are referred to as the covariance structure parameters in the CFA model. But this simple enumeration of parameters does not take the so-called *identification* issue into account. Simply put, because the latent factors are not observed, their scales (as measured by factor variances here) can be arbitrarily defined. Consequently, when estimating the model parameters from data, an infinite number of equivalent solutions can be generated by different sets of arbitrary estimates of factor variances. An identification problem thus occurs when there are more than one set of estimates that can fit the model equally well. In this situation, one must find a way to uniquely identify a single set of estimates as the solution for interpreting the model.

To deal with such an identification problem, the scales of the latent factors must be somehow fixed. Two equivalent types of identification constraints are commonly used to fix (or identify) the scales of latent factors:

- Fix factor variances to 1
- Fix one loading for each factor to 1

Technically, the constant value of 1 is not required—any other positive constants can serve the identification purpose.

When all latent factors are exogenous in a model (such as the CFA model), both types of identification constraints are equally easy to specify in SEM software. However, in general structural equation modeling where latent factors can be endogenous, factor variances of endogenous factors are no longer independent parameters. Consequently, fixing these endogenous factor variances to 1, although theoretically possible, might not be readily available in SEM software. Therefore, for convenience, the second type of identification constraints is used in most applications.

In fact, software packages such as *lavaan* would automatically fix the first specified loading of each factor to 1. This means that if we input our CFA model into *lavaan* following the same order specified in Equation (2.1), b_1, b_4, and b_7 would be fixed to 1 automatically by *lavaan*. This way the scale of each factor is anchored with one observed variable so that the factor variances would become identified and estimable.

The addition of identification constraints reduces the number of covariance structure parameters in a CFA model (and in a general strcutural equation model). For example, the total number of *independent* parameters in the current CFA model is 21 $(= 9 + 6 + 9 - 3)$, reflecting a reduction of three fixed factor loadings.

Once the identification problem is dealt with, parameters of a CFA model can be estimated by analyzing a sample data. In a CFA, there are three main type of estimation results of interest:

- The model fit indices, which you use to study whether the factors hypothesized in the model can capture the relationships among the manifest (observed) variables reasonably well.

- The loading estimates and their statistical significance, which you use to examine the relationships between the manifest (observed) variables and the latent factors.

- The estimates of factor covariances (or correlations) and their statistical significance, which you use to examine the relationships among the latent factors in the model.

Examples in later sections demonstrate the interpretations of these estimation results.

2.2.3   Covariance Structure Parameters and Model Matrices

The preceding section describes three types of parameters in a CFA. In general, these parameters are collected in a parameter vector θ, which contains:

- Factor loadings b_{ik}'s, which are parameters collected in the factor loading matrix B, where the first index i is for observed variables (row labels of B) and the second index k is for latent factors (column labels of B)
- Factor variances and covariances ϕ_{kl}'s, which are elements collected in the factor covariance matrix Φ
- Error variances $\sigma_{e_i}^2$'s for observed variables, which are collected as the diagonal elements of the error covariance matrix Ψ

For example, for the running example with nine observed variables and three factors, the parameters are defined in the following model matrices:

$$
B = \begin{pmatrix}
b_1 & 0 & 0 \\
b_2 & 0 & 0 \\
b_3 & 0 & 0 \\
0 & b_4 & 0 \\
0 & b_5 & 0 \\
0 & b_6 & 0 \\
0 & 0 & b_7 \\
0 & 0 & b_8 \\
0 & 0 & b_9
\end{pmatrix}
$$

$$
\Phi = \begin{pmatrix}
\phi_{11} & \phi_{12} & \phi_{13} \\
\phi_{21} & \phi_{22} & \phi_{23} \\
\phi_{31} & \phi_{32} & \phi_{33}
\end{pmatrix}
$$

$$\Psi = \begin{pmatrix} \sigma^2_{e_1} & 0 & 0 & 0 & 0 & 0 & 0 & 0 & 0 \\ 0 & \sigma^2_{e_2} & 0 & 0 & 0 & 0 & 0 & 0 & 0 \\ 0 & 0 & \sigma^2_{e_3} & 0 & 0 & 0 & 0 & 0 & 0 \\ 0 & 0 & 0 & \sigma^2_{e_4} & 0 & 0 & 0 & 0 & 0 \\ 0 & 0 & 0 & 0 & \sigma^2_{e_5} & 0 & 0 & 0 & 0 \\ 0 & 0 & 0 & 0 & 0 & \sigma^2_{e_6} & 0 & 0 & 0 \\ 0 & 0 & 0 & 0 & 0 & 0 & \sigma^2_{e_7} & 0 & 0 \\ 0 & 0 & 0 & 0 & 0 & 0 & 0 & \sigma^2_{e_8} & 0 \\ 0 & 0 & 0 & 0 & 0 & 0 & 0 & 0 & \sigma^2_{e_9} \end{pmatrix} \tag{2.2}$$

Note that because the factor loading matrix B in the current example has a *simple* structure, the notation for the factor loadings has been simplified in our presentations. For example, b_2 in the matrix B would have been b_{21} if it is desirable to make explicit that it is a loading on the first factor.

2.2.4 Covariance Structures of the CFA Model and Its Estimation

Recall from Chapter 1 that the null hypothesis of covariance structure analysis is of the form:

$$H_0: \quad \Sigma = \Sigma(\theta) \tag{2.3}$$

For the CFA model, $\Sigma(\theta)$ can be written explicitly as a matrix expression so that the null hypothesis becomes:

$$H_0: \quad \Sigma = B\Phi B' + \Psi \tag{2.4}$$

Using the model matrix definitions in Equation (2.2) for the running CFA example of the *HolzingerSwineford1939* data, the covariance structures in matrix Equation (2.4) are equivalent to the following set of elementwise algebraic expressions of covariance structures:

$$\Sigma_{ij} = b_i b_j \phi_{kk} \; (i \neq j \text{ and both variables } i \text{ and } j \text{ load on the same factor } k)$$
$$\Sigma_{ij} = b_i b_j \phi_{kl} \; (i \neq j \text{ and variables } i \text{ and } j \text{ load on distinct factors } k \text{ and } l,$$
$$\text{respectively})$$
$$\Sigma_{ii} = b_i^2 \phi_{kk} + \sigma^2_{e_i} \; (\text{variable } i \text{ loads on factor } k) \tag{2.5}$$

For identification purposes, b_1, b_4, and b_7 are fixed to 1 in Equations (2.2)–(2.5). Hence, θ is a parameter vector that contains 21 mathematically independent elements—that is:

$$\theta = (b_2, b_3, b_5, b_6, b_8, b_9, \phi_{11}, \phi_{21}, \phi_{22}, \phi_{31}, \phi_{32}, \phi_{33},$$
$$\sigma^2_{e_1}, \sigma^2_{e_2}, \sigma^2_{e_3}, \sigma^2_{e_4}, \sigma^2_{e_5}, \sigma^2_{e_6}, \sigma^2_{e_7}, \sigma^2_{e_8}, \sigma^2_{e_9}) \tag{2.6}$$

In the population, the null hypothesis states that the elements of Σ (observed variances and covariances) in a CFA model can be explained or expressed as known functions of θ; and these functions are defined either by Equation (2.4) (matrix form) or by (2.5) (elementwise form).

Given the sample data with covariance matrix S, you would like to find the *best* value for θ to construct an estimate of $\Sigma(\theta)$ (by using Equation (2.4) or (2.5)) such that it reproduces the sample covariance matrix S as much as possible. This *best* value for θ is obtained by optimizing a particular discrepancy function. Assuming a multivariate distribution of the p observed variables, the following maximum likelihood (ML) discrepancy function is minimized by finding the optimal value for θ given the sample covariance matrix S and the covariance structure model:

$$F_{ML}\left(S,\Sigma(\theta)\right) = \log|\Sigma(\theta)| - \log|S| + \text{trace}\left(S\Sigma(\theta)^{-1}\right) - p \qquad (2.7)$$

where $|S|$ and $|\Sigma(\theta)|$ are the determinants of S and $\Sigma(\theta)$, respectively. The so-obtained optimal value for θ is the maximum likelihood estimate.

The ML discrepancy function was mentioned in Chapter 1 and a saturated model was fitted to illustrate the estimation of parameters. For the current CFA model, however, there are some distinctive features to consider:

1. In general, the CFA model is not saturated.
2. Because the CFA model is not saturated, you cannot estimate θ by solving mathematical equations analytically (say, by solving the system of equation in Equation (2.5)).
3. It would now be interesting to test or interpret the model fit of a nonsaturated model to see if the data can be well explained by the hypothesized CFA model.

To illustrate, consider our running example with nine observed variables and three factors. The number of distinct elements in Σ or S is 45 ($=9 \times 10/2$) and the number of parameters in θ is 21. This means that if you expand Equation (2.5) to all distinct covariance elements Σ_{ij} for the nine observed variables, there would be 45 equations for solving for the 21 unknowns in θ. Hence, the model is not saturated but is *over-identified*. A discrepancy function like F_{ML} must then be minimized numerically and iteratively to get an optimal estimate of θ. The degrees of freedom of the model are computed as the difference between the number of distinct elements in Σ and the number of independent parameters in the CFA model. For example, the degrees of freedom for the running CFA example is 24 ($= 45 - 21$), which is positive, indicating an over-identified (or simply, identified) model. The chi-square model fit statistics and the computations of many model fit indices would be based on this model degrees of freedom.

2.2.5   Estimation of Mean and Covariance Structure Parameters

So far, we have assumed that the observed variables were *centered* at zeros in previous sections. What would happen if we relax this assumption? A short answer to this is that the mean structures would now be analyzed simultaneously with the covariance structures. This section explains some basic ideas about mean and covariance structure analysis.

To model the mean structures of non-centered observed variables, we need to add the intercepts $a's$ to the CFA model. For our running example, the set of equations for the model becomes:

$$
\begin{aligned}
x1 &= a_1 + b_1 \times \text{Visual} + e_1 \\
x2 &= a_2 + b_2 \times \text{Visual} + e_2 \\
x3 &= a_3 + b_3 \times \text{Visual} + e_3 \\
x4 &= a_4 + b_4 \times \text{Textual} + e_4 \\
x5 &= a_5 + b_5 \times \text{Textual} + e_5 \\
x6 &= a_6 + b_6 \times \text{Textual} + e_6 \\
x7 &= a_7 + b_7 \times \text{Speed} + e_7 \\
x8 &= a_8 + b_8 \times \text{Speed} + e_8 \\
x9 &= a_9 + b_9 \times \text{Speed} + e_9
\end{aligned}
\tag{2.8}
$$

In addition to the intercept parameters, the means of latent factors also become model parameters in the CFA. In structural equation modeling, estimating a model with mean structure parameters together with covariance structures is called a mean and covariance structure analysis. In the current example, the nine intercepts and three factor means are mean structure parameters that are analyzed on top of those covariance structure parameters described previously. Hence, the new parameter vector θ has to include these new mean structure parameters, and the null hypothesis is of the following extended form:

$$
H_0: \quad \Sigma = \Sigma(\theta), \quad \mu = \mu(\theta)
\tag{2.9}
$$

Again, as factors are not observed, their locations are arbitrary (much like the fact that their scales are also arbitrary, as discussed previously). The factor means can take on any values without affecting the model fit. Therefore, for the identification of the locations of factors, factor means are usually fixed to zero in CFA (except for multiple-group analysis—see later Chapters). Consequently, only the intercept parameters are added to the parameter vector θ. For the running example, the nine intercepts a_i's are added to θ so that:

$$
\begin{aligned}
\theta = (&b_2, b_3, b_5, b_6, b_8, b_9, \phi_{11}, \phi_{21}, \phi_{22}, ..., \\
&\sigma_{e6}^2, \sigma_{e7}^2, \sigma_{e8}^2, \sigma_{e9}^2, a_1, a_2, a_3, a_4, a_5, a_6, a_7, a_8, a_9)
\end{aligned}
\tag{2.10}
$$

Because the factor means are fixed zeros, the mean and covariance structures specified in the null hypothesis can be expressed in the following elementwise form:

$$\Sigma_{ij} = b_i b_j \phi_{kk} \quad (i \neq j \text{ and both variables } i \text{ and } j \text{ load on the same factor } k)$$
$$\Sigma_{ij} = b_i b_j \phi_{kl} \quad (i \neq j \text{ and variables } i \text{ and } j \text{ load on factors } k \text{ and } l,$$
$$\text{respectively})$$
$$\Sigma_{ii} = b_i^2 \phi_{kk} + \sigma_{e_i}^2 \quad (\text{variable } i \text{ load on factor } k)$$
$$\mu_i = a_i \tag{2.11}$$

Now, the ML discrepancy function would need to include a term (at the back) for the estimation of the mean structure parameters. That is,

$$F_{ML}\left((S,\bar{x}),(\Sigma(\theta),\mu(\theta))\right) = \log|\Sigma(\theta)| - \log|S| + \text{trace}\left(S\Sigma(\theta)^{-1}\right) - p$$
$$+ (\bar{x} - \mu(\theta))'\Sigma(\theta)^{-1}(\bar{x} - \mu(\theta)) \tag{2.12}$$

where $\bar{x}$ is the sample mean vector.

How would this new discrepancy function affect the estimation results of the current CFA model, as compared with those estimation results by fitting the covariance structures only? The short answer is "not much." The parameter estimates for loadings, factor variance and covariances, and the error variances would be the same for the two types of analyses. The χ^2 model fit statistic and most fit indices would be identical. A notable addition is that the analysis of mean and covariance structures would include the estimates for the intercepts in the result.

The reason for observing mostly the same estimation results for the two types of analysis is due to the fact that the mean structures in the null hypothesis (Equation 2.11) is saturated. That is, the additional mean elements (μ_i's) to fit is the same number as the number of added intercept parameters (a_i's). For estimation in samples, each a_i would then be estimated by the sample mean $\bar{x}_i$ when the maximum likelihood method is used. The fit of the mean structures would thus be perfect (i.e., $\mu(\hat{\theta}) = \bar{x}$) so that the last term of the ML discrepancy function for mean and covariance structures is zero. The saturated mean structures also imply that the model degrees of freedom of the mean and covariance structure model would not change from that of the analysis of covariance structures alone. In the current example, the degrees of freedom with mean and covariance structure is still 24 (=(45+9)−(21+9)), which is the same degrees of freedom with covariance structure alone.

Because of the aforementioned properties of saturated mean structures, you could elect to fit a covariance structure model without the intercept parameters if you are interested in the covariance structure model parameters only. However, nonsaturated mean structures do occur in some setting of structural equation modeling (e.g., multiple-group analysis or models with item invariance—see later chapters). For this reason, some software would default to analyze mean and covariance structures when you input the raw data for analysis.

2.2.6 CFA Model for MBI Data

We now illustrate the CFA model depicted in Figure 2.1. Some mathematical details are ignored because the ideas behind these details are the same as those of previous sections.

2.2.6.1 Measurement Model for Latent Variable EE

Since the latent variable EE is measured by the nine items, the *measurement* model for EE can be written as follows:

$$
\begin{aligned}
I1 &= a_1 + b_1 \times \text{EE} + e_1 \\
I2 &= a_2 + b_2 \times \text{EE} + e_2 \\
I3 &= a_3 + b_3 \times \text{EE} + e_3 \\
I6 &= a_6 + b_6 \times \text{EE} + e_6 \\
I8 &= a_8 + b_8 \times \text{EE} + e_8 \\
I13 &= a_{13} + b_{13} \times \text{EE} + e_{13} \\
I14 &= a_{14} + b_{14} \times \text{EE} + e_{14} \\
I16 &= a_{16} + b_{16} \times \text{EE} + e_{16} \\
I20 &= a_{20} + b_{20} \times \text{EE} + e_{20}
\end{aligned}
\tag{2.13}
$$

In this component, there are nine intercept parameters (i.e., the nine a's), nine slope (or factor loading) parameters (i.e., the nine b's) and nine variances associated with the errors in the nine e's. Therefore, the total number of parameters to be estimated are 27 in this component.

2.2.6.2 Measurement Model for Latent Variable DP

Since the latent variable DP is measured by the five items, the *measurement* model for DP can be written as follows:

$$
\begin{aligned}
I5 &= a_5 + b_5 \times \text{DP} + e_5 \\
I10 &= a_{10} + b_{10} \times \text{DP} + e_{10} \\
I11 &= a_{11} + b_{11} \times \text{DP} + e_{11} \\
I15 &= a_{15} + b_{15} \times \text{DP} + e_{15} \\
I22 &= a_{22} + b_{22} \times \text{DP} + e_{22}
\end{aligned}
\tag{2.14}
$$

In this component, there are five intercept parameters (i.e., the five a's), five slope (or factor loading) parameters (i.e., the five b's) and five variances associated with the errors in the five e's. Therefore, the total number of parameters to be estimated is 15 in this component.

2.2.6.3   Measurement Model for Latent Variable PA

Since the latent variable PA is measured by the eight items, the *measurement model* for PA can be written as follows:

$$I4 = a_4 + b_4 \times \text{PA} + e_4$$
$$I7 = a_7 + b_7 \times \text{PA} + e_7$$
$$I9 = a_9 + b_9 \times \text{PA} + e_9$$
$$I12 = a_{12} + b_{12} \times \text{PA} + e_{12}$$
$$I17 = a_{17} + b_{17} \times \text{PA} + e_{17}$$
$$I18 = a_{18} + b_{18} \times \text{PA} + e_{18}$$
$$I19 = a_{19} + b_{19} \times \text{PA} + e_{19}$$
$$I21 = a_{21} + b_{21} \times \text{PA} + e_{21} \tag{2.15}$$

In this component, there are eight intercept parameters (i.e., the eight a's), eight slopes (or factor loading) parameters (i.e., the eight b's) and eight variances associated with the errors in the eight e's. Therefore, the total number of parameters to be estimated are 24 in this component.

2.2.6.4   Covariance among the Latent Variables

Along with the three measurement models, there are three additional covariance (or correlation) parameters to be estimated as follows:

$$\text{EE} \sim\sim \text{DP} \tag{2.16}$$
$$\text{EE} \sim\sim \text{PA} \tag{2.17}$$
$$\text{DP} \sim\sim \text{PA} \tag{2.18}$$

2.2.6.5   Means and Variances for the Latent Variables

There are three means and three variances for the three latent variables in the current CFA model. Note that some SEM software might characterize the factor means as *intercepts*, which can be a more general concept for the location parameter that applies to endogenous as well as exogenous factors. However, in the current CFA model, because all latent factors are exogenous, it would be more specific to call these location parameters *factor means*.

By simple enumeration, the total number of parameters in the current model is 75 (i.e., 27 in *EE*, 15 in *DP*, 24 in *PA*, 3 covariances, 3 means and 3 variances for the three latent variables). However, as discussed previously about the arbitrary locations and scales of latent factors, the CFA model parameterized this way is not estimable unless some identification constraints are imposed. Similar to the strategies discussed in Sections 2.2.2 and 2.2.5, we fix the first factor loading in each component measurement model to one (as shown in Figure 2.1) and we set the three latent factor means to zero for identification purposes. Thus, due to these six identification constraints, the number of parameters to estimate in the current CFA model is 69 ($= 75 - 6$). The model degrees of freedom (DF) is the difference between the total number of elements in Σ and μ (which is $275 = 22 \times 25/2$) and the number of model parameters. Hence, DF $= 275 - 69 = 206$.

2.3 CFA with R *lavaan*

2.3.1 Data Analysis for MBI

In this section, we continue the CFA modeling with the *MBI* data. We first fit a basic CFA model and then proceed to improve the model fit by utilizing the *modification indices*.

2.3.1.1 Fitting the Original Model

To use *lavaan* for the CFA model, we first specify the model as follows:

```
CFA4MBI <- '
# latent variables
EE =~ I1+I2+I3+I6+I8+I13+I14+I16+I20
DP =~ I5+I10+I11+I15+I22
PA =~ I4+I7+I9+I12+I17+I18+I19+I21

# covariance
EE~~DP
EE~~PA
DP~~PA

# Intercepts for items
I1  ~ 1; I2  ~ 1; I3  ~ 1; I4  ~ 1; I5  ~ 1
I6  ~ 1; I7  ~ 1; I8  ~ 1; I9  ~ 1; I10 ~ 1
I11 ~ 1; I12 ~ 1; I13 ~ 1; I14 ~ 1; I15 ~ 1
I16 ~ 1; I17 ~ 1; I18 ~ 1; I19 ~ 1; I20 ~ 1
I21 ~ 1; I22 ~ 1
'
```

As seen from the above model specification, *lavaan* uses operator =~ to specify *measured by* for measurement models. Also *lavaan* uses operator ~ ~ to specify variances and covariances. The convention in *lavaan* for this notation is that it is *variances* if both sides is the same (such as PA ~ ~ PA) and *covariance* if the left-side is different from the right-side (for example EE ~ ~ PA). The specification for *intercept* is ~.

With the model specification, we can fit the CFA model by calling the *lavaan* function *cfa* as follows:

```
# fit the CFA model
fitMBI = cfa(CFA4MBI, data=dMBI)
```

For model fitting, we can call *summary* with option *fit.measures=TRUE* to output the fitted measures as follows:

```
# Print the model summary with the fitting measures
summary(fitMBI, fit.measures=TRUE)
```

```
## lavaan 0.6-12 ended normally after 46 iterations
##
##    Estimator                                         ML
##    Optimization method                           NLMINB
##    Number of model parameters                        69
##
##    Number of observations                           372
##
## Model Test User Model:
##
##    Test statistic                               695.719
##    Degrees of freedom                               206
##    P-value (Chi-square)                           0.000
##
## Model Test Baseline Model:
##
##    Test statistic                              3452.269
##    Degrees of freedom                               231
##    P-value                                        0.000
##
## User Model versus Baseline Model:
##
##    Comparative Fit Index (CFI)                    0.848
##    Tucker-Lewis Index (TLI)                       0.830
##
## Loglikelihood and Information Criteria:
##
##    Loglikelihood user model (H0)             -12811.043
##    Loglikelihood unrestricted model (H1)     -12463.184
##
##    Akaike (AIC)                               25760.087
##    Bayesian (BIC)                             26030.490
##    Sample-size adjusted Bayesian (BIC)        25811.575
##
## Root Mean Square Error of Approximation:
##
```

```
## RMSEA                                            0.080
## 90 Percent confidence interval - lower           0.073
## 90 Percent confidence interval - upper           0.087
## P-value RMSEA <= 0.05                             0.000
##
## Standardized Root Mean Square Residual:
##
## SRMR                                             0.070
##
## Parameter Estimates:
##
## Standard errors                              Standard
## Information                                  Expected
## Information saturated (h1) model          Structured
##
## Latent Variables:
##                  Estimate  Std.Err  z-value  P(>|z|)
## EE =~
##   I1               1.000
##   I2               0.887    0.061   14.621    0.000
##   I3               1.021    0.068   15.085    0.000
##   I6               0.764    0.064   12.013    0.000
##   I8               1.143    0.066   17.299    0.000
##   I13              1.017    0.065   15.544    0.000
##   I14              0.848    0.069   12.251    0.000
##   I16              0.715    0.058   12.410    0.000
##   I20              0.753    0.056   13.410    0.000
## DP =~
##   I5               1.000
##   I10              1.142    0.127    8.986    0.000
##   I11              1.353    0.142    9.511    0.000
##   I15              0.905    0.109    8.318    0.000
##   I22              0.768    0.121    6.361    0.000
## PA =~
##   I4               1.000
##   I7               0.970    0.150    6.482    0.000
##   I9               1.780    0.254    7.007    0.000
##   I12              1.499    0.221    6.769    0.000
##   I17              1.348    0.181    7.463    0.000
##   I18              1.918    0.262    7.329    0.000
##   I19              1.716    0.238    7.205    0.000
##   I21              1.356    0.218    6.219    0.000
##
## Covariances:
##                  Estimate  Std.Err  z-value  P(>|z|)
```

```
## EE ~~
## DP              0.701    0.099    7.061    0.000
## PA             -0.192    0.042   -4.537    0.000
## DP ~~
## PA             -0.172    0.035   -4.850    0.000
##
## Intercepts:
##              Estimate  Std.Err  z-value  P(>|z|)
## .I1            4.366    0.086   50.743    0.000
## .I2            4.868    0.080   60.823    0.000
## .I3            3.527    0.090   39.296    0.000
## .I4            6.298    0.052  121.827    0.000
## .I5            2.199    0.077   28.544    0.000
## .I6            2.707    0.082   33.002    0.000
## .I7            6.312    0.043  145.108    0.000
## .I8            3.043    0.089   34.018    0.000
## .I9            6.035    0.068   88.565    0.000
## .I10           2.204    0.075   29.427    0.000
## .I11           2.239    0.079   28.279    0.000
## .I12           5.699    0.062   92.237    0.000
## .I13           3.586    0.087   41.163    0.000
## .I14           4.027    0.089   45.049    0.000
## .I15           1.769    0.067   26.296    0.000
## .I16           2.473    0.075   33.182    0.000
## .I17           6.406    0.044  145.103    0.000
## .I18           5.702    0.066   86.454    0.000
## .I19           5.946    0.062   96.547    0.000
## .I20           2.245    0.073   30.639    0.000
## .I21           5.852    0.066   89.258    0.000
## .I22           2.581    0.082   31.538    0.000
## EE             0.000
## DP             0.000
## PA             0.000
##
## Variances:
##              Estimate  Std.Err  z-value  P(>|z|)
## .I1            1.128    0.095   11.861    0.000
## .I2            1.105    0.090   12.214    0.000
## .I3            1.301    0.108   12.031    0.000
## .I6            1.553    0.121   12.888    0.000
## .I8            0.852    0.081   10.553    0.000
## .I13           1.142    0.097   11.821    0.000
## .I14           1.804    0.140   12.844    0.000
## .I16           1.235    0.096   12.812    0.000
## .I20           1.075    0.085   12.585    0.000
```

##					
##	.I5	1.503	0.125	12.026	0.000
##	.I10	1.169	0.107	10.901	0.000
##	.I11	1.044	0.112	9.330	0.000
##	.I15	1.106	0.093	11.838	0.000
##	.I22	2.076	0.160	12.958	0.000
##	.I4	0.802	0.062	12.901	0.000
##	.I7	0.523	0.042	12.572	0.000
##	.I9	1.117	0.093	11.952	0.000
##	.I12	0.987	0.080	12.287	0.000
##	.I17	0.375	0.035	10.739	0.000
##	.I18	0.909	0.081	11.224	0.000
##	.I19	0.844	0.073	11.557	0.000
##	.I21	1.245	0.098	12.764	0.000
##	EE	1.625	0.190	8.551	0.000
##	DP	0.705	0.132	5.321	0.000
##	PA	0.193	0.048	4.047	0.000

As seen from the output, the model fitting with *lavaan* ended normally with 46 iterations and the number of free parameters is 69.

Although all parameters are significant and corresponding to the CFA model, the χ^2-test statistic is 695.719 with degrees of freedom of 206 and a p-value of 0.000, which indicated that the model fitting is not satisfactory. This is further confirmed by *Comparative Fit Index (CFI)* of 0.848, *Tucker-Lewis Index (TLI)* of 0.830, and *Root Mean Square Error of Approximation* of 0.080.

2.3.1.2   Modification Indices and Model Refitting

To examine the potential model lack-of-fit, *lavaan* suggests a list of potential model *modifications* using a R function *modindices* on model *modification indices*. For the aforementioned model fitting, this can be printed as follows:

```
modindices(fitMBI, minimum.value=3, sort=TRUE)
```

##	lhs	op	rhs	mi	epc	sepc.lv	sepc.all	sepc.nox
## 183	I6	~~	I16	91.28	0.733	0.733	0.529	0.529
## 120	I1	~~	I2	82.45	0.613	0.613	0.549	0.549
## 84	EE	=~	I12	41.52	-0.313	-0.400	-0.335	-0.335
## 285	I10	~~	I11	38.08	0.580	0.580	0.525	0.525
## 335	I7	~~	I21	33.53	0.263	0.263	0.326	0.326
## 323	I4	~~	I7	33.43	0.209	0.209	0.324	0.324
## 106	PA	=~	I1	28.73	0.872	0.383	0.231	0.231
## 348	I18	~~	I19	18.61	0.250	0.250	0.285	0.285
## 185	I6	~~	I5	17.19	0.354	0.354	0.232	0.232
## 275	I5	~~	I15	15.58	0.313	0.313	0.243	0.243
## 175	I3	~~	I12	15.51	-0.255	-0.255	-0.225	-0.225

```
## 201  I8  ~~ I20 14.21  0.230   0.230   0.240   0.240
## 101  DP  =~ I12 14.17 -0.329  -0.276  -0.232  -0.232
## 329  I4  ~~ I21 13.10  0.201   0.201   0.201   0.201
## 217 I13  ~~ I20 13.07  0.237   0.237   0.214   0.214
## 107  PA  =~  I2 12.69  0.565   0.248   0.161   0.161
## 111  PA  =~ I13 12.66 -0.583  -0.256  -0.152  -0.152
## 333  I7  ~~ I18 11.81 -0.145  -0.145  -0.211  -0.211
## 182  I6  ~~ I14 11.33 -0.311  -0.311  -0.186  -0.186
## 144  I2  ~~ I13 10.34 -0.219  -0.219  -0.195  -0.195
## 124  I1  ~~ I13 10.26 -0.225  -0.225  -0.199  -0.199
## 328  I4  ~~ I19  9.93 -0.153  -0.153  -0.185  -0.185
## 121  I1  ~~  I3  9.73  0.231   0.231   0.191   0.191
## 90   DP  =~  I2  9.57 -0.357  -0.300  -0.194  -0.194
## 339  I9  ~~ I19  9.53  0.188   0.188   0.194   0.194
## 147  I2  ~~ I20  9.40 -0.194  -0.194  -0.178  -0.178
## 145  I2  ~~ I14  9.14  0.244   0.244   0.173   0.173
## 126  I1  ~~ I16  8.97 -0.206  -0.206  -0.174  -0.174
## 350 I19  ~~ I21  8.66 -0.179  -0.179  -0.175  -0.175
## 226 I13  ~~ I12  8.21  0.175   0.175   0.165   0.165
## 225 I13  ~~  I9  8.19 -0.188  -0.188  -0.167  -0.167
## 274  I5  ~~ I11  8.10 -0.264  -0.264  -0.211  -0.211
## 146  I2  ~~ I16  7.86 -0.187  -0.187  -0.160  -0.160
## 327  I4  ~~ I18  7.61 -0.141  -0.141  -0.165  -0.165
## 89   DP  =~  I1  7.31 -0.321  -0.269  -0.162  -0.162
## 96   DP  =~ I16  7.20  0.319   0.268   0.186   0.186
## 113  PA  =~ I16  7.07 -0.434  -0.191  -0.133  -0.133
## 80   EE  =~ I22  7.05  0.254   0.324   0.205   0.205
## 210  I8  ~~ I12  6.91 -0.146  -0.146  -0.159  -0.159
## 247 I16  ~~  I5  6.88  0.200   0.200   0.147   0.147
## 280  I5  ~~ I12  6.68  0.179   0.179   0.147   0.147
## 155  I2  ~~  I9  6.57  0.164   0.164   0.147   0.147
## 194  I6  ~~ I17  6.43 -0.113  -0.113  -0.148  -0.148
## 176  I3  ~~ I17  6.19  0.105   0.105   0.150   0.150
## 286 I10  ~~ I15  6.15 -0.188  -0.188  -0.165  -0.165
## 331  I7  ~~ I12  5.92 -0.101  -0.101  -0.141  -0.141
## 287 I10  ~~ I22  5.82 -0.228  -0.228  -0.146  -0.146
## 199  I8  ~~ I14  5.73 -0.185  -0.185  -0.149  -0.149
## 349 I18  ~~ I21  5.65 -0.153  -0.153  -0.144  -0.144
## 77   EE  =~ I10  5.64 -0.202  -0.257  -0.178  -0.178
## 135  I1  ~~  I9  5.59  0.155   0.155   0.138   0.138
## 296 I11  ~~ I15  5.18 -0.185  -0.185  -0.172  -0.172
## 109  PA  =~  I6  5.12 -0.413  -0.181  -0.115  -0.115
## 165  I3  ~~ I16  5.05 -0.165  -0.165  -0.130  -0.130
## 313 I15  ~~ I19  4.88 -0.125  -0.125  -0.130  -0.130
## 81   EE  =~  I4  4.82  0.094   0.119   0.120   0.120
```

```
## 127  I1  ~~ I20  4.71 -0.141  -0.141     -0.128    -0.128
## 112  PA  =~ I14  4.70  0.427   0.188      0.109     0.109
## 122  I1  ~~  I6  4.48 -0.162  -0.162     -0.123    -0.123
## 229  I13 ~~ I19  4.47 -0.123  -0.123     -0.125    -0.125
## 98   DP  =~  I4  4.39  0.160   0.135      0.135     0.135
## 289  I10 ~~  I7  4.20  0.094   0.094      0.120     0.120
## 308  I15 ~~  I7  4.15 -0.088  -0.088     -0.115    -0.115
## 258  I16 ~~ I19  3.83 -0.114  -0.114     -0.111    -0.111
## 110  PA  =~  I8  3.76 -0.292  -0.128     -0.074    -0.074
## 345  I17 ~~ I18  3.61  0.078   0.078      0.134     0.134
## 166  I3  ~~ I20  3.51 -0.130  -0.130     -0.110    -0.110
## 92   DP  =~  I6  3.46  0.247   0.207      0.131     0.131
## 234  I14 ~~ I10  3.45 -0.157  -0.157     -0.108    -0.108
## 125  I1  ~~ I14  3.33  0.151   0.151      0.106     0.106
## 215  I13 ~~ I14  3.30  0.152   0.152      0.106     0.106
## 346  I17 ~~ I19  3.24 -0.069  -0.069     -0.123    -0.123
## 117  PA  =~ I11  3.22  0.378   0.166      0.109     0.109
## 315  I22 ~~  I4  3.21  0.125   0.125      0.097     0.097
## 228  I13 ~~ I18  3.08 -0.107  -0.107     -0.105    -0.105
## 142  I2  ~~  I6  3.06 -0.131  -0.131     -0.100    -0.100
## 152  I2  ~~ I22  3.06  0.147   0.147      0.097     0.097
```

As seen from this output, we used *sort=TRUE* to sort the *mi* (i.e., modification indices) from the largest to lowest. If we add the error covariance I6~ ~ I16 into the model with $mi = 91.282$, we would reduce the χ^2-statistic by about 91.282 points. We can examine this by re-fitting the model as follows:

```
# Make another model
CFA4MBI2 <- '
# latent variables
EE =~ I1+I2+I3+I6+I8+I13+I14+I16+I20
DP =~ I5+I10+I11+I15+I22
PA =~ I4+I7+I9+I12+I17+I18+I19+I21

# add correlation between I6 and I16
I6 ~~ I16

# covariance
EE~~DP
EE~~PA
DP~~PA

# Intercepts for items
I1 ~ 1; I2 ~ 1; I3 ~ 1; I4 ~ 1; I5 ~ 1
I6 ~ 1; I7 ~ 1; I8 ~ 1; I9 ~ 1; I10 ~ 1
```

```
I11 ~ 1; I12 ~ 1; I13 ~ 1; I14 ~ 1; I15 ~ 1
I16 ~ 1; I17 ~ 1; I18 ~ 1; I19 ~ 1; I20 ~ 1
I21 ~ 1; I22 ~ 1
'
```

With the modified model *CFA4MBI2*, we can re-fit this model as follows:

```
# fit the CFA model
fitMBI2 = cfa(CFA4MBI2, data=dMBI)
```

With this refitted model, we can call *summary* with option *fit.measures=TRUE* to output the fitted measures as follows:

```
# Print the model summary with the fitting measures
summary(fitMBI2, fit.measures=TRUE)
```

```
## lavaan 0.6-12 ended normally after 48 iterations
##
##    Estimator                                         ML
##    Optimization method                           NLMINB
##    Number of model parameters                        70
##
##    Number of observations                           372
##
## Model Test User Model:
##
##    Test statistic                               597.731
##    Degrees of freedom                               205
##    P-value (Chi-square)                           0.000
##
## Model Test Baseline Model:
##
##    Test statistic                              3452.269
##    Degrees of freedom                               231
##    P-value                                        0.000
##
## User Model versus Baseline Model:
##
##    Comparative Fit Index (CFI)                    0.878
##    Tucker-Lewis Index (TLI)                       0.863
##
## Loglikelihood and Information Criteria:
##
##    Loglikelihood user model (H0)             -12762.049
##    Loglikelihood unrestricted model (H1)     -12463.184
```

```
##
##   Akaike (AIC)                                      25664.098
##   Bayesian (BIC)                                    25938.421
##   Sample-size adjusted Bayesian (BIC)               25716.333
##
## Root Mean Square Error of Approximation:
##
##   RMSEA                                                 0.072
##   90 Percent confidence interval - lower                0.065
##   90 Percent confidence interval - upper                0.078
##   P-value RMSEA <= 0.05                                 0.000
##
## Standardized Root Mean Square Residual:
##
##   SRMR                                                  0.069
##
## Parameter Estimates:
##
##   Standard errors                                    Standard
##   Information                                        Expected
##   Information saturated (h1) model                 Structured
##
## Latent Variables:
##                    Estimate  Std.Err  z-value  P(>|z|)
##   EE =~
##     I1              1.000
##     I2              0.887    0.059    14.954    0.000
##     I3              1.015    0.066    15.332    0.000
##     I6              0.715    0.063    11.339    0.000
##     I8              1.133    0.064    17.567    0.000
##     I13             1.002    0.064    15.631    0.000
##     I14             0.847    0.068    12.473    0.000
##     I16             0.672    0.057    11.777    0.000
##     I20             0.746    0.055    13.552    0.000
##   DP =~
##     I5              1.000
##     I10             1.151    0.129     8.954    0.000
##     I11             1.363    0.144     9.465    0.000
##     I15             0.909    0.110     8.273    0.000
##     I22             0.771    0.122     6.338    0.000
##   PA =~
##     I4              1.000
##     I7              0.969    0.149     6.486    0.000
##     I9              1.779    0.254     7.013    0.000
##     I12             1.496    0.221     6.768    0.000
```

```
##      I17                1.347    0.180    7.468    0.000
##      I18                1.917    0.261    7.334    0.000
##      I19                1.714    0.238    7.209    0.000
##      I21                1.356    0.218    6.224    0.000
##
## Covariances:
##                       Estimate  Std.Err  z-value  P(>|z|)
## .I6 ~~
##    .I16                0.733    0.091    8.057    0.000
## EE ~~
##     DP                 0.697    0.099    7.018    0.000
##     PA                -0.188    0.042   -4.448    0.000
## DP ~~
##     PA                -0.171    0.035   -4.842    0.000
##
## Intercepts:
##                       Estimate  Std.Err  z-value  P(>|z|)
##    .I1                 4.366    0.086   50.743    0.000
##    .I2                 4.868    0.080   60.823    0.000
##    .I3                 3.527    0.090   39.296    0.000
##    .I4                 6.298    0.052  121.827    0.000
##    .I5                 2.199    0.077   28.544    0.000
##    .I6                 2.707    0.082   33.002    0.000
##    .I7                 6.312    0.043  145.107    0.000
##    .I8                 3.043    0.089   34.018    0.000
##    .I9                 6.035    0.068   88.565    0.000
##    .I10                2.204    0.075   29.427    0.000
##    .I11                2.239    0.079   28.279    0.000
##    .I12                5.699    0.062   92.237    0.000
##    .I13                3.586    0.087   41.163    0.000
##    .I14                4.027    0.089   45.049    0.000
##    .I15                1.769    0.067   26.296    0.000
##    .I16                2.473    0.075   33.182    0.000
##    .I17                6.406    0.044  145.104    0.000
##    .I18                5.702    0.066   86.454    0.000
##    .I19                5.946    0.062   96.546    0.000
##    .I20                2.245    0.073   30.639    0.000
##    .I21                5.852    0.066   89.258    0.000
##    .I22                2.581    0.082   31.538    0.000
##     EE                 0.000
##     DP                 0.000
##     PA                 0.000
##
## Variances:
##                       Estimate  Std.Err  z-value  P(>|z|)
```

##	.I1	1.091	0.093	11.690	0.000
##	.I2	1.076	0.089	12.088	0.000
##	.I3	1.283	0.108	11.932	0.000
##	.I6	1.654	0.127	12.974	0.000
##	.I8	0.844	0.081	10.398	0.000
##	.I13	1.156	0.098	11.794	0.000
##	.I14	1.780	0.139	12.786	0.000
##	.I16	1.317	0.102	12.904	0.000
##	.I20	1.071	0.085	12.538	0.000
##	.I5	1.511	0.125	12.047	0.000
##	.I10	1.164	0.107	10.864	0.000
##	.I11	1.038	0.112	9.272	0.000
##	.I15	1.108	0.094	11.842	0.000
##	.I22	2.077	0.160	12.957	0.000
##	.I4	0.801	0.062	12.898	0.000
##	.I7	0.523	0.042	12.570	0.000
##	.I9	1.116	0.093	11.947	0.000
##	.I12	0.988	0.080	12.291	0.000
##	.I17	0.375	0.035	10.739	0.000
##	.I18	0.909	0.081	11.220	0.000
##	.I19	0.844	0.073	11.557	0.000
##	.I21	1.244	0.098	12.762	0.000
##	EE	1.662	0.191	8.679	0.000
##	DP	0.697	0.132	5.285	0.000
##	PA	0.193	0.048	4.050	0.000

It can be seen from this refitted model, the χ^2-statistic reduced to 597.731 from 695.719 in the original CFA model. The difference, $97.988 = 695.719 - 597.731$, is close to the $mi = 91.282$, which was *predicted* from the original analysis without refitting the model.

We can iterate this model re-fitting process multiple times until we find a satisfactory model for the data. Of course, we need to be mindful that the modified *components* are substantively meaningful in the application and this modification should be highly theory-driven. We leave this to the interested readers for further investigation.

2.3.1.3   Final CFA Model

Based on the above *modification indices*, we now include three more terms, which are the item correlations of I6 $\sim \sim$ I16, I1 $\sim \sim$ I2, and I10 $\sim \sim$ I11 and the cross-loading from EE to *Item12* as EE = ˜ I12, in a modified model as follows:

```
# Make another model
CFA4MBI3 <- '
# Latent variables
```

```
EE =~ I1+I2+I3+I6+I8+I13+I14+I16+I20
DP =~ I5+I10+I11+I15+I22
PA =~ I4+I7+I9+I12+I17+I18+I19+I21

# Additional item correlations based on MIs
I6   ~~ I16
I1   ~~ I2
I10  ~~ I11
# Cross-loading from EE to Item12 based on MI
EE =~ I12

# Covariance
EE ~~ DP ; EE ~~ PA; DP ~~ PA

# Intercepts for items
I1   ~ 1; I2   ~ 1; I3   ~ 1; I4   ~ 1; I5   ~ 1
I6   ~ 1; I7   ~ 1; I8   ~ 1; I9   ~ 1; I10  ~ 1
I11  ~ 1; I12  ~ 1; I13  ~ 1; I14  ~ 1; I15  ~ 1
I16  ~ 1; I17  ~ 1; I18  ~ 1; I19  ~ 1; I20  ~ 1
I21  ~ 1; I22  ~ 1
'
```

With the modified model *CFA4MBI3*, we can refit this model as follows:

```
# fit the CFA model
fitMBI3 = cfa(CFA4MBI3, data=dMBI)
```

With this refitted model, we can call *summary* with option *fit.measures=TRUE* to output the fitted measures as follows:

```
# Print the model summary with the fitting measures
summary(fitMBI3, fit.measures=TRUE)
```

```
## lavaan 0.6-12 ended normally after 52 iterations
##
##    Estimator                                         ML
##    Optimization method                           NLMINB
##    Number of model parameters                        73
##
##    Number of observations                           372
##
## Model Test User Model:
##
##    Test statistic                               446.419
##    Degrees of freedom                               202
```

```
## P-value (Chi-square)                              0.000
##
## Model Test Baseline Model:
##
##    Test statistic                             3452.269
##    Degrees of freedom                              231
##    P-value                                       0.000
##
## User Model versus Baseline Model:
##
##    Comparative Fit Index (CFI)                   0.924
##    Tucker-Lewis Index (TLI)                      0.913
##
## Loglikelihood and Information Criteria:
##
##    Loglikelihood user model (H0)            -12686.394
##    Loglikelihood unrestricted model (H1)    -12463.184
##
##    Akaike (AIC)                             25518.787
##    Bayesian (BIC)                           25804.867
##    Sample-size adjusted Bayesian (BIC)      25573.260
##
## Root Mean Square Error of Approximation:
##
##    RMSEA                                         0.057
##    90 Percent confidence interval - lower        0.050
##    90 Percent confidence interval - upper        0.064
##    P-value RMSEA <= 0.05                         0.052
##
## Standardized Root Mean Square Residual:
##
##    SRMR                                          0.054
##
## Parameter Estimates:
##
##    Standard errors                            Standard
##    Information                                Expected
##    Information saturated (h1) model         Structured
##
## Latent Variables:
##                   Estimate  Std.Err  z-value  P(>|z|)
##    EE =~
##      I1             1.000
##      I2             0.878    0.049    17.969    0.000
##      I3             1.073    0.075    14.303    0.000
```

```
##     I6                0.764    0.069    11.011   0.000
##     I8                1.215    0.075    16.299   0.000
##     I13               1.072    0.073    14.739   0.000
##     I14               0.880    0.075    11.673   0.000
##     I16               0.727    0.063    11.556   0.000
##     I20               0.806    0.061    13.113   0.000
##   DP =~
##     I5                1.000
##     I10               0.889    0.114     7.829   0.000
##     I11               1.105    0.125     8.819   0.000
##     I15               0.921    0.105     8.791   0.000
##     I22               0.776    0.115     6.730   0.000
##   PA =~
##     I4                1.000
##     I7                0.973    0.147     6.611   0.000
##     I9                1.763    0.248     7.105   0.000
##     I12               1.131    0.188     6.022   0.000
##     I17               1.327    0.176     7.555   0.000
##     I18               1.890    0.255     7.421   0.000
##     I19               1.695    0.232     7.300   0.000
##     I21               1.342    0.213     6.290   0.000
##   EE =~
##     I12              -0.316    0.050    -6.366   0.000
##
## Covariances:
##                    Estimate  Std.Err  z-value  P(>|z|)
##  .I6 ~~
##    .I16              0.706    0.090     7.869   0.000
##  .I1 ~~
##    .I2               0.588    0.082     7.131   0.000
##  .I10 ~~
##    .I11              0.517    0.101     5.106   0.000
##   EE ~~
##     DP               0.747    0.104     7.197   0.000
##     PA              -0.167    0.040    -4.168   0.000
##   DP ~~
##     PA              -0.181    0.038    -4.768   0.000
##
## Intercepts:
##                    Estimate  Std.Err  z-value  P(>|z|)
##    .I1              4.366    0.086    50.743    0.000
##    .I2              4.868    0.080    60.823    0.000
##    .I3              3.527    0.090    39.296    0.000
##    .I4              6.298    0.052   121.827    0.000
##    .I5              2.199    0.077    28.544    0.000
```

```
##    .I6            2.707    0.082   33.002     0.000
##    .I7            6.312    0.043  145.108     0.000
##    .I8            3.043    0.089   34.018     0.000
##    .I9            6.035    0.068   88.566     0.000
##    .I10           2.204    0.075   29.427     0.000
##    .I11           2.239    0.079   28.279     0.000
##    .I12           5.699    0.062   92.237     0.000
##    .I13           3.586    0.087   41.163     0.000
##    .I14           4.027    0.089   45.049     0.000
##    .I15           1.769    0.067   26.297     0.000
##    .I16           2.473    0.075   33.182     0.000
##    .I17           6.406    0.044  145.103     0.000
##    .I18           5.702    0.066   86.454     0.000
##    .I19           5.946    0.062   96.547     0.000
##    .I20           2.245    0.073   30.639     0.000
##    .I21           5.852    0.066   89.258     0.000
##    .I22           2.581    0.082   31.538     0.000
##    EE             0.000
##    DP             0.000
##    PA             0.000
##
## Variances:
##                 Estimate  Std.Err  z-value   P(>|z|)
##    .I1            1.268    0.106   11.992     0.000
##    .I2            1.238    0.101   12.310     0.000
##    .I3            1.285    0.109   11.814     0.000
##    .I6            1.636    0.127   12.902     0.000
##    .I8            0.783    0.080    9.788     0.000
##    .I13           1.115    0.097   11.536     0.000
##    .I14           1.822    0.143   12.781     0.000
##    .I16           1.281    0.100   12.793     0.000
##    .I20           1.031    0.083   12.361     0.000
##    .I5            1.407    0.126   11.211     0.000
##    .I10           1.455    0.126   11.510     0.000
##    .I11           1.355    0.130   10.397     0.000
##    .I15           1.004    0.094   10.735     0.000
##    .I22           2.008    0.159   12.631     0.000
##    .I4            0.795    0.062   12.846     0.000
##    .I7            0.515    0.041   12.477     0.000
##    .I9            1.108    0.093   11.860     0.000
##    .I12           0.898    0.072   12.545     0.000
##    .I17           0.374    0.035   10.642     0.000
##    .I18           0.906    0.081   11.130     0.000
##    .I19           0.838    0.073   11.459     0.000
##    .I21           1.240    0.097   12.723     0.000
```

##	EE	1.486	0.186	7.982	0.000
##	DP	0.800	0.144	5.567	0.000
##	PA	0.199	0.048	4.112	0.000

With *CFA4MBI3*, the number of parameters to be estimated is 73 and the *lavaan* was ended normally after 52 iterations with maximum likelihood estimation.

As seen from the output, the χ^2-statistic is 446.419 and the degrees of freedom of 202 with the *p*-value of 0.000, indicating poor model fitting. However, all the alternative model fitting indices are satisfactory where the *Comparative Fit Index (CFI)* is 0.924, the *Tucker-Lewis Index (TLI)* is 0.913, the *Root Mean Square Error of Approximation (RMSEA)* is 0.057, and the *Standardized Root Mean Square Residual (SRMR)* is 0.054, respectively.

With this satisfactory model fitting, we can now examine the model parameters. As seen from the output, all the 73 parameters (the original 69 along with the 4 new parameters suggested by the model modification indices *MI*) are statistically significant.

This model can be graphically illustrated in Figure 2.6. As seen from this figure, we have included three more item correlations (in *brown* color) along with a cross-loading from *EE* to *item12* (in *brown* color) in comparison with Figure 2.1.

2.3.2   Data Analysis for HolzingerSwineford1939

Similar to the CFA modeling with *MBI* data, we first fit a basic three-factor CFA model to the *HS* data corresponding to Figure 2.5, and then proceed to improve fit by using the *modification indices*.

2.3.2.1   Fitting the Original Three-Factor CFA Model

The basic *HS* model in Figure 2.5 can be specified as following:

```
# CFA Model Specification
HSModel <- '
visual  =~ x1+x2+x3
textual =~ x4+x5+x6
speed   =~ x7+x8+x9
'
```

With this specified CFA model, we can now call *cfa* to fit the above model as follows:

```
# Call cfa to fit the model
fitHS = cfa(HSModel, data=dHS)
# Print the model summary
summary(fitHS, fit.measures=TRUE)
```

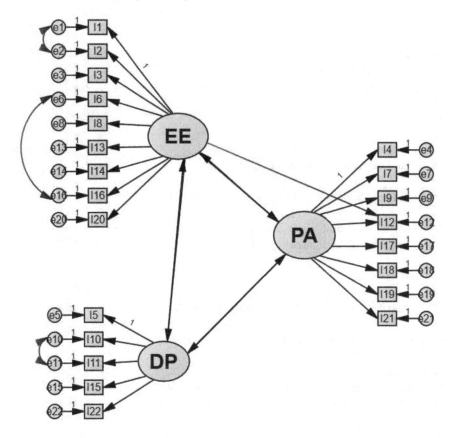

FIGURE 2.6
Final CFA Model among Latent Variables EE, DP and PA.

```
## lavaan 0.6-12 ended normally after 35 iterations
##
##   Estimator                                      ML
##   Optimization method                        NLMINB
##   Number of model parameters                     21
##
##   Number of observations                        301
##
## Model Test User Model:
##
##   Test statistic                             85.306
##   Degrees of freedom                             24
##   P-value (Chi-square)                        0.000
##
```

```
## Model Test Baseline Model:
##
##   Test statistic                              918.852
##   Degrees of freedom                               36
##   P-value                                       0.000
##
## User Model versus Baseline Model:
##
##   Comparative Fit Index (CFI)                   0.931
##   Tucker-Lewis Index (TLI)                      0.896
##
## Loglikelihood and Information Criteria:
##
##   Loglikelihood user model (H0)             -3737.745
##   Loglikelihood unrestricted model (H1)     -3695.092
##
##   Akaike (AIC)                               7517.490
##   Bayesian (BIC)                             7595.339
##   Sample-size adjusted Bayesian (BIC)        7528.739
##
## Root Mean Square Error of Approximation:
##
##   RMSEA                                         0.092
##   90 Percent confidence interval - lower        0.071
##   90 Percent confidence interval - upper        0.114
##   P-value RMSEA <= 0.05                         0.001
##
## Standardized Root Mean Square Residual:
##
##   SRMR                                          0.065
##
## Parameter Estimates:
##
##   Standard errors                            Standard
##   Information                                Expected
##   Information saturated (h1) model         Structured
##
## Latent Variables:
##                    Estimate  Std.Err  z-value  P(>|z|)
##   visual =~
##     x1                1.000
##     x2                0.554    0.100    5.554    0.000
##     x3                0.729    0.109    6.685    0.000
##   textual =~
##     x4                1.000
```

```
##       x5                1.113   0.065   17.014   0.000
##       x6                0.926   0.055   16.703   0.000
##    speed =~
##       x7                1.000
##       x8                1.180   0.165   7.152    0.000
##       x9                1.082   0.151   7.155    0.000
##
## Covariances:
##                       Estimate Std.Err z-value  P(>|z|)
##    visual ~~
##       textual           0.408   0.074   5.552    0.000
##       speed             0.262   0.056   4.660    0.000
##    textual ~~
##       speed             0.173   0.049   3.518    0.000
##
## Variances:
##                       Estimate Std.Err z-value  P(>|z|)
##      .x1               0.549   0.114   4.833    0.000
##      .x2               1.134   0.102   11.146   0.000
##      .x3               0.844   0.091   9.317    0.000
##      .x4               0.371   0.048   7.779    0.000
##      .x5               0.446   0.058   7.642    0.000
##      .x6               0.356   0.043   8.277    0.000
##      .x7               0.799   0.081   9.823    0.000
##      .x8               0.488   0.074   6.573    0.000
##      .x9               0.566   0.071   8.003    0.000
##       visual           0.809   0.145   5.564    0.000
##       textual          0.979   0.112   8.737    0.000
##       speed            0.384   0.086   4.451    0.000
```

As seen from the output, the *lavaan* ended normally after 35 iterations for this CFA model with 21 parameters to be estimated from 301 observations.

The χ^2-statistic is 85.306 with the degrees of freedom of 24 and p-value of 0.000, indicating poor model fitting based on the χ^2-statistic. Although the *Comparative Fit Index (CFI)* of 0.931 is satisfactory, other fitting indices such as *Tucker-Lewis Index (TLI)* of 0.896 and *Rooted Mean Square Error of Approximation (RMSEA)* of 0.092 do not indicate a good model fit.

2.3.2.2   Modification Indices and Model Re-fitting

For model refitting, we make use of the *modification indices* in *lavaan* to see which theoretically sounded paths or links could be included to improve the model fitting. The *lavaan* function *modindices* can print a lengthy list ranking from the best to least. We will only print the first six using R function *head* with the following R code chunk:

```
# only the first 6 rows using 'head' function in R
head(modindices(fitHS, minimum.value = 3, sort=TRUE))
```

```
##            lhs op rhs    mi    epc sepc.lv sepc.all
## 30     visual =~  x9 36.41  0.577   0.519    0.515
## 76         x7 ~~  x8 34.15  0.536   0.536    0.859
## 28     visual =~  x7 18.63 -0.422  -0.380   -0.349
## 78         x8 ~~  x9 14.95 -0.423  -0.423   -0.805
## 33    textual =~  x3  9.15 -0.272  -0.269   -0.238
## 55         x2 ~~  x7  8.92 -0.183  -0.183   -0.192
##     sepc.nox
## 30     0.515
## 76     0.859
## 28    -0.349
## 78    -0.805
## 33    -0.238
## 55    -0.192
```

As seen from the output, the model fit χ^2-statistic could be reduced by about 36.411 and 18.631 by including the cross-loadings from *visual* to x9 and x7, respectively. Similarly, the χ^2-statistic could be reduced by about 34.145 and 14.946, respectively, by including the **residual item covariances** (error covariances) between x7 and x8 or x8 and x9. Note that these reductions are not additive, i.e., the total reduction of χ^2-statistic by adding these *MI*s together in general is not equal to the sum of the component reductions.

Examining item definitions among x1 to x9, it makes more theoretical sense to include the residual item correlation between x7 and x8 because x7 is to describe the *Speeded addition* and x8 is to describe the *Speeded counting of dots*. We can include this **residual item correlation** with a CFA model refitting as follows:

```
# CFA Model Specification
HSModel2 <- '
# Measurement Model
visual  =~ x1+x2+x3
textual =~ x4+x5+x6
speed   =~ x7+x8+x9

# Add residual item correlation based on MI
x7 ~~ x8
'
```

With this specified CFA model, we can now call *cfa* to fit the above model as follows:

```
# Call cfa to fit the model
fitHS2 = cfa(HSModel2, data=dHS)
# Print the model summary
summary(fitHS2, fit.measures=TRUE)
```

```
## lavaan 0.6-12 ended normally after 43 iterations
##
##   Estimator                                      ML
##   Optimization method                        NLMINB
##   Number of model parameters                     22
##
##   Number of observations                        301
##
## Model Test User Model:
##
##   Test statistic                             53.272
##   Degrees of freedom                             23
##   P-value (Chi-square)                        0.000
##
## Model Test Baseline Model:
##
##   Test statistic                            918.852
##   Degrees of freedom                             36
##   P-value                                     0.000
##
## User Model versus Baseline Model:
##
##   Comparative Fit Index (CFI)                 0.966
##   Tucker-Lewis Index (TLI)                    0.946
##
## Loglikelihood and Information Criteria:
##
##   Loglikelihood user model (H0)           -3721.728
##   Loglikelihood unrestricted model (H1)   -3695.092
##
##   Akaike (AIC)                             7487.457
##   Bayesian (BIC)                           7569.013
##   Sample-size adjusted Bayesian (BIC)      7499.242
##
```

```
## Root Mean Square Error of Approximation:
##
##   RMSEA                                              0.066
##   90 Percent confidence interval - lower             0.043
##   90 Percent confidence interval - upper             0.090
##   P-value RMSEA <= 0.05                              0.118
##
## Standardized Root Mean Square Residual:
##
##   SRMR                                               0.047
##
## Parameter Estimates:
##
##   Standard errors                               Standard
##   Information                                   Expected
##   Information saturated (h1) model            Structured
##
## Latent Variables:
##                     Estimate  Std.Err  z-value  P(>|z|)
##   visual =~
##     x1               1.000
##     x2               0.576    0.098    5.898    0.000
##     x3               0.752    0.103    7.289    0.000
##   textual =~
##     x4               1.000
##     x5               1.115    0.066    17.015   0.000
##     x6               0.926    0.056    16.682   0.000
##   speed =~
##     x7               1.000
##     x8               1.244    0.194    6.414    0.000
##     x9               2.515    0.641    3.924    0.000
##
## Covariances:
##                     Estimate  Std.Err  z-value  P(>|z|)
##  .x7 ~~
##    .x8               0.353    0.067    5.239    0.000
##   visual ~~
##     textual          0.400    0.073    5.511    0.000
##     speed            0.184    0.054    3.423    0.001
##   textual ~~
##     speed            0.102    0.036    2.854    0.004
##
## Variances:
##                     Estimate  Std.Err  z-value  P(>|z|)
##    .x1               0.576    0.101    5.678    0.000
```

##					
##	.x2	1.122	0.100	11.171	0.000
##	.x3	0.832	0.087	9.552	0.000
##	.x4	0.372	0.048	7.791	0.000
##	.x5	0.444	0.058	7.600	0.000
##	.x6	0.357	0.043	8.287	0.000
##	.x7	1.036	0.090	11.501	0.000
##	.x8	0.795	0.080	9.988	0.000
##	.x9	0.088	0.188	0.466	0.641
##	visual	0.783	0.135	5.810	0.000
##	textual	0.978	0.112	8.729	0.000
##	speed	0.147	0.056	2.615	0.009

With this added item correlation based on *MI*, the *Comparative Fit Index (CFI)* is increased to 0.966, and the *Tucker-Lewis Index (TLI)* to 0.946. The *Root Mean Square Error of Approximation (RMSEA)* is decreased to 0.066 with 90% confidence interval of $(0.043, 0.090)$. The p-value of RMSEA $<= 0.05$ (testing close fit) is 0.118, supporting a good model fit. In addition, the *Standardized Root Mean Square Residual (SRMR)* is 0.047. All these alternative fit indices are satisfactory.

With this satisfactory CFA model, we can see all the factor loadings are statistically significant, as well as all the covariances and variances (except the error variance of x9).

This would serve as the final CFA model for this data. With the added residual item correlation between x7 and x8, this CFA model can be shown graphically described in Figure 2.7.

2.4 CFA with SAS

The *MBI* data is used here to illustrate the CFA modeling by SAS. The following code reads in the csv file into a SAS data set:

```
proc import datafile='c:\ELEMmbi.csv'
   out  = dMBI
   dbms = csv
   replace;
run;

libname U "c:\";

data u.dEleMBI;
   set work.dEleMBI;
run;
```

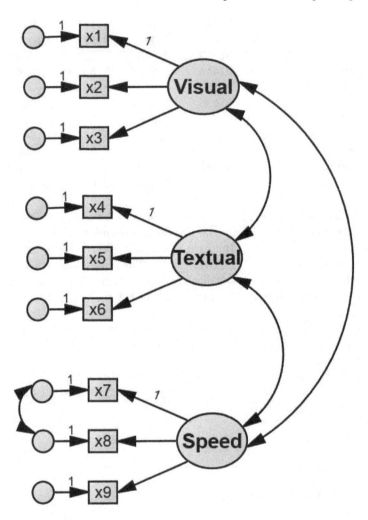

FIGURE 2.7
Final CFA Model for the Holzinger & Swineford Data.

The CFA model for the data set can be specified by the PATH statement
of PROC CALIS as follows:

```
proc calis 'c:\dEleMBI';
    path
        EE ==> I1 I2 I3 I6 I8 I13 I14 I16 I20 = 1,
        PA ==> I4 I7 I9 I12 I17 I18 I19 I21 = 1,
        DP ==> I5 I10 I11 I15 I22 = 1;
run;
```

In the PATH statement, each of the *EE, PA,* and *DP* factors are reflected by the corresponding set of items. Unlike *lavaan,* PROC CALIS does not automatically fix any loading values for identification purposes. Instead, you can choose which loadings to be fixed values. In the current specification, a fixed value of 1 after the equal sign applies to the factor loading for the *first* item in each of the three factor-variables relations. Other factor loadings in the specification are free parameters.

By default, PROC CALIS treats error variance, factor variances and factor covariances as free parameters and therefore there is no need to specify them when you specify such a basic CFA model.

Like many other structural equation modeling software, PROC CALIS can produce a lot of output. We illustrate some important output here. The following table shows the variables in the model, and it classifies variables according to whether they are endogenous or exogenous, and whether they are manifest (observed) or latent (unobserved). It is clear from the table that all observed variables in the CFA model are endogenous and all latent factors are exogenous. This table would be useful to check whether your CFA model has been specified as intended.

Variables in the Model		
Endogenous	Manifest	I1 I10 I11 I12 I13 I14 I15 I16 I17 I18 I19 I2 I20 I21 I22 I3 I4 I5 I6 I7 I8 I9
	Latent	
Exogenous	Manifest	
	Latent	DP EE PA
Number of Endogenous Variables = 22		
Number of Exogenous Variables = 3		

The next table shows the optimization results. PROC CALIS used ten iterations to get to a converged solution. The table also shows other technical details in the optimization.

Optimization Results			
Iterations	10	Function Calls	25
Jacobian Calls	12	Active Constraints	0
Objective Function	1.8702123894	Max Abs Gradient Element	0.0000199598
Lambda	0	Actual Over Pred Change	1.3911154412
Radius	0.0006034975		

Convergence criterion (GCONV=1E-8) satisfied.

As can be seen from the fit summary results below, PROC CALIS produces a large number of fit statistics, including many of those we have seen in the *lavaan* output: model fit chi-square = 693.8488 (df = 206, $p<.0001$); CFI = 0.848 (shown as Bentler Comparative Fit Index), TLI = 0.830 (shown as Bentler-Bonett Non-normed Index), RMSEA = 0.080, SRMR = 0.0729.

You can customize the output of fit indices by some options. These options will be shown in a later example.

Fit Summary		
Modeling Info	Number of Observations	372
	Number of Variables	22
	Number of Moments	253
	Number of Parameters	47
	Number of Active Constraints	0
	Baseline Model Function Value	9.2803
	Baseline Model Chi-Square	3442.9884
	Baseline Model Chi-Square DF	231
	Pr > Baseline Model Chi-Square	<.0001
Absolute Index	Fit Function	1.8702
	Chi-Square	693.8488
	Chi-Square DF	206
	Pr > Chi-Square	<.0001
	Z-Test of Wilson & Hilferty	15.2255
	Hoelter Critical N	129
	Root Mean Square Residual (RMR)	0.1412
	Standardized RMR (SRMR)	0.0729
	Goodness of Fit Index (GFI)	0.8491

Parsimony Index	Adjusted GFI (AGFI)	0.8147
	Parsimonious GFI	0.7572
	RMSEA Estimate	0.0799
	RMSEA Lower 90% Confidence Limit	0.0734
	RMSEA Upper 90% Confidence Limit	0.0865
	Probability of Close Fit	<.0001
	ECVI Estimate	2.1403
	ECVI Lower 90% Confidence Limit	1.9299
	ECVI Upper 90% Confidence Limit	2.3726
	Akaike Information Criterion	787.8488
	Bozdogan CAIC	1019.0368
	Schwarz Bayesian Criterion	972.0368
	McDonald Centrality	0.5191
Incremental Index	Bentler Comparative Fit Index	0.8481
	Bentler-Bonett NFI	0.7985
	Bentler-Bonett Non-normed Index	0.8297
	Bollen Normed Index Rho1	0.7740
	Bollen Non-normed Index Delta2	0.8493
	James et al. Parsimonious NFI	0.7121

The loading estimates are shown next.

PATH List						
Path			Parameter	Estimate	Standard Error	t Value	Pr > |t|
EE ===>	I1			1.00000			
EE ===>	I2	_Parm01	0.88682	0.06073	14.6015	<.0001
EE ===>	I3	_Parm02	1.02142	0.06780	15.0643	<.0001
EE ===>	I6	_Parm03	0.76444	0.06372	11.9973	<.0001
EE ===>	I8	_Parm04	1.14339	0.06618	17.2758	<.0001
EE ===>	I13	_Parm05	1.01713	0.06552	15.5229	<.0001
EE ===>	I14	_Parm06	0.84786	0.06930	12.2350	<.0001
EE ===>	I16	_Parm07	0.71514	0.05770	12.3937	<.0001
EE ===>	I20	_Parm08	0.75290	0.05622	13.3923	<.0001
PA ===>	I4		1.00000			
PA ===>	I7	_Parm09	0.96967	0.14980	6.4733	<.0001
PA ===>	I9	_Parm10	1.77960	0.25431	6.9976	<.0001
PA ===>	I12	_Parm11	1.49902	0.22176	6.7598	<.0001
PA ===>	I17	_Parm12	1.34841	0.18091	7.4535	<.0001
PA ===>	I18	_Parm13	1.91762	0.26200	7.3192	<.0001
PA ===>	I19	_Parm14	1.71608	0.23849	7.1956	<.0001
PA ===>	I21	_Parm15	1.35623	0.21836	6.2109	<.0001
DP ===>	I5		1.00000			
DP ===>	I10	_Parm16	1.14170	0.12723	8.9734	<.0001
DP ===>	I11	_Parm17	1.35252	0.14239	9.4985	<.0001
DP ===>	I15	_Parm18	0.90523	0.10897	8.3073	<.0001
DP ===>	I22	_Parm19	0.76767	0.12084	6.3527	<.0001

The factor loadings for items 1, 4, and 5 have been fixed to 1 as intended and therefore there are no standard errors or significance tests for these loadings. All other loading estimates are statistically significant.

Estimates of factor variances and error variances are shown in the following table. They are all significantly different from zero.

Variance Parameters						
Variance Type	Variable	Parameter	Estimate	Standard Error	t Value	Pr > |t|
Exogenous	EE	_Add01	1.62961	0.19083	8.5397	<.0001
	PA	_Add02	0.19320	0.04780	4.0416	<.0001
	DP	_Add03	0.70647	0.13295	5.3139	<.0001
Error	I1	_Add04	1.13125	0.09551	11.8445	<.0001
	I2	_Add05	1.10801	0.09084	12.1971	<.0001
	I3	_Add06	1.30449	0.10857	12.0151	<.0001
	I4	_Add07	0.80377	0.06239	12.8832	<.0001
	I5	_Add08	1.50722	0.12550	12.0096	<.0001
	I6	_Add09	1.55731	0.12100	12.8707	<.0001
	I7	_Add10	0.52407	0.04174	12.5553	<.0001
	I8	_Add11	0.85420	0.08105	10.5391	<.0001
	I9	_Add12	1.12005	0.09384	11.9356	<.0001
	I10	_Add13	1.17205	0.10766	10.8863	<.0001
	I11	_Add14	1.04647	0.11232	9.3172	<.0001
	I12	_Add15	0.98979	0.08067	12.2703	<.0001
	I13	_Add16	1.14494	0.09699	11.8048	<.0001
	I14	_Add17	1.80894	0.14103	12.8263	<.0001
	I15	_Add18	1.10873	0.09378	11.8221	<.0001
	I16	_Add19	1.23863	0.09680	12.7952	<.0001
	I17	_Add20	0.37568	0.03503	10.7248	<.0001
	I18	_Add21	0.91185	0.08135	11.2093	<.0001
	I19	_Add22	0.84592	0.07329	11.5417	<.0001
	I20	_Add23	1.07823	0.08579	12.5683	<.0001
	I21	_Add24	1.24805	0.09791	12.7465	<.0001
	I22	_Add25	2.08118	0.16083	12.9401	<.0001

The final set of estimates are for the covariances among factors, as shown in the following table:

Covariances Among Exogenous Variables						
Var1	Var2	Parameter	Estimate	Standard Error	t Value	Pr > |t|
PA	EE	_Add26	-0.19260	0.04250	-4.5313	<.0001
DP	EE	_Add27	0.70298	0.09969	7.0515	<.0001
DP	PA	_Add28	-0.17235	0.03558	-4.8433	<.0001

All the covariance estimates are significantly different from zero. PROC CALIS also outputs the standardized version of these covariance estimates—that is, the correlations of factors are shown in the following table.

Standardized Results for Covariances Among Exogenous Variables						
Var1	Var2	Parameter	Estimate	Standard Error	t Value	Pr > |t|
PA	EE	_Add26	-0.34325	0.05434	-6.3166	<.0001
DP	EE	_Add27	0.65517	0.04081	16.0543	<.0001
DP	PA	_Add28	-0.46650	0.05532	-8.4321	<.0001

A confirmatory factor analysis with both mean and covariance structures can also be fitted by PROC CALIS. By default, PROC CALIS analyzes covariance structures only. If the mean structures are also of interest in the current CFA model, the quickest way to specify the default mean structures is to add the MEANSTR option in the PROC CALIS statement, as shown in the following specification:

```
proc calis 'c:\dEleMBI'  meanstr;
   path
      EE ==> I1 I2 I3 I6 I8 I13 I14 I16 I20 = 1,
      PA ==> I4 I7 I9 I12 I17 I18 I19 I21 = 1,
      DP ==> I5 I10 I11 I15 I22 = 1;
run;
```

With the MEANSTR option, PROC CALIS sets the mean structure parameters by default—that is, the intercepts of all endogenous observed variables and the means of all exogenous observed variables would be free parameters in the model. However, the intercepts of all endogenous latent variables and the means of all exogenous latent variables would be fixed zeros by default. Options to override these defaults are available, as will be illustrated in later chapters.

Estimation results of the current mean and covariance structure analysis are the same as that of the covariance structure analysis. In addition, the current results include estimates of the intercepts of the observed variables, as shown in the following table.

Means and Intercepts						
Type	Variable	Parameter	Estimate	Standard Error	t Value	Pr > |t|
Intercept	I1	_Add29	4.36559	0.08627	50.6067	<.0001
	I2	_Add30	4.86828	0.08026	60.6595	<.0001
	I3	_Add31	3.52688	0.08999	39.1904	<.0001
	I4	_Add32	6.29839	0.05184	121.5	<.0001
	I5	_Add33	2.19892	0.07725	28.4668	<.0001
	I6	_Add34	2.70699	0.08225	32.9133	<.0001
	I7	_Add35	6.31183	0.04361	144.7	<.0001
	I8	_Add36	3.04301	0.08969	33.9268	<.0001
	I9	_Add37	6.03495	0.06832	88.3274	<.0001
	I10	_Add38	2.20430	0.07511	29.3482	<.0001
	I11	_Add39	2.23925	0.07940	28.2026	<.0001
	I12	_Add40	5.69892	0.06195	91.9888	<.0001
	I13	_Add41	3.58602	0.08735	41.0526	<.0001
	I14	_Add42	4.02688	0.08963	44.9281	<.0001
	I15	_Add43	1.76882	0.06745	26.2258	<.0001
	I16	_Add44	2.47312	0.07473	33.0927	<.0001
	I17	_Add45	6.40591	0.04427	144.7	<.0001
	I18	_Add46	5.70161	0.06613	86.2218	<.0001
	I19	_Add47	5.94624	0.06176	96.2870	<.0001
	I20	_Add48	2.24462	0.07346	30.5562	<.0001
	I21	_Add49	5.85215	0.06574	89.0182	<.0001
	I22	_Add50	2.58065	0.08205	31.4529	<.0001

You can request model modification indices by adding the MODIFICA-TION option in the PROC CALIS statement, as shown in the following specification:

```
proc calis 'c:\dEleMBI' modification;
   path
      EE ==> I1 I2 I3 I6 I8 I13 I14 I16 I20 = 1,
      PA ==> I4 I7 I9 I12 I17 I18 I19 I21 = 1,
      DP ==> I5 I10 I11 I15 I22 = 1;
run;
```

The main model modification results are shown in the following tables. Note that the *LM stat* in the PROC CALIS output is labeled as *mi* in *lavaan.*

Rank Order of the 10 Largest LM Stat for Error Variances and Covariances				
Error of	Error of	LM Stat	Pr > ChiSq	Parm Change
I6	I16	91.03675	<.0001	0.73458
I2	I1	82.22657	<.0001	0.61437
I11	I10	37.98163	<.0001	0.58171
I7	I21	33.43863	<.0001	0.26387
I7	I4	33.34202	<.0001	0.20999
I19	I18	18.55762	<.0001	0.25047
I6	I5	17.14619	<.0001	0.35516
I5	I15	15.54121	<.0001	0.31348
I3	I12	15.46928	<.0001	-0.25610
I8	I20	14.17288	0.0002	0.23055

Rank Order of the 10 Largest LM Stat for Path Relations				
To	From	LM Stat	Pr > ChiSq	Parm Change
I16	I6	91.03694	<.0001	0.47169
I6	I16	91.03685	<.0001	0.59306
I1	I2	82.22604	<.0001	0.55448
I2	I1	82.22599	<.0001	0.54309
I12	I3	48.95255	<.0001	-0.22902
I12	EE	41.40548	<.0001	-0.31338
I12	I8	40.59895	<.0001	-0.21233
I10	I11	37.98383	<.0001	0.55590
I11	I10	37.98289	<.0001	0.49633
EE	I12	35.24600	<.0001	-0.36338

As argued previously when fitting the same model by *lavaan*, three residual covariances (error covariances) for items are suggested at the top three rows in the first table. After including these covariances, it make sense to add the I12 ← EE path that is shown in the second table. Adding those item-to-item paths (i.e., the first five rows in the second table) are either redundant to the added error covariances or destructive to the factor-variable relations in a CFA model.

You can add the three suggested error covariances in the PCOV statement and the I12 ← EE path in the PATH statement to the original model, as shown in the following revised specification:

```
proc calis 'c:\dEleMBI';
    path
        EE ==> I1 I2 I3 I6 I8 I13 I14 I16 I20 = 1,
        PA ==> I4 I7 I9 I12 I17 I18 I19 I21 = 1,
        DP ==> I5 I10 I11 I15 I22 = 1,
        EE ==> I12;
    pcov I6 I16, I1 I2, I10 I11;
    fitindex on(only)=[chisq df probchi RMSEA PROBCLFIT SRMR CFI
                       BENTLERNNFI] noindextype;
run;
```

Note that the FITINDEX statement is also added to simplify the output so that the output shows only those specified fit indices. The following table shows that the model fit improves quite a lot and the revised model now has an acceptable fit.

Fit Summary	
Chi-Square	445.2194
Chi-Square DF	202
Pr > Chi-Square	<.0001
Standardized RMR (SRMR)	0.0566
RMSEA Estimate	0.0570
Probability of Close Fit	0.0542
Bentler Comparative Fit Index	0.9243
Bentler-Bonett Non-normed Index	0.9134

Fitting the CFA model to the *HolzingerSwineford1939* data using *PROC CALIS* is left to readers as an exercise.

2.5 Conclusions and Discussion

This chapter discusses the latent variable modeling with a CFA model. We illustrated the CFA modeling with two numerical examples. For each example, we first fit the original proposed CFA model and found that the CFA model could not fit the data satisfactorily. To improve the model fitting, we made use of the *modification indices (MI)* in *lavaan* (or the *LM statistics* in *PROC CALIS*) to search for a better CFA model. These modification indices are recommended as an initial guidance for building satisfactory CFA models. Perhaps even more importantly, SEM practitioners should use these indices along with their substantive knowledge about the data for their own research and CFA modeling.

The CFA model illustrated in this chapter is for first-order CFA model only where in the *MBI* example, the three factors of *EE*, *DP* and *PA* are measured in the first-order by their corresponding 22 items, and in *HS* example, the three factors of *visual*, *textual* and *speed* are measured by the 9 items. For further understanding of these CFA analysis, we recommend interested readers to review Chapter 4 in Byrne (2012) for *MBI* data and *lavaan* manual for *HS* data.

Higher-order CFA can be examined and tested in a similar way. For example in the MBI example, we can formulate a second-order CFA with a higher level denoted by *burnout*, which is measured by these three latent factors of EE, *DP* and *PA*. In this case, we can easily make a change to the CFA model in CFA4MBI to include this second-order factor. The specification of the covariances EE $\sim \sim$ DP; EE $\sim \sim$ PA; DP $\sim \sim$ PA would not be necessary because these covariances are now implied by the second-order CFA model.

The second-order CFA model is specified as follows:

```
# Make Second-Order CFA model
CFA4MBI4 <- '
# Latent variables at the first-order
EE =~ I1+I2+I3+I6+I8+I13+I14+I16+I20
DP =~ I5+I10+I11+I15+I22
PA =~ I4+I7+I9+I12+I17+I18+I19+I21

# Make the second-order
burnout =~ EE + DP + PA

# Intercepts for items
I1  ~ 1; I2  ~ 1; I3  ~ 1; I4  ~ 1; I5  ~ 1
I6  ~ 1; I7  ~ 1; I8  ~ 1; I9  ~ 1; I10 ~ 1
I11 ~ 1; I12 ~ 1; I13 ~ 1; I14 ~ 1; I15 ~ 1
I16 ~ 1; I17 ~ 1; I18 ~ 1; I19 ~ 1; I20 ~ 1
I21 ~ 1; I22 ~ 1
'
```

With this second-order CFA model, we can fit the CFA model by calling *cfa* similarly. We leave this second-order CFA model as an exercise for interested readers to practice further.

2.6 Exercises

1. Fit the second-order CFA model for *MBI* as discussed in the Section *Conclusions and Discussions*, perform *MI* to identify the best CFA model for this data.

2. Using the data from Li et al. (2020) to verify the CFA model proposed in Figure 1 in Li et al. (2020) on organizational climate, job

satisfaction, and turnover in voluntary child welfare workers. Here
are some of the data explanations and the rest can be found in
the paper:

- *i1*: "I plan to stay in child welfare practice as long as possible
  (Intent 1),"

- *i2*: "Under no circumstance will I voluntarily leave child welfare
  (Intent 2),"

- *i3*: "I plan to leave child welfare as soon as possible (Intent 3)."

- *i4*: "How often have you looked in the paper for a new job?"
  (Intent 4);

- *i5*: "How often you have looked in professional journals for a new
  job?" (Intent 5);

- *i6*: "How many phone inquiries have you made about other jobs?"
  (Intent 6);

- *i7*: "How many resumes have you sent out?" (Intent 7);

- *i8*: "How often do you search the internet for jobs?" (Intent 8);

- *i9*: "How many interviews have you had?" (Intent 9).

Data can be read into *R* using the following code chunk:

```
# Read the data into R session using 'read.csv'
dTurnover = read.csv("data/Li-Turnover.dat",
                     na.strings ="-9999", header=F)
colnames(dTurnover) = c("i1","i2","i3","i4","i5","i6","i7","i8",
                        "i9","role","job","org","sup","jspay",
                        "jspro","jsben","sex","age","race",
                        "edu","sal")
# Check the data dimension
#dim(dTurnover)
# Print the data summary
#summary(dTurnover)
```

3

Mediation Analysis

Mediation analysis is a statistical method frequently used by social and behavioral scientists to understand the complex relationships among study variables. Researchers can apply it in observational studies to sharpen their understanding of an observed relationship between predictor and outcome variables by adding mediators. Several books, such as Hayes (2018); Jose (2013); MacKinnon (2008), and numerous published articles, such as Baron and Kenny (1986), have described mediation analysis in detail.

We will illustrate mediation analysis using the famous *Industrialization and Political Democracy* data set used by Bollen (1989) on structural equation modeling with *R* package *lavaan* and the *CALIS* procedure of *SAS* in this chapter. Most of the illustrations can be found in *R* package *lavaan* by Rosseel (2012) and the *lavaan* webpage at https://www.lavaan.ugent.be/. The illustration in this chapter is then a further promotion of this *R* package for interested researchers and students.

Note to readers: We will use *R* package *lavaan* (i.e., *latent variable analysis*) in this chapter. Remember to install this *R* package to your computer using *install.packages("lavaan")* and load this package into *R* session using *library(lavaan)* before running all the *R* programs in this chapter as follows.

```
# Load the lavaan package into R session
library(lavaan)
```

3.1 Industrialization and Political Democracy Data Set

Used throughout Bollen's 1989 book (Bollen, 1989), this data set contains various measures of political democracy and industrialization in developing countries. This data set is included with **lavaan** as a built-in data set named *PoliticalDemocracy*, which is a data frame with 75 observations from 11 variables defined as follows:

- y1 = Expert ratings of the freedom of the press in 1960
- y2 = The freedom of political opposition in 1960
- y3 = The fairness of elections in 1960

DOI: 10.1201/9781003365860-3

- y4 = The effectiveness of the elected legislature in 1960

- y5 = Expert ratings of the freedom of the press in 1965

- y6 = The freedom of political opposition in 1965

- y7 = The fairness of elections in 1965

- y8 = The effectiveness of the elected legislature in 1965

- x1 = The gross national product (GNP) per capita in 1960

- x2 = The inanimate energy consumption per capita in 1960

- x3 = The percentage of the labor force in industry in 1960

We can see the dimension of this data frame along with the first six observations using the following *R* chunk:

```
# The dimension of the data
dim(PoliticalDemocracy)
```

```
## [1] 75 11
```

```
# Print the first 6 observation for illustration
round(head(PoliticalDemocracy),3)
```

```
##      y1   y2    y3   y4   y5   y6    y7    y8    x1
## 1  2.50 0.00  3.33 0.00 1.25 0.00  3.73 3.333 4.44
## 2  1.25 0.00  3.33 0.00 6.25 1.10  6.67 0.737 5.38
## 3  7.50 8.80 10.00 9.20 8.75 8.09 10.00 8.212 5.96
## 4  8.90 8.80 10.00 9.20 8.91 8.13 10.00 4.615 6.29
## 5 10.00 3.33 10.00 6.67 7.50 3.33 10.00 6.667 5.86
## 6  7.50 3.33  6.67 6.67 6.25 1.10  6.67 0.368 5.53
##      x2   x3
## 1 3.64 2.56
## 2 5.06 3.57
## 3 6.26 5.22
## 4 7.57 6.27
## 5 6.82 4.57
## 6 5.14 3.89
```

As seen from the *R* output, this data set contains 75 observations and 11 columns. These 11 columns correspond to the 11 observed items, which is graphically described in Figure 3.1.

As seen in this figure, the classical model described in Bollen (1989) for this data is illustrated with three latent variables with *ind60* measured by x1, x2 and x3, *dem60* measured by y1, y2, y3, and y4, and *dem65* measured by y5, y6, y7, and y8. The focus of the current analysis is to investigate the impact of *ind60* on *dem65* with *dem60* as a mediator between *ind60* and *dem65*.

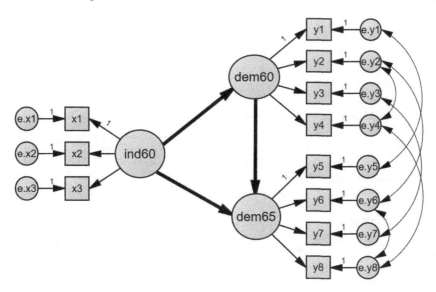

FIGURE 3.1
Relationship among Three Latent Variables: ind60, dem60, and dem65.

3.2 Mediation Model and Statistical Inference

In this section, we describe the basic mediation model corresponding to the structural equation model in Figure 3.1 and discuss the parameter estimation and the associated statistical inference.

3.2.1 Basic Mediation Model

There are many types of mediation models. The basic mediation model is to describe the mediation effect between an independent (or a predictor) variable (denoted by X) and an dependent (or outcome) variable (denoted by Y) which is mediated by a third variable, which is called a mediator (denoted by M), as shown in the top diagram of Figure 3.2.

To describe the concepts of mediation effects further, it would be useful to first go back to examine the simple relationship between the independent and dependent variables, as depicted in the bottom diagram of Figure 3.2. When we hypothesize an effect of X on Y in a statistical model, we might not assume such an effect occurs instantly. The effect might be realized through a process or mechanism. For example, parental encouragement (independent variable) might first enhance children's achievement motivation (mediator), which in turn boosts the academic performance (dependent/outcome variable) of children. Meanwhile, parental encouragement might also first increase children's

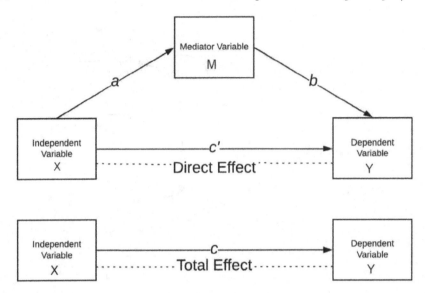

FIGURE 3.2
Mediation model.

anxiety level (another mediator), which in turn hinders the academic performance of children. We could even break down each component process further by including many other intermediate variables in the process.

Nonetheless, if we are not interested in the processes or mechanisms involved, we can just treat them collectively like a *black box*. In this case, the diagram at the bottom of Figure 3.2 represents the *total* or overall effect of X on Y, without knowing what might have happened in the black box.

When the process or mechanism is of interest, we open the black box and characterize the targeted mechanism by a mediator variable. Continuing with the previous example of parental encouragement effect on academic performance, if we hypothesize that the mediation process through enhancement of achievement motivation is a prevalent one, a model that includes such a mediator variable is depicted by the top diagram of Figure 3.2. The pathway from X to Y through M represents the *indirect* effect of X through the specified mediator M. The indirect effect is also called the *mediated* effect. The direct pathway from X to Y represents the *direct* effect. More accurately, the *direct* effect should be interpreted as the effect of X on Y that is not through the mediation process of M. This so-called *direct* effect might in fact involve other processes (excluding that of M) that are treated collectively as a black box.

As mentioned, there could be many types of mediation models. The focus of this chapter is the basic one depicted as the upper diagram in Figure 3.2. The next section presents some quantitative analysis of this basic mediation model. Later parts of this chapter fit the mediation model represented in Figure 3.1, where *dem60* is the mediator variable between *ind60* on *dem65*.

3.2.2 Direct, Indirect, and Total Effects: Statistical Model and Definitions

In the previous section, we explained the basic mediation model in Figure 3.1 conceptually by describing the direct and indirect pathways of effects. We now continue to cast the corresponding statistical mediation model into linear regression equations as follows:

$$M = a_0 + a \times X + e_M$$
$$Y = b_0 + b \times M + c' \times X + e_Y \tag{3.1}$$

In the first linear equation, M is regressed on X with intercept a_0, slope a, and error term e_M. This equation represents the $X \to M$ path with a as the effect of X on M in Figure 3.2. In the second equation, Y is regressed on M and X with intercept b_0, slopes b and c', and error term e_Y. This equation represents the two paths $M \to Y$ and $X \to Y$ with b as the effect of M on Y and c' as the direct effect of X on Y in Figure 3.2.

Substituting the first equation into the second equation yields an equation that regresses Y on X:

$$Y = (b_0 + b \times a_0) + (ba + c') \times X + (b \times e_M + e_Y) \tag{3.2}$$

Had we treated the mediation process as a black box and used the bottom diagram in Figure 3.1 to model Y, the linear regression model would have been:

$$Y = d + c \times X + e \tag{3.3}$$

where d is the intercept, c is the slope, and e the error term in this "black box" model. In Equation (3.3), c corresponds to the *total* or overall effect of X on Y in Figure 3.2.

Equating the coefficients in Equations (3.2) and (3.3), we have

$$\text{Total effect} = c = ba + c' \tag{3.4}$$

Because c' represents the direct effect, the indirect effect can be computed by:

$$\text{Indirect effect} = \text{Total effect} - \text{Direct effect} = c - c' = ba \tag{3.5}$$

Hence, Equation (3.5) shows two equivalent ways to compute the indirect effect:

- Difference method $(c-c')$: To use this method, the linear regression model in Equation (3.3) and the second linear regression model in Equation (3.1) must be fit.

- Product method (ba): To use this method, the two linear regression models in Equation (3.1) must be fit.

In structural equation modeling, specifying component paths that correspond to models in Equation (3.1) is a common practice and the fitting of the regression model (3.3) is unnecessary. Therefore, in structural equation modeling, the product method of computing direct, indirect, and total effects is most common, as summarized in the following formulas:

$$\text{Direct effect} = c'$$
$$\text{Indirect effect} = b \times a$$
$$\text{Total effect} = c' + b \times a \tag{3.6}$$

Because the *indirect* effect is the product of b and a, it can be positive (i.e., the higher values on X, the higher impact on Y) when a and b are either both positive or both negative, whereas it can be negative (i.e., the higher values on X, the lower impact on Y) when either a or b, but not both, is negative. This complicates the interpretation of the *indirect* effect and some cautions should be taken for specific application with substantive knowledge.

3.2.3   Estimation of Direct, Indirect and Total Effects

Structural equation modeling techniques can estimate all parameters (a, b, c', and so on) simultaneously in Equation (3.1) given the model for the data. Hence, we can use the following "plug-in" formulas to estimate of the direct, indirect, and total effects from the data:

$$\text{Direct effect} = \hat{c}'$$
$$\text{Indirect effect} = \hat{b} \times \hat{a}$$
$$\text{Total effect} = \hat{c}' + \hat{b} \times \hat{a} \tag{3.7}$$

where $\hat{a}$ is an estimate of a, and so on. Estimated standard errors of these effects can then be computed by standard asymptotic method (i.e., the so-called "delta-method"), which is supported by most, if not all, SEM software. Alternatively, bootstrap methods can be used to estimate standard errors.

3.2.4   General Structural Equation Model and Its Implied Structured Covariance Matrix

In Chapter 1, we discussed the covariance structures of a regression model that involves only observed variables. In Chapter 2, we examined confirmatory factor models and extended the discussion of covariance structures to situations that involves relationships between observed variables and latent factors. Both of these covariance structures are derived without specifying the functional relationships among latent factors.

The mediation model in this chapter is an example that fills this gap by positing functional relationships among latent factors. That is, in our mediation example, the latent factors *ind60*, *dem60*, and *dem65* are functionally related and they are also reflected by observed variables. By extending the covariance structures to those among latent factors, this section demonstrates the construction of a general structural equation model and its implied structured covariance matrix for the observed variables.

The LISREL (linear structural relations) model by Jöreskog (1970) and his colleagues is one of the earliest, if not the earliest, formulation of a general structural equation model. It is certainly of historical interests to describe the LISREL-type model. Using the current mediation model as an example, Bollen (1989) has provided a detailed description of the LISREL model. Here we recapitulate the essential elements of the LISREL-type model. To broaden our scope of general structural equation modeling, we will compare the LISREL formulation with another formulation, the RAM formulation by McArdle and McDonald (1984).

3.2.4.1 LISREL-Type Model

There are three components in a LISREL-type model: (1) Structural model for describing the functional relationships among latent constructs or factors; (2) Measurement model for the x-variables, which are the measurement variables for the exogenous (or indepndent) factors, denoted as ξ; (3) Measurement model for the y-variables, which are the measurement variables for the endogenous (or dependent) latent factors, denoted as η.

The structural model is represented as:

$$\eta = B\eta + \Gamma\xi + \zeta \tag{3.8}$$

Latent factors ξ and η and the error term ζ are treated as random variables in the model equation. All these variables are centered (i.e., with expected values 0) and so there is no need to include intercept terms in Equation (3.8). In addition, the error variables ζ are independent of ξ.

In our running example of mediation analysis, there are three latent factors in the model: *ind60* is an exogenous (independent) latent factor and both *dem60* and *dem65* are endogenous (dependent) latent factors. Hence, in this case, ξ is a 1×1 vector that represents the *ind60* factor, η is a 2×1 vector for the *dem60* and *dem65* factors, and ζ is a 2×1 vector for error terms.

Because η appears in both sides of of the matrix equation (3.8), the equation might look strange at the first glance. However, this formulation allows for the specification of the functional relationships among endogenous factors themselves. For example, the coefficient matrices for the mediation model are specified as:

$$B = \begin{pmatrix} 0 & 0 \\ \beta_{21} & 0 \end{pmatrix}, \Gamma = \begin{pmatrix} \gamma_{11} \\ \gamma_{21} \end{pmatrix} \tag{3.9}$$

which is just a matrix form of the following two linear equations:

$$\text{dem60} = \gamma_{11} \times \text{ind60} + \zeta_1$$
$$\text{dem65} = \beta_{21} \times \text{dem60} + \gamma_{21} \times \text{ind60} + \zeta_2 \tag{3.10}$$

The structural model as specified in Equation (3.8) involves only latent variables and therefore it cannot be analyzed empirically without measuring the observed indicators of the latent factors. So now enter the measurement models that relate observed variables x and y to the latent factors:

$$x = \Lambda_x \xi + \delta$$
$$y = \Lambda_y \eta + \epsilon \tag{3.11}$$

In Equation (3.11), x and y are vectors of observed variables that are reflective of the latent factors ξ and η, respectively, and δ and ϵ are error terms that have zero expected values and are independent of the latent factors. In addition, without loss of generality, x and y are assumed to have zero expected values so that there is no need to include intercepts in these measurement models.

In our mediation example, x is a 3×1 vector of measurement variables for ξ and y is a 8×1 vector of measurement variables for η. Correspondingly, δ is a 3×1 vector of random errors for x and ϵ is a 8×1 vector of random errors for y. According to the model representation in Figure 3.1, the loading matrices Λ_x and Λ_y are of the following forms:

$$\Lambda_x = \begin{pmatrix} \lambda_{x1} \\ \lambda_{x2} \\ \lambda_{x3} \end{pmatrix}, \Lambda_y = \begin{pmatrix} \lambda_{y1} & 0 \\ \lambda_{y2} & 0 \\ \lambda_{y3} & 0 \\ \lambda_{y4} & 0 \\ 0 & \lambda_{y5} \\ 0 & \lambda_{y6} \\ 0 & \lambda_{y7} \\ 0 & \lambda_{y8} \end{pmatrix} \tag{3.12}$$

For the identification of the scales of latent factors, λ_{x1}, λ_{y1}, and λ_{y5} are all set to the fixed value 1.

In the structural model (Equation 3.8) and the measurement models (Equation 3.11), ξ, ζ, δ, and ϵ are exogenous variables and their covariance matrices are denoted as Φ, Ψ, Θ_δ, and Θ_ϵ, respectively. None of these covariance matrices is required to be diagonal, although the error covariance matrices Ψ, Θ_δ, and Θ_ϵ are usually set to be diagonal in initial modeling when there are no prior knowledge about correlated errors. The non-constant elements of these covariance matrices are model parameters. For example, corresponding to the path diagram specified in Figure 3.1, various covariance matrices in our mediation model are specified as follows:

$$\Phi = (\phi), \Psi = \begin{pmatrix} \psi_{11} & 0 \\ 0 & \psi_{22} \end{pmatrix}, \Theta_\delta = \begin{pmatrix} \theta_{x1,x1} & 0 & 0 \\ 0 & \theta_{x2,x2} & 0 \\ 0 & 0 & \theta_{x3,x3} \end{pmatrix}$$

$$\Theta_\epsilon = \begin{pmatrix} \theta_{y1,y1} & 0 & 0 & 0 & \theta_{y1,y5} & 0 & 0 & 0 \\ 0 & \theta_{y2,y2} & 0 & \theta_{y2,y4} & 0 & \theta_{y2,y6} & 0 & 0 \\ 0 & 0 & \theta_{y3,y3} & 0 & 0 & 0 & \theta_{y3,y7} & 0 \\ 0 & \theta_{y2,y4} & 0 & \theta_{y4,y4} & 0 & 0 & 0 & \theta_{y4,y8} \\ \theta_{y1,y5} & 0 & 0 & 0 & \theta_{y5,y5} & 0 & 0 & 0 \\ 0 & \theta_{y2,y6} & 0 & 0 & 0 & \theta_{y6,y6} & 0 & \theta_{y6,y8} \\ 0 & 0 & \theta_{y3,y7} & 0 & 0 & 0 & \theta_{y7,y7} & 0 \\ 0 & 0 & 0 & \theta_{y4,y8} & 0 & \theta_{y6,y8} & 0 & \theta_{y8,y8} \end{pmatrix}$$

$$(3.13)$$

To derive the covariance structures of the observed variables under the LISREL-type model, we first combine the measurement models in Equation (3.11) into a single matrix equation

$$\begin{pmatrix} x \\ y \end{pmatrix} = \begin{pmatrix} \Lambda_x & 0 \\ 0 & \Lambda_y \end{pmatrix} \begin{pmatrix} \xi \\ \eta \end{pmatrix} + \begin{pmatrix} \delta \\ \epsilon \end{pmatrix} \tag{3.14}$$

The implied covariance structures of the observed variables x and y is then derived as:

$$\begin{aligned} \Sigma &= \begin{pmatrix} \Sigma_{xx} & \Sigma_{xy} \\ \Sigma_{yx} & \Sigma_{yy} \end{pmatrix} \\ &= \begin{pmatrix} \Lambda_x & 0 \\ 0 & \Lambda_y \end{pmatrix} \begin{pmatrix} \text{Cov}(\xi, \xi') & \text{Cov}(\xi, \eta') \\ \text{Cov}(\eta, \xi') & \text{Cov}(\eta, \eta') \end{pmatrix} \begin{pmatrix} \Lambda_x' & 0 \\ 0 & \Lambda_y' \end{pmatrix} \\ &\quad + \begin{pmatrix} \text{Cov}(\delta, \delta') & 0 \\ 0 & \text{Cov}(\epsilon, \epsilon') \end{pmatrix} \end{aligned} \tag{3.15}$$

After some laborious matrix algebra and assuming that $(I-B)$ is invertible, it can be shown that:

$$\Sigma = \begin{pmatrix} \Lambda_x \Phi \Lambda_x' + \Theta_\delta & \Lambda_x \Phi \Gamma' (I-B)^{-1'} \Lambda_y' \\ \Lambda_y (I-B)^{-1} \Gamma \Phi \Lambda_x' & \Lambda_y (I-B)^{-1} (\Gamma \Phi \Gamma' + \Psi)(I-B)^{-1'} \Lambda_y' + \Theta_\epsilon \end{pmatrix}$$

$$(3.16)$$

The right side of Equation (3.16) is the implied covariance structures that is being tested under the null hypothesis. Notice that the submatrix pertaining to the covariance structures of the x-variable (i.e., $\Lambda_x \Phi \Lambda_x' + \Theta_\delta$) is of the same form as that of a confirmatory factor model, which has been discussed in Chapter 2.

Although the expression for the implied covariance structures looks formidable, the computations of the covariance structures when doing model fitting and testing are quite straightforward. That is, given the model specification such as the graphical form shown in Figure 3.1 or any other equivalent syntactic forms, a computer program represents the eight LISREL model matrices B, Γ, Λ_x, Λ_y, Φ, Ψ, Θ_δ, and Θ_ϵ accordingly and then it carries out the model computations routinely. The computation of covariance structures using Equation (3.16) is not harder (though slightly more time-consuming) than those for simpler covariance structures in Chapter 1 or 2. The model fitting and parameter estimation of the LISREL-type models are done by minimizing a discrepancy function (such as the ML discrepancy function), which has been described in Chapters 1 and 2.

3.2.4.2 RAM-Type Model

The LISREL-type model distinguishes structural models from measurement models. A motivation of such a distinction is to address the measurement error problems so that the functional relationships studied in the structural model are based on the "purified" latent constructs that are free of measurement errors. However, there are other formulations of structural equation models that do not make such a distinction explicitly. One of these formulations is the RAM (reticular action model) that is proposed by McArdle and McDonald (1984). To broaden our perspective about covariance structure models, this section describes the RAM formulation.

The model equation of a RAM-type model is simply

$$v = Av + e \qquad (3.17)$$

where v is a $k \times 1$ vector of variables (observed as well as latent) in the model and e represents the unique source of variation in v that is not due to the path effects from any other variables in v.

Because of such a simple model in Equation (3.17), there are considerably fewer model matrices in a RAM-type model than in an *equivalent* LISREL-type model, which considers eight model matrices for a covariance structure analysis. Indeed, a RAM-type model needs to specify only two model matrices: a $k \times k$ square (nonsymmetric) matrix A for indicating the paths from the column variables to the row variables and a $k \times k$ symmetric matrix P for specifying the covariance matrix of e.

Depending on whether a variable in v is exogenous or endogenous, the corresponding diagonal element in P is interpreted differently. For an exogenous variable in the model, the corresponding diagonal element in P is the *variance* of the variable itself. For an endogenous variable in the model, the corresponding diagonal element in P is the *error variance* of the variable.

With the formulation in Equation (3.17) and the invertibility of $(I - A)$, the model equation can be rewritten as:

$$v = (I - A)^{-1}e \qquad (3.18)$$

Then the covariance structures of v is derived simply as:

$$\Sigma = (I - A)^{-1}P(I - A)^{-1'} \tag{3.19}$$

Notice that the covariance matrix Σ in Equation (3.19) includes variances and covariances of all observed and latent variables in v. In model fitting and parameter estimation, however, only the covariance structures pertaining to the observed variables are used.

The RAM-type models are clearly simpler in terms of model formulation than that of the LISREL-type models, but do these two formulations lead to different estimation results? The answer is "no" if the same estimation method is used for these formulations. With some mathematical maneuver, the equivalence of the LISREL and RAM models can be established. However, instead of giving a rigorous proof of equivalence, here we will just illustrate the idea.

First, define v as a vector of variables that collects x, y, ξ, and η variables in the LISREL-type model. Both the rows and columns of the RAM model matrix A are then indexed by this set of variables v. Because only ξ is exogenous in the model, the rows and columns of the RAM model matrix P are indexed by δ, ϵ, ξ, and ζ. That is, only ξ is not replaced by an error term for indexing the P matrix. Now, with these variable indices or labels, the RAM model matrices can be defined in the following supermatrix forms that contain the LISREL model matrices as submatrices:

$$A = \begin{pmatrix} 0 & 0 & \Lambda_x & 0 \\ 0 & 0 & 0 & \Lambda_y \\ 0 & 0 & 0 & 0 \\ 0 & 0 & \Gamma & B \end{pmatrix}, P = \begin{pmatrix} \Theta_\delta & 0 & 0 & 0 \\ 0 & \Theta_\epsilon & 0 & 0 \\ 0 & 0 & \Phi & 0 \\ 0 & 0 & 0 & \Psi \end{pmatrix} \tag{3.20}$$

Equation (3.20) essentially embeds all the parameters in the eight LISREL model matrices in the RAM model models. Hence, this implies that the RAM formulation can represent any LISREL-type model. The unstructured covariance matrix in Equation (3.19) for the RAM covariance structures is organized as follows:

$$\Sigma = \begin{pmatrix} \Sigma_{xx} & \Sigma_{xy} & \Sigma_{x\xi} & \Sigma_{x\eta} \\ \Sigma_{yx} & \Sigma_{yy} & \Sigma_{y\xi} & \Sigma_{y\eta} \\ \Sigma_{\xi x} & \Sigma_{\xi y} & \Sigma_{\xi\xi} & \Sigma_{\xi\eta} \\ \Sigma_{\eta x} & \Sigma_{\eta y} & \Sigma_{\eta\xi} & \Sigma_{\eta\eta} \end{pmatrix} \tag{3.21}$$

With some laborious matrix algebra, which is not shown here, the part of Σ in Equation (3.21) that pertains to x and y have exactly the same covariance structures (which are specified in the corresponding submatrix on the right-hand side of Equation (3.19)) as that of the LISREL covariance structures derived in Equation (3.16). Therefore, the same covariance structures of observed variables are being analyzed in either of the LISREL- and RAM-type models.

In addition to the LISREL and RAM formulations, there are other formulations of structural equation models, including those by Bentler and Weeks (1982) and McDonald (1978). While it might be debatable which formulation is theoretically the simplest, conceptually the most useful, or computationally the most convenient, the most important message here is that despite the differences in appearances in all these formulations, they would yield the same estimation results given the same estimation method applied to the data.

3.2.5 Statistical Inference on Mediation Effect

In order to determine the statistical significance of the mediation effect (i.e., *indirect* effect), a statistic based on the *indirect* effect must be compared to its distribution under null hypothesis. Since the mediation effect (i.e., the *indirect* effect) is a product of two estimates of $\hat{a}$ and $\hat{b}$, there are typically two ways to calculate the standard error for statistical testing with the classical Sobel test in Sobel (1982) and another with bootstrap resampling approach.

3.2.5.1 Sobel Test

The Sobel test is based on the so-called *delta* method (Sobel, 1982). The Sobel test is formulated as the ratio of the estimated *indirect* effect to its estimated standard error of measurement to derive a z-statistic as

$$z = \frac{\hat{a} \times \hat{b}}{\hat{SE}\left(\hat{a} \times \hat{b}\right)} \qquad (3.22)$$

where the $\hat{SE}\left(\hat{a} \times \hat{b}\right)$ is the estimated standard error of $\hat{a} \times \hat{b}$, which is calculated as $\hat{SE}\left(\hat{a} \times \hat{b}\right) = \sqrt{\hat{a}^2\hat{\sigma}_b^2 + \hat{b}^2\hat{\sigma}_a^2}$

This z-statistic is compared with the normal distribution to determine its significance, assuming that $\hat{a} \times \hat{b}$ is normally distributed. However, the normality assumption is true only asymptotically (i.e., at large sample sizes), which means that at small sample sizes the standard error estimate might not be unbiased and the *p*-value might not be accurate.

3.2.5.2 Bootstrap Resampling Approach

A better method for statistical significance is to use the bootstrap resampling approach. Bootstrap resampling is a general nonparametric approach that does not make any assumption about the distribution of the data. It also does not assume a particular form of sampling distribution of the statistic of interest. In fact, it establishes the sampling distribution through resamplng.

In bootstrap resampling, the sample data are treated as a representation of the population from which the observations are originally drawn. Now treating this sample data as a pseudo-population for bootstrap resampling, the observations in the pseudo-population are *resampled* randomly with replacement to form a bootstrap sample, which is of the same size as the original data. Repeat the bootstrap resampling many times, say 1000 times, to obtain 1000 bootstrap samples. You will use these 1000 samples to make statistical inferences. For example, in each bootstrap sample, you can fit the same structural equation model and obtain an indirect effect estimate. These 1000 bootstrapped estimates form an empirical sampling distribution. The bootstrap percentile method constructs confidence limits of the indirect effect from this empirical distribution. If the confidence limits include the zero point, you fail to reject the null hypothesis of no indirect effect. Otherwise, you conclude that the indirect effect estimate is statistically significant. You can also compute the standard error of the estimated indirect effect as the standard deviation of the 1000 bootstrap estimates. The standard error obtained this way does not assume any distribution form of the population and can be used to compute the corresponding test statistic and confidence interval.

In fact, the bootstrap resmapling approach is not limited to the testing of indirect effects, it can also be applied to other estimates in the model, the *total* effect estimates, and any other parametric functions as well. See Section 3.3.2 for an application of the bootstrap.

3.2.5.3 Full Mediation versus Partial Mediation

Recall that in the basic mediation model, the impact of variable X on an outcome variable Y through a *mediator* variable M is known as the *indirect* effect, while the direct impact of X on Y is known as the *direct* effect.

The mediator variable M can account for either all or some of the relationship between X and Y. *Full mediation* occurs when the direct effect in the mediation model vanishes (i.e., $c' = 0$). However, in *partial mediation*, the mediator M accounts for some, but not all, of the relationship between the X and Y. In this partial mediation situation, there is not only a non-null relationship through the mediator M to the outcome variable Y (i.e., $a \times b \neq 0$), but also some direct relationship between X and Y (i.e., $c' \neq 0$).

To summarize, to validate *full mediation* between X to Y through M from the sample data, we need to verify the following two criteria:

- Variable X does not have a significant *direct* effect on outcome variable Y—that is, being able to *retain* the null hypothesis $H_0 : c' = 0$

- Variable X does have a significant impact on moderator M, which also has a significant impact on response variable Y—that is, being able to reject null hypotheses $H_0 : a = 0$ and $H_0 : b = 0$. In this case, the *indirect* effect estimate would be statistically significantly different from zero.

In contrast, to validate the *partial mediation* between X to Y through M, we need to verify the following two criteria:

- Variable X has a significant *direct* effect on outcome variable Y—that is, being able to reject the null hypothesis $H_0 : c' = 0$

- Variable X has a significant impact on moderator M, which has a significant impact on response variable Y—that is, being able to reject null hypotheses $H_0 : a = 0$ and $H_0 : b = 0$.

3.3 Data Analysis Using R

In this section, we first describe the general structural equation modeling corresponding to the model described in Figure 3.1 and then go to mediation analysis to estimate and test for direct and indirect effects.

3.3.1 General Structural Equation Modeling

To fit the model described in Figure 3.1 using R *lavaan*, we can specify the three components needed in *lavaan*:

- Component 1 is the *measurement model* to describe the three latent variables of *ind60*, *dem60* and *dem65* with the 11 observed variables of x1 to x3, and y1 to y8, which is implemented in R *lavaan* convention using $=\tilde{}$.

- Component 2 is the *regression model* to describe the regression relationship among the three latent variables, which is implemented in R *lavaan* convention using $\tilde{}$.

- Component 3 is the *residual covariance* to describe the residual covariances among the 11 observed variables defined in Figure 3.1, which is implemented in R *lavaan* convention using $\tilde{}\tilde{}$.

The R implementation is as follows:

```
modelPD <- '
  # Component 1: Measurement model
    ind60 =~ x1 + x2 + x3
    dem60 =~ y1 + y2 + y3 + y4
    dem65 =~ y5 + y6 + y7 + y8
  # Component 2: Regression model
    dem60 ~ ind60
    dem65 ~ ind60 + dem60
  # Component 3: Residual covariances
    y1 ~~ y5
    y2 ~~ y4 + y6
    y3 ~~ y7
    y4 ~~ y8
    y6 ~~ y8'
```

With this specified *modelPD*, we can call *sem* to fit this model to the data frame *PoliticalDemocracy*. After the model is fitted, we can call *summary* to print the fitted measures (i.e., *fit.measures=TRUE*) as follows:

```
# Model fitting
fit.PD <- sem(modelPD, data = PoliticalDemocracy)
# Print the model fitting
summary(fit.PD, fit.measures=TRUE)
```

```
## lavaan 0.6-12 ended normally after 68 iterations
##
##   Estimator                                         ML
##   Optimization method                           NLMINB
##   Number of model parameters                        31
##
##   Number of observations                            75
##
## Model Test User Model:
##
##   Test statistic                                38.125
##   Degrees of freedom                                35
##   P-value (Chi-square)                           0.329
##
## Model Test Baseline Model:
##
##   Test statistic                               730.654
##   Degrees of freedom                                55
##   P-value                                        0.000
##
## User Model versus Baseline Model:
##
##   Comparative Fit Index (CFI)                    0.995
##   Tucker-Lewis Index (TLI)                       0.993
##
## Loglikelihood and Information Criteria:
##
##   Loglikelihood user model (H0)              -1547.791
##   Loglikelihood unrestricted model (H1)      -1528.728
##
##   Akaike (AIC)                                3157.582
##   Bayesian (BIC)                              3229.424
##   Sample-size adjusted Bayesian (BIC)         3131.720
##
```

```
## Root Mean Square Error of Approximation:
##
##    RMSEA                                                    0.035
##    90 Percent confidence interval - lower                   0.000
##    90 Percent confidence interval - upper                   0.092
##    P-value RMSEA <= 0.05                                     0.611
##
## Standardized Root Mean Square Residual:
##
##    SRMR                                                     0.044
##
## Parameter Estimates:
##
##    Standard errors                                      Standard
##    Information                                          Expected
##    Information saturated (h1) model                   Structured
##
## Latent Variables:
##                    Estimate  Std.Err  z-value  P(>|z|)
##    ind60 =~
##      x1              1.000
##      x2              2.180    0.139   15.742    0.000
##      x3              1.819    0.152   11.967    0.000
##    dem60 =~
##      y1              1.000
##      y2              1.257    0.182    6.889    0.000
##      y3              1.058    0.151    6.987    0.000
##      y4              1.265    0.145    8.722    0.000
##    dem65 =~
##      y5              1.000
##      y6              1.186    0.169    7.024    0.000
##      y7              1.280    0.160    8.002    0.000
##      y8              1.266    0.158    8.007    0.000
##
## Regressions:
##                    Estimate  Std.Err  z-value  P(>|z|)
##    dem60 ~
##      ind60           1.483    0.399    3.715    0.000
##    dem65 ~
##      ind60           0.572    0.221    2.586    0.010
##      dem60           0.837    0.098    8.514    0.000
##
## Covariances:
##                    Estimate  Std.Err  z-value  P(>|z|)
##    .y1 ~~
```

```
##     .y5                0.624    0.358    1.741    0.082
##   .y2 ~~
##     .y4                1.313    0.702    1.871    0.061
##     .y6                2.153    0.734    2.934    0.003
##   .y3 ~~
##     .y7                0.795    0.608    1.308    0.191
##   .y4 ~~
##     .y8                0.348    0.442    0.787    0.431
##   .y6 ~~
##     .y8                1.356    0.568    2.386    0.017
##
## Variances:
##                     Estimate  Std.Err  z-value  P(>|z|)
##     .x1                0.082    0.019    4.184    0.000
##     .x2                0.120    0.070    1.718    0.086
##     .x3                0.467    0.090    5.177    0.000
##     .y1                1.891    0.444    4.256    0.000
##     .y2                7.373    1.374    5.366    0.000
##     .y3                5.067    0.952    5.324    0.000
##     .y4                3.148    0.739    4.261    0.000
##     .y5                2.351    0.480    4.895    0.000
##     .y6                4.954    0.914    5.419    0.000
##     .y7                3.431    0.713    4.814    0.000
##     .y8                3.254    0.695    4.685    0.000
##      ind60             0.448    0.087    5.173    0.000
##     .dem60             3.956    0.921    4.295    0.000
##     .dem65             0.172    0.215    0.803    0.422
```

As seen from the output, we have achieved a very satisfactory model fitting as indicated by:

- Chi-square test with p-value of 0.329 from the test-statistic of 38.125 with the degrees of freedom of 35

- Comparative Fit Index (CFI) of 0.995

- Tucker-Lewis Index (TLI) of 0.993

- Root Mean Square Error of Approximation (RMSEA) of 0.035 with 90% confidence interval of (0.000, 0.092) and the p-value of 0.611 for RMSEA ≤ 0.05

- Standardized Root Mean Square Residual (SRMR) of 0.044

With the satisfactory model fitting, we can examine the results from the three components. As seen from the output, all the observed variables are significantly loaded to their corresponding latent variables in the *measurement model*. Therefore in this *measurement model*, the *ind60* (i.e., industrialization at 1960) is significantly measured by the three indicators of x1, (i.e., the gross national product per capita in 1960), x2 (i.e., the inanimate energy consumption per capita in 1960), and x3 (i.e., the percentage of the labor force in industry in 1960). Similarly, the *dem60* (i.e., democracy at 1960) is statistically measured by the four indicators of y1 (i.e., expert ratings of the freedom of the press in 1960), y2 (i.e., the freedom of political opposition in 1960), y3 (i.e., the fairness of elections in 1960), y4 (i.e., the effectiveness of the elected legislature in 1960). Furthermore, the *dem65* (i.e., democracy at 1965) is statistically measured by the four indicators of y5 (i.e., expert ratings of the freedom of the press in 1965), y6 (i.e., the freedom of political opposition in 1965), y7 (i.e., the fairness of elections in 1965), y8 (i.e., the effectiveness of the elected legislature in 1965).

In the second component of *regression model*, all three regression paths are also statistically significant, i.e., *dem60* ~ *ind60* (est = 1.483 with p-value = 0.000), *dem65* ~ *ind60* (est = 0.572 with p-value = 0.010), and *dem65* ~ *dem60* (est = 0.837 with p-value = 0.000). These regression parameters are typically used to measure the *direct effects* in mediation analysis. This indicates that both the *dem65* and *dem60* are statistically significant related to *ind60*. In addition, *dem65* is significantly related to *dem60*. These three *direct effects* are statistically significant.

However, in the third component of *residual covariance* model, only two out of the six covariances are statistically significant under the significance level of 0.05 and they are the covariances between y2 ~~ y6 and y6 ~~ y8. This indicates that four of the six covariances are not statistically significant and we could probably remove the nonsignificant residual covariances to form a simplified model.

3.3.2   Mediation Analysis

Further interest in mediation analysis is to estimate the *indirect* and the *total* effects beside the *direct* effects. As explained in Section 3.2 in this chapter, the *indirect effect* from *ind60* to *dem65* is calculated by the product of two regression coefficients between *dem60* ~ *ind60* and *dem65* ~ *dem60*, which is 1.241 (i.e., 1.483 times 0.837). Similarly, the *total effect* from *ind60* to *dem65* is calculated by the sum of *indirect effect* and *direct effect* between *ind60* and *dem65*, which is 1.813 (i.e., 0.572 + 1.483 times 0.837).

To make the calculation systematical, we can make use of *lavaan* to name the each parameter (i.e., parameter *labels*, an important feature in *lavaan* as described below) for further calculations.

```
# Mediation model
MedMod <- '
  # Component 1: Measurement model
```

```
    ind60 =~ x1 + x2 + x3
    dem60 =~ y1 + y2 + y3 + y4
    dem65 =~ y5 + y6 + y7 + y8
# Component 2: Regression model with parameter labels
    # 2.1: path between ind60 and dem65
    dem65 ~ cprm*ind60
    # 2.2: paths through mediator dem60
    dem60 ~ a*ind60
    dem65 ~ b*dem60
    # 2.3: indirect effect (a*b)
    indirect := a*b
    # 2.4: total effect
    total := cprm + (a*b)
# Component 3: Residual covariances
    y1 ~~ y5
    y2 ~~ y4 + y6
    y3 ~~ y7
    y4 ~~ y8
    y6 ~~ y8'
```

Note in the R code chunk, we *label*ed the path between *ind60* and *dem65* as *cprm* (i.e., dem65 $\sim c * $ind60) and the other two paths through mediator *dem60* as a and b (i.e., dem60 $\sim a * $ind60 and dem65 $\sim b * $dem60) corresponding to Figure 3.1. We further used a new operator from *lavaan* := to *denote* and define the *indirect* and the *total* effects.

The model fitting will give the same fitting measures, and therefore, we use *fit.measures=FALSE* to suppress the output. The R code to fit the mediation model is then as follows:

```
# Model fitting
fit.MedMod <- sem(MedMod, data = PoliticalDemocracy)
# Print the model fitting
summary(fit.MedMod, fit.measures=FALSE)
```

```
## lavaan 0.6-12 ended normally after 68 iterations
##
##    Estimator                                         ML
##    Optimization method                           NLMINB
##    Number of model parameters                        31
##
##    Number of observations                            75
##
## Model Test User Model:
##
```

```
##    Test statistic                               38.125
##    Degrees of freedom                               35
##    P-value (Chi-square)                          0.329
##
## Parameter Estimates:
##
##    Standard errors                             Standard
##    Information                                 Expected
##    Information saturated (h1) model          Structured
##
## Latent Variables:
##                     Estimate  Std.Err  z-value  P(>|z|)
##    ind60 =~
##      x1                1.000
##      x2                2.180    0.139   15.742    0.000
##      x3                1.819    0.152   11.967    0.000
##    dem60 =~
##      y1                1.000
##      y2                1.257    0.182    6.889    0.000
##      y3                1.058    0.151    6.987    0.000
##      y4                1.265    0.145    8.722    0.000
##    dem65 =~
##      y5                1.000
##      y6                1.186    0.169    7.024    0.000
##      y7                1.280    0.160    8.002    0.000
##      y8                1.266    0.158    8.007    0.000
##
## Regressions:
##                     Estimate  Std.Err  z-value  P(>|z|)
##    dem65 ~
##      ind60  (cprm)    0.572    0.221    2.586    0.010
##    dem60 ~
##      ind60    (a)     1.483    0.399    3.715    0.000
##    dem65 ~
##      dem60    (b)     0.837    0.098    8.514    0.000
##
## Covariances:
##                     Estimate  Std.Err  z-value  P(>|z|)
##    .y1 ~~
##      .y5              0.624    0.358    1.741    0.082
##    .y2 ~~
##      .y4              1.313    0.702    1.871    0.061
##      .y6              2.153    0.734    2.934    0.003
##    .y3 ~~
##      .y7              0.795    0.608    1.308    0.191
```

```
##    .y4 ~~
##       .y8              0.348     0.442     0.787     0.431
##    .y6 ~~
##       .y8              1.356     0.568     2.386     0.017
##
## Variances:
##                     Estimate   Std.Err   z-value   P(>|z|)
##    .x1                 0.082     0.019     4.184     0.000
##    .x2                 0.120     0.070     1.718     0.086
##    .x3                 0.467     0.090     5.177     0.000
##    .y1                 1.891     0.444     4.256     0.000
##    .y2                 7.373     1.374     5.366     0.000
##    .y3                 5.067     0.952     5.324     0.000
##    .y4                 3.148     0.739     4.261     0.000
##    .y5                 2.351     0.480     4.895     0.000
##    .y6                 4.954     0.914     5.419     0.000
##    .y7                 3.431     0.713     4.814     0.000
##    .y8                 3.254     0.695     4.685     0.000
##     ind60              0.448     0.087     5.173     0.000
##    .dem60              3.956     0.921     4.295     0.000
##    .dem65              0.172     0.215     0.803     0.422
##
## Defined Parameters:
##                     Estimate   Std.Err   z-value   P(>|z|)
##     indirect           1.242     0.355     3.494     0.000
##     total              1.814     0.374     4.856     0.000
```

As seen from the output, the parameter estimation results are the same except for an extra part of *Defined Parameters*, where the estimated *indirect* effect is 1.242 (SE = 0.355, z-value = 3.494, p-value <0.000), which is statistically significant. The *direct* effect is represented in *dem65* ∼ *ind60* with parameter c' (cprm), which is estimated at 0.572 with SE = 0.221, z-value = 2.586 and p-value = 0.010, which is also statistically significant. Furthermore, the *total* effect is 1.814 (SE = 0.374, z-value = 4.856, p-value < 0.000). This is the case of *partial mediation* where the *total* effect size of 1.814 is composed of significant *direct* effect estimate of 0.572 and signficant mediation effect estimate of 1.242.

Note that by default, the standard errors for these defined parameters (i.e., *indirect* and *total* effects) are calculated by the *delta* method (Sobel, 1982). These standard errors could be biased and inefficient. A better calculation can be obtained using the *bootstrap* standard errors by specifying *se = "bootstrap"* along with the number of bootstraps (such as, *bootstrap = 1000* for 1,000 bootstraps) in the *sem* fitting function. This *bootstrap* inference can be implemented in *R sem* in the following *R* code chunk:

```
# Model fitting
fit.MedMod <- sem(MedMod, data = PoliticalDemocracy,
                  se='bootstrap',bootstrap=1000)

## Warning in lav_model_nvcov_bootstrap(lavmodel =
## lavmodel, lavsamplestats = lavsamplestats, : lavaan
## WARNING: 321 bootstrap runs resulted in nonadmissible
## solutions.

# Print the model fitting
summary(fit.MedMod, fit.measures=FALSE)

## lavaan 0.6-12 ended normally after 68 iterations
##
##   Estimator                                         ML
##   Optimization method                           NLMINB
##   Number of model parameters                        31
##
##   Number of observations                            75
##
## Model Test User Model:
##
##   Test statistic                                38.125
##   Degrees of freedom                                35
##   P-value (Chi-square)                           0.329
##
## Parameter Estimates:
##
##   Standard errors                            Bootstrap
##   Number of requested bootstrap draws             1000
##   Number of successful bootstrap draws            1000
##
## Latent Variables:
##                    Estimate  Std.Err  z-value  P(>|z|)
##   ind60 =~
##     x1                1.000
##     x2                2.180    0.145   14.988    0.000
##     x3                1.819    0.137   13.246    0.000
##   dem60 =~
##     y1                1.000
##     y2                1.257    0.156    8.043    0.000
##     y3                1.058    0.140    7.535    0.000
##     y4                1.265    0.157    8.069    0.000
```

```
##    dem65 =~
##     y5                      1.000
##     y6                      1.186    0.193    6.149    0.000
##     y7                      1.280    0.186    6.875    0.000
##     y8                      1.266    0.205    6.179    0.000
##
## Regressions:
##                           Estimate  Std.Err  z-value  P(>|z|)
##    dem65 ~
##      ind60     (cprm)       0.572    0.236    2.422    0.015
##    dem60 ~
##      ind60     (a)          1.483    0.355    4.174    0.000
##    dem65 ~
##      dem60     (b)          0.837    0.089    9.381    0.000
##
## Covariances:
##                           Estimate  Std.Err  z-value  P(>|z|)
##   .y1 ~~
##     .y5                     0.624    0.461    1.353    0.176
##   .y2 ~~
##     .y4                     1.313    0.754    1.741    0.082
##     .y6                     2.153    0.865    2.489    0.013
##   .y3 ~~
##     .y7                     0.795    0.630    1.262    0.207
##   .y4 ~~
##     .y8                     0.348    0.446    0.780    0.435
##   .y6 ~~
##     .y8                     1.356    0.734    1.847    0.065
##
## Variances:
##                           Estimate  Std.Err  z-value  P(>|z|)
##   .x1                       0.082    0.019    4.405    0.000
##   .x2                       0.120    0.074    1.613    0.107
##   .x3                       0.467    0.081    5.733    0.000
##   .y1                       1.891    0.487    3.881    0.000
##   .y2                       7.373    1.280    5.759    0.000
##   .y3                       5.067    1.016    4.987    0.000
##   .y4                       3.148    0.801    3.931    0.000
##   .y5                       2.351    0.605    3.884    0.000
##   .y6                       4.954    0.951    5.209    0.000
##   .y7                       3.431    0.647    5.302    0.000
##   .y8                       3.254    0.901    3.612    0.000
##    ind60                    0.448    0.070    6.386    0.000
##   .dem60                    3.956    0.917    4.313    0.000
##   .dem65                    0.172    0.246    0.702    0.482
```

```
##
## Defined Parameters:
##                      Estimate  Std.Err  z-value  P(>|z|)
##      indirect          1.242    0.357    3.482    0.000
##      total             1.814    0.410    4.423    0.000
```

As can be seen, there are some slight differences in estimating the standard
errors between the *delta* method and the *bootstrap* methods. In both methods,
we can conclude that both the *indirect* and *total* effects are statistically significant, indicating that the effect from *ind60* to *dem65* is partially mediated
by the mediator *dem60*.

3.4   Data Analysis Using SAS

To use *SAS* to analyze the same mediation model, a *SAS* data set *Polictical-
Democracy* is read in by the following code:

```
data PoliticalDemocracy;
  set "c:\PoliticalDemocracy";
  y1 = PRESS60; y2 = FREOP60; y3 = FREEL60; y4 = LEGIS60;
  y5 = PRESS65; y6 = FREOP65; y7 = FREEL65; y8 = LEGIS65;
  x1 = GNPPC60; x2 = ENPC60;  x3 = INDLF60;
run;
```

To use the same variable names as that of the previous *lavaan* analysis,
variables y1–y8 and x1–x3 are created in the DATA step. The following CALIS
code specifies the model and the analysis:

```
proc calis data=PoliticalDemocracy;
  path
      ind60 ==> x1-x3 = 1,
      dem60 ==> y1-y4 = 1,
      dem65 ==> y5-y8 = 1,
      ind60 ==> dem60 dem65,
      dem60 ==> dem65;
  pcov y1 y5, y2 y6, y3 y7, y4 y8,
       y2 y4, y6 y8;
  fitindex on(only)=[chisq df probchi RMSEA PROBCLFIT
          SRMR CFI BENTLERNNFI] noindextype;
  effpart ind60 ==> dem65;
run;
```

In the PATH statement, the first three specifications are for the measurement models—that is, latent constructs *ind60*, *dem60*, and *dem65* are reflected by the respective measurement variables. The fixed values after the equal signs are for the first path coefficient (or loading) in each of the specifications. That is, the path effects for x1, y1, and y5 from their respective latent factors are all fixed to 1. The next two path specifications in the PATH statement defines the structural model for the latent constructs *ind60*, *dem60*, and *dem65*. In the PCOV statement, six error covariances among the observed variables are specified. In the FITINDEX statement, you instruct the output to contain only those specified fit indices. Finally, you specify the EFFPART statement to study the direct-indirect effect decomposition of *ind60* on *dem65*.

The SAS *PROC CALIS* output is as follows:

Fit Summary	
Chi-Square	37.6169
Chi-Square DF	35
Pr > Chi-Square	0.3503
Standardized RMR (SRMR)	0.0444
RMSEA Estimate	0.0318
Probability of Close Fit	0.6287
Bentler Comparative Fit Index	0.9961
Bentler-Bonett Non-normed Index	0.9938

Except for some minor numerical differences, the fit summary table from *PROC CALIS* displays more or less the same set of fit statistics as those that were previously shown in the *lavaan* results. The parameter estimates from *PROC CALIS* are also very similar to that of *lavaan* and therefore are not shown here.

The estimated total, direct, and indirect effects, and their standard error estimates match closely to that of *lavaan* as shown below. The p-values indicate that all these effects are all statistically significant.

Effects of ind60			
Effect / Std Error / t Value / p Value			
	Total	Direct	Indirect
dem65	1.8141	0.5723	1.2418
	0.3761	0.2228	0.3578
	4.8234	2.5687	3.4705
	<.0001	0.0102	0.000520

Instead of using the EFFPART statement to specify the effect decomposition of interest, *PROC CALIS* also supports parameter definitions for *direct* and *indirect* effects and their estimation by specifying the PARAMETER statement. This approach is similar to that of the previous *lavaan* specifications. For example, you can use the following code to define *total*, *direct*, and *indirect* effects:

```
proc calis data=PoliticalDemocracy;
   path
      ind60 ==> x1-x3 = 1,
      dem60 ==> y1-y4 = 1,
      dem65 ==> y5-y8 = 1,
      ind60 ==> dem60 dem65 = a cprm,
      dem60 ==> dem65 = b;
   pcov y1 y5, y2 y6, y3 y7, y4 y8,
        y2 y4, y6 y8;
   parameter total direct indirect;
      indirect = a*b;
      direct   = cprm;
      total    = direct + indirect;
run;
```

In the PATH statement, parameters *a*, *b*, and *cprm* are specified at the intended locations for the path coefficients. Then, in the PARAMETER statement and the three statements that follow, *indirect*, *direct*, and *total* effects are defined as functions of the model parameters *a*, *b*, and *cprm*. The following output displays the estimation of these effects:

Additional Parameters					
Type	Parameter	Estimate	Standard Error	t Value	Pr > |t|
Dependent	total	1.81410	0.37610	4.8234	<.0001
	direct	0.57231	0.22280	2.5687	0.0102
	indirect	1.24179	0.35782	3.4705	0.0005

These results are exactly the same as that by using the EFFPART statement specification.

3.5 Discussions

Mediation analysis is widely used in social and behavioral sciences to understand the mediation effect between independent and outcome variables by adding mediators. In this chapter, we illustrated the mediation analysis with *R* and *SAS* using the famous *Industrialization and Political Democracy* data set used by Bollen (1989). In this data set, only one mediator (i.e., *dem60*) was included in the analysis of this data, which is the most basic mediation analysis with one target effect variable (X) and one outcome variable (Y)in the mediation analysis.

There are situations that a mediation model can include multiple Xs and multiple Ys. In some other applications, there might also be more than one mediators in the process. Most commonly, these mediators are all specified as either parallel or serial pathways that mediates the effect of X on Y. But there could be more complicated cases where these mediators are specified as combined parallel and serial pathways between X on Y. For more discussion on this, see MacKinnon (2008), Jose (2013), and Hayes (2018).

The R and SAS implementation to these multiple mediators, multiple independent and dependent variables can be extended. We leave this to interested readers for their explorations.

3.6 Exercises

Figure 2 in Li et al. (2020) presented a general SEM model on organizational climate, job satisfaction, and turnover in voluntary child welfare workers. The data from this paper were used in Chapter 2 as an exercise to validate the CFA model. The following exercises are to review and verify the results presented in this paper for the general SEM:

1. Fit a mediation model to test whether *Job Satisfaction* is a significant mediator between *Organization Climate* and *Intent to Stay*.

2. Fit a mediation model to test whether *Job Satisfaction* is a significant mediator between *Organization Climate* and *Intent to Leave*.

Hint: Data can be read into R using the following code chunk:

```
# Read the data into R session using 'read.csv'
dTurnover = read.csv("data/Li-Turnover.dat",
                     na.strings ="-9999", header=F)
colnames(dTurnover) = c("i1","i2","i3","i4","i5","i6","i7","i8",
                        "i9","role","job","org","sup","jspay",
                        "jspro","jsben","sex","age","race",
                        "edu","sal")
# Check the data dimension: not run
#dim(dTurnover)
# Print the data summary: not run
#summary(dTurnover)
```

4

Structural Equation Modeling with Non-Normal Data

A critical distributional assumption of the maximum likelihood estimation in structural equation modeling is that the data are drawn from a joint multivariate normal distribution.

So far for all the illustrations in structural equation modeling, such a multivariate normal distribution has been implicitly assumed. However, this normality assumption might not be practically possible. In this chapter, we explore the structural equation modeling for non-normal data using the famous *Industrialization and Political Democracy* data by Bollen (1989) used in Chapter 3.

Note to readers: We will use *R* package *lavaan* (i.e., *latent variable analysis*) in this chapter. Remember to install this *R* package to your computer using *install.packages("lavaan")* and load this package into *R* session using *library(lavaan)* before running all the *R* programs in this chapter as follows.

```
# Load the lavaan package into R session
library(lavaan)
```

For general inquiries about SAS analytic software products, readers can go to the webpage: http://support.sas.com/ and for the documentation of the CALIS procedure: https://support.sas.com/rnd/app/stat/procedures/calis.html.

4.1 Industrialization and Political Democracy Data

As illustrated in Chapter 3, the *PoliticalDemocracy* data contains 75 observations and 11 columns x1 to x3 and y1 to y8. These 11 columns correspond to the 11 observed items, which is graphically described in Figure 4.1.

The classical structural equation model described in Bollen (1989) is constructed with three latent variables with *ind60* measured by x1, x2 and x3, *dem60* measured by y1, y2, y3, and y4, and *dem65* measured by y5, y6, y7, and y8. The focus of the analysis is to investigate the impact of *ind60* on *dem65* with *dem60* as a mediator between *ind60* and *dem65*. In Chapter 3, we

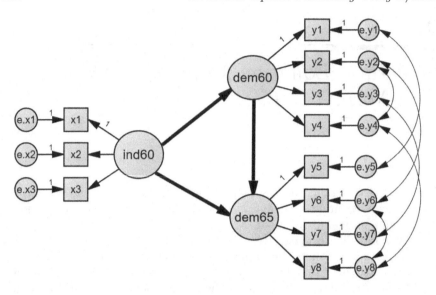

FIGURE 4.1
Relationship among Three Latent Variables ind60, dem60 and dem65.

fit this model to the data assuming that the 11 observed variables are jointly distributed as multivariate normal distribution with the observed covariance matrix S calculated as follows:

```
round(cov(PoliticalDemocracy),3)
```

```
##          y1      y2      y3     y4   y5      y6      y7     y8
## y1 6.879   6.251   5.839   6.09 5.06   5.746   5.812   5.67
## y2 6.251  15.580   5.839   9.51 5.60   9.386   7.535   7.76
## y3 5.839   5.839  10.764   6.69 4.94   4.727   7.006   5.64
## y4 6.089   9.509   6.688  11.22 5.70   7.442   7.488   8.01
## y5 5.064   5.603   4.939   5.70 6.83   4.977   5.821   5.34
## y6 5.746   9.386   4.727   7.44 4.98  11.375   6.748   8.25
## y7 5.812   7.535   7.006   7.49 5.82   6.748  10.799   7.59
## y8 5.671   7.758   5.639   8.01 5.34   8.247   7.592  10.53
## x1 0.734   0.619   0.787   1.15 1.08   0.853   0.937   1.10
## x2 1.273   1.491   1.552   2.24 2.06   1.805   1.996   2.23
## x3 0.911   1.170   1.039   1.84 1.58   1.572   1.626   1.69
##          x1    x2     x3
## y1 0.734 1.27 0.911
## y2 0.619 1.49 1.170
## y3 0.787 1.55 1.039
## y4 1.150 2.24 1.838
## y5 1.082 2.06 1.583
```

```
## y6 0.853 1.80 1.572
## y7 0.937 2.00 1.626
## y8 1.103 2.23 1.692
## x1 0.537 0.99 0.823
## x2 0.990 2.28 1.806
## x3 0.823 1.81 1.976
```

where the observed correlation matrix is

```
round(cor(PoliticalDemocracy),3)
```

```
##        y1    y2    y3    y4    y5    y6    y7    y8
## y1 1.000 0.604 0.679 0.693 0.739 0.650 0.674 0.666
## y2 0.604 1.000 0.451 0.719 0.543 0.705 0.581 0.606
## y3 0.679 0.451 1.000 0.609 0.576 0.427 0.650 0.530
## y4 0.693 0.719 0.609 1.000 0.652 0.659 0.680 0.737
## y5 0.739 0.543 0.576 0.652 1.000 0.565 0.678 0.630
## y6 0.650 0.705 0.427 0.659 0.565 1.000 0.609 0.753
## y7 0.674 0.581 0.650 0.680 0.678 0.609 1.000 0.712
## y8 0.666 0.606 0.530 0.737 0.630 0.753 0.712 1.000
## x1 0.382 0.214 0.327 0.469 0.565 0.345 0.389 0.464
## x2 0.321 0.250 0.313 0.443 0.523 0.354 0.402 0.456
## x3 0.247 0.211 0.225 0.390 0.431 0.332 0.352 0.371
##        x1    x2    x3
## y1 0.382 0.321 0.247
## y2 0.214 0.250 0.211
## y3 0.327 0.313 0.225
## y4 0.469 0.443 0.390
## y5 0.565 0.523 0.431
## y6 0.345 0.354 0.332
## y7 0.389 0.402 0.352
## y8 0.464 0.456 0.371
## x1 1.000 0.894 0.799
## x2 0.894 1.000 0.851
## x3 0.799 0.851 1.000
```

and the variances in the diagonals as follows:

```
round(apply(PoliticalDemocracy,2,var),3)
```

```
##      y1     y2     y3     y4     y5     y6     y7
##   6.879 15.580 10.764 11.219  6.826 11.375 10.799
##      y8     x1     x2     x3
##  10.534  0.537  2.282  1.976
```

The first question would be whether this joint multivariate normal distribution assumption is tenable. If not tenable, the next question would be what the options are to correct the model fitting and estimation.

Let us first examine the assumption of normal distribution of the observed variables. For this purpose, we can either graphically plot the data distribution or statistically test whether the null hypothesis of normal distribution can be rejected.

For graphical examination, we make use of the *R* package *tidyverse* to reshape the data into a new dataframe *dPD* as follows:

```
# Call tidyverse to make the data
library(tidyverse)
# Stack the data with the 11 variables
dPD <- PoliticalDemocracy %>%
  gather("x1","x2","x3","y1","y2","y3","y4","y5","y6","y7","y8",
         key="Vars", value="Value")
# Dimension of the new data *dPD*
dim(dPD)
```

```
## [1] 825    2
```

This *reshaped* the original data *PoliticalDemocracy* with 75 observations and 11 variables into a new dataframe *dPD* with 825 observation and 2 variables. With this new dataframe, we can call *ggplot* to plot the distributions of these 11 variables using *geom_histogram* as follows:

```
# Call ggplot
library(ggplot2)
# Make the histogram plot
ggplot(dPD, aes(Value))+
  geom_histogram()+
  facet_wrap(~factor(Vars), nrow=4, scales="free")+
  labs(x=" ", y="Histogram Counts")
```

```
## `stat_bin()` using `bins = 30`. Pick better value
## with `binwidth`.
```

As illustrated by the empirical distributions that are shown in Figure 4.2, the assumption that the observed variables are normally distributed is hardly defensible, especially for the eight observed variables y1 to y8. For example, the distribution of y2 peaks at the end points and y3 has a heavy point mass at the upper end of the distribution—both are indications of nonnormal distributions.

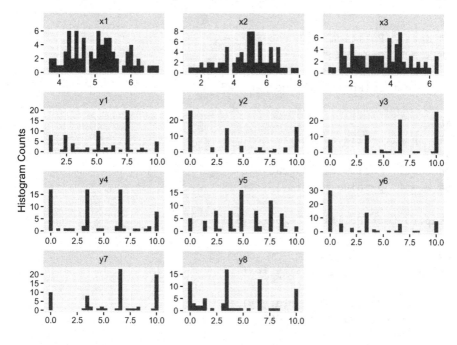

FIGURE 4.2
The Histograms for the 11 Observed Variables.

We can further call *Shapiro* test to statistically test whether these observed variables are normally distributed. The *Shapiro* test is implemented in *R* function *shapiro.test*, which can be called as follows:

```
shapiro.test(PoliticalDemocracy$x1)
```

```
##
##  Shapiro-Wilk normality test
##
## data:  PoliticalDemocracy$x1
## W = 1, p-value = 0.1
```

```
shapiro.test(PoliticalDemocracy$x2)
```

```
##
##  Shapiro-Wilk normality test
```

```
##
## data:  PoliticalDemocracy$x2
## W = 1, p-value = 0.2

shapiro.test(PoliticalDemocracy$x3)

##
##  Shapiro-Wilk normality test
##
## data:  PoliticalDemocracy$x3
## W = 1, p-value = 0.1

shapiro.test(PoliticalDemocracy$y1)

##
##  Shapiro-Wilk normality test
##
## data:  PoliticalDemocracy$y1
## W = 0.9, p-value = 0.0006

shapiro.test(PoliticalDemocracy$y2)

##
##  Shapiro-Wilk normality test
##
## data:  PoliticalDemocracy$y2
## W = 0.8, p-value = 0.00000008

shapiro.test(PoliticalDemocracy$y3)

##
##  Shapiro-Wilk normality test
##
## data:  PoliticalDemocracy$y3
## W = 0.9, p-value = 0.0000005

shapiro.test(PoliticalDemocracy$y4)

##
##  Shapiro-Wilk normality test
##
## data:  PoliticalDemocracy$y4
## W = 0.9, p-value = 0.00002
```

```
shapiro.test(PoliticalDemocracy$y5)
```

```
##
##   Shapiro-Wilk normality test
##
## data:   PoliticalDemocracy$y5
## W = 1, p-value = 0.02
```

```
shapiro.test(PoliticalDemocracy$y6)
```

```
##
##   Shapiro-Wilk normality test
##
## data:   PoliticalDemocracy$y6
## W = 0.8, p-value = 0.00000002
```

```
shapiro.test(PoliticalDemocracy$y7)
```

```
##
##   Shapiro-Wilk normality test
##
## data:   PoliticalDemocracy$y7
## W = 0.9, p-value = 0.000002
```

```
shapiro.test(PoliticalDemocracy$y8)
```

```
##
##   Shapiro-Wilk normality test
##
## data:   PoliticalDemocracy$y8
## W = 0.9, p-value = 0.00002
```

These tests further confirmed the observation from Figure 4.2 that the distributions of these observed variables might not be normally distributed, let alone the joint multivariate normality of these variables. Hence, the validity of using the maximum likelihood method for model estimation might be in doubt.

4.2   Methods for Non-Normal Data

4.2.1   Parameter Estimations

As discussed previously, the *default* estimation in almost all SEM software is the maximum likelihood (ML) estimation based on the multivariate normal likelihood function of the data, which is equivalently defined by the following *ML* discrepancy function :

$$F_{\text{ML}}\left(S, \Sigma(\theta)\right) = \log|\Sigma(\theta)| - \log|S| + \text{trace}\left(\Sigma(\theta)^{-1}S\right) - p \qquad (4.1)$$

where $|S|$ and $|\Sigma(\theta)|$ are the determinants of S and $\Sigma(\theta)$, respectively, and p is the dimension of S or $\Sigma(\theta)$.

There are several other estimators that have been developed to fit the SEM, which are:

1. GLS: the *Generalized Least Squares(GLS)* or simply *GLS* estimator for complete data only, whose discrepancy function is formulated as follows:

$$F_{\text{GLS}}\left(S, \Sigma(\theta)\right) = \frac{1}{2} \times \text{trace}\left[S^{-1}(S - \Sigma(\theta))\right]^2. \qquad (4.2)$$

2. WLS: the *Weighted Least Squares* estimator, or *Asymptotically Distribution-Free (ADF)* estimator (Browne, 1984), whose discrepancy function is formulated as follows:

$$F_{\text{ADF}}\left(S, \Sigma(\theta)\right) = \text{vecs}(S - \Sigma(\theta))'W\text{vecs}(S - \Sigma(\theta)) \qquad (4.3)$$

where vecs(.) extracts the lower triangular elements of the square matrix in the argument and puts the elements into a vector, and W is symmetric weight matrix. Because S is of dimension p, vecs(S) is a vector of order $p^* = p(p+1)/2$ and so W is a $p^* \times p^*$ symmetric matrix. For optimality reasons, W is usually computed as a consistent estimate of the asymptotic covariance matrix of vecs($\sqrt{n}S$) in applications.

3. DWLS: the *Diagonally Weight Least Squares* estimator, which can be viewed of a special case of *WLS* by using a diagonal weight matrix for W in the discrepancy function. That is,

$$F_{\text{DWLS}}\left(S, \Sigma(\theta)\right) = \text{vecs}(S - \Sigma(\theta))'D\text{vecs}(S - \Sigma(\theta)) \qquad (4.4)$$

where D is symmetric weight matrix. Similar to the case for the *ADF* discrepancy function, in applications a consistent estimate of the asymptotic covariance matrix of vecs($\sqrt{n}S$) is first computed. Then, D is formed by retaining only the diagonal elements of the consistent estimate of the asymptotic covariance matrix.

4. ULS: the *Unweighted Least Squares* estimator, which can also be viewed of a special case of *WLS* by using the identity matrix for W in the discrepancy function. That is,

$$F_{\text{ULS}}\left(S, \Sigma(\theta)\right) = \text{vecs}(S - \Sigma(\theta))'\text{vecs}(S - \Sigma(\theta)) \qquad (4.5)$$

Many other variants of discrepancy functions have been proposed, but the ones described here are the most basic and popular ones. Although the *ADF* estimation is suitable for any arbitrary distribution and hence theoretically more general than the *ML* estimation in data applications, simulation results (Hu et al., 1992) show that it needs to have an extremely large sample size (e.g., $N > 5000$) to achieve desirable statistical properties. Thus, the maximum likelihood estimator with the discrepancy function F_{ML} is still the most popular one in applications. Nowadays, instead of using the *ADF* estimation to relax the multivariate normality assumption, the *ML* estimation with adjustment of nonnormality made to the chi-square model fit test and the standard error estimates is perhaps the main statistical strategy for fitting SEM models with nonnormal data.

4.2.2 Statistical Inferences and Testing by ML Estimation with Adjustments

When the observed variables are jointly multivariate normally distributed, the maximum likelihood estimation by minimizing F_{ML} is well-known to be the optimal estimator. The ML estimation under this situation is asymptotically the most efficient in the sense that the standard errors of the ML estimates attain the minimum possible among all consistent estimators when the sample size gets large. The model test statistic is distributed as a χ^2 variate asymptotically.

When the observed variables are not multivariate normally distributed, the ML discrepancy function F_{ML} is mis-specified. Although consistent estimates can still be obtained by minimizing F_{ML} under this situation, the asymptotic efficiency can no longer be claimed. Indeed, formulas for computing standard errors and the model test statistic would need to be adjusted for non-normality. In the SEM literature, this kind of adjustment to ML estimation is usually coined as a *robust* version of *ML*. Several *robust* versions are described as follows:

- MLM: the maximum likelihood estimation with robust standard errors and a Satorra-Bentler scaled test statistic for complete data only.

- MLMVS: the maximum likelihood estimation with robust standard errors and a mean- and variance-adjusted test statistic (aka the *Satterthwaite* approach) for complete data only.

- MLMV: the maximum likelihood estimation with robust standard errors and a mean- and variance-adjusted test statistic (using a scale-shifted approach) for complete data only.

- MLF: the maximum likelihood estimation with standard errors based on the first-order derivatives for both complete and incomplete data.

- MLR: the maximum likelihood estimation with robust (Huber-White) standard errors and a scaled test statistic that is (asymptotically) equal to the Yuan-Bentler test statistic for both complete and incomplete data.

Refer to Rosseel (2012) for detailed discussions on these estimations and various *robust* options.

4.3   Data Analysis Using *R*

We still use the SEM in Figure 4.1 to illustrate the difference between the *default* estimation assuming that the data are jointly normally distributed with the *robust* estimation where data can be non-normally distributed, which is the case for this *PoliticalDemocracy* data.

The SEM model was implemented in Chapter 3 as *modelPD*:

```
# The SEM for Political Democracy
modelPD <- '
  # Component 1: Measurement model
    ind60 =~ x1 + x2 + x3
    dem60 =~ y1 + y2 + y3 + y4
    dem65 =~ y5 + y6 + y7 + y8
  # Component 2: Regression model
    dem60 ~ ind60
    dem65 ~ ind60 + dem60
  # Component 3: Residual covariances
    y1 ~~ y5
    y2 ~~ y4 + y6
    y3 ~~ y7
    y4 ~~ y8
    y6 ~~ y8'
```

With this specified *modelPD*, we can call *sem* to fit this model to the data with *robust* estimator *estimator="MLR"* (i.e., the maximum likelihood estimation with robust Huber White standard errors and a scaled test statistic that is asymptotically equal to the Yuan-Bentler test statistic). After the model is fitted, we examine the model fitting using *fit.measures = TRUE* from the *summary* to print the model fitting measures without the parameter estimates (i.e., *estimates=FALSE*). This is implemented as follows in *R*:

```
# Model fitting with Robust ML
fit.PD.MLR <- sem(modelPD,data= PoliticalDemocracy,
                  estimator="MLR")
```

```
# Print the model fitting without estimates
summary(fit.PD.MLR, fit.measures=TRUE, estimates=FALSE)
```

```
## lavaan 0.6-12 ended normally after 68 iterations
##
##    Estimator                                          ML
##    Optimization method                            NLMINB
##    Number of model parameters                         31
##
##    Number of observations                             75
##
## Model Test User Model:
##                                             Standard     Robust
##    Test Statistic                            38.125     41.401
##    Degrees of freedom                            35         35
##    P-value (Chi-square)                       0.329      0.211
##    Scaling correction factor                             0.921
##       Yuan-Bentler correction (Mplus variant)
##
## Model Test Baseline Model:
##
##    Test statistic                           730.654    702.841
##    Degrees of freedom                            55         55
##    P-value                                    0.000      0.000
##    Scaling correction factor                             1.040
##
## User Model versus Baseline Model:
##
##    Comparative Fit Index (CFI)                0.995      0.990
##    Tucker-Lewis Index (TLI)                   0.993      0.984
##
##    Robust Comparative Fit Index (CFI)                    0.991
##    Robust Tucker-Lewis Index (TLI)                       0.986
##
## Loglikelihood and Information Criteria:
##
##    Loglikelihood user model (H0)          -1547.791  -1547.791
##    Scaling correction factor                             1.012
##       for the MLR correction
##    Loglikelihood unrestricted model (H1)  -1528.728  -1528.728
##    Scaling correction factor                             0.964
##       for the MLR correction
##
##    Akaike (AIC)                            3157.582   3157.582
##    Bayesian (BIC)                          3229.424   3229.424
##    Sample-size adjusted Bayesian (BIC)     3131.720   3131.720
```

```
##
## Root Mean Square Error of Approximation:
##
##    RMSEA                                          0.035      0.049
##    90 Percent confidence interval - lower         0.000      0.000
##    90 Percent confidence interval - upper         0.092      0.103
##    P-value RMSEA <= 0.05                          0.611      0.474
##
##    Robust RMSEA                                              0.047
##    90 Percent confidence interval - lower                   0.000
##    90 Percent confidence interval - upper                   0.097
##
## Standardized Root Mean Square Residual:
##
##    SRMR                                           0.044      0.044
```

There are two columns in the above output, where the first column is the output for the *Standard* maximum likelihood estimation as seen in Chapter 3 and the second column is the output for the *robust* maximum likelihood estimation with *scaling correction factor*.

The *scaling correction factor* is estimated as 0.921 which is used to scale the *standard* Chi-squared test-statistic of 38.125 and resulted the *robust* Chi-squared test-statistic of 41.401 ($= 38.125/0.921$). With this scaled Chi-square test-statistic, the associated p-value is now changed from 0.329 to 0.211. Other changes on the values of *loglikelihood* as well as goodness-of-fitting measures of *CFI*, *TLI*, *RMSEA* and *SRMR* can be observed due to the scaling factor.

With this *robust* maximum likelihood estimation with the *scaling correction factor*, the values of the parameter estimations can be also examined. In the following *R* code chunk, we first extract the parameter estimates from both fittings using *parameterEstimates*. We then combine these two model fittings using *R* function *cbind* to combine them together side-by-side.

```
# Extract all the tables related to the parameter estimates
Est.ML  = parameterEstimates(fit.PD)
Est.MLR = parameterEstimates(fit.PD.MLR)
# Combine the two tables
ML2MLR = cbind(Est.ML[, c("est","se","z","pvalue")],
               Est.MLR[,c("est","se","z","pvalue")])
# Rename the combined table
colnames(ML2MLR) = c("ML.est","ML.se","ML.z","ML.pvalue",
                     "MLR.est","MLR.se","MLR.z","MLR.pvalue")
rownames(ML2MLR) = paste(Est.ML[,"lhs"],Est.ML[,"op"],
                         Est.ML[,"rhs"],sep="")
# Print the combined table
round(ML2MLR,3)
```

```
##              ML.est ML.se    ML.z ML.pvalue MLR.est
## ind60=~x1     1.000 0.000      NA        NA   1.000
## ind60=~x2     2.180 0.139  15.742     0.000   2.180
## ind60=~x3     1.819 0.152  11.967     0.000   1.819
## dem60=~y1     1.000 0.000      NA        NA   1.000
## dem60=~y2     1.257 0.182   6.889     0.000   1.257
## dem60=~y3     1.058 0.151   6.987     0.000   1.058
## dem60=~y4     1.265 0.145   8.722     0.000   1.265
## dem65=~y5     1.000 0.000      NA        NA   1.000
## dem65=~y6     1.186 0.169   7.024     0.000   1.186
## dem65=~y7     1.280 0.160   8.002     0.000   1.280
## dem65=~y8     1.266 0.158   8.007     0.000   1.266
## dem60~ind60   1.483 0.399   3.715     0.000   1.483
## dem65~ind60   0.572 0.221   2.586     0.010   0.572
## dem65~dem60   0.837 0.098   8.514     0.000   0.837
## y1~~y5        0.624 0.358   1.741     0.082   0.624
## y2~~y4        1.313 0.702   1.871     0.061   1.313
## y2~~y6        2.153 0.734   2.934     0.003   2.153
## y3~~y7        0.795 0.608   1.308     0.191   0.795
## y4~~y8        0.348 0.442   0.787     0.431   0.348
## y6~~y8        1.356 0.568   2.386     0.017   1.356
## x1~~x1        0.082 0.019   4.184     0.000   0.082
## x2~~x2        0.120 0.070   1.718     0.086   0.120
## x3~~x3        0.467 0.090   5.177     0.000   0.467
## y1~~y1        1.891 0.444   4.256     0.000   1.891
## y2~~y2        7.373 1.374   5.366     0.000   7.373
## y3~~y3        5.067 0.952   5.324     0.000   5.067
## y4~~y4        3.148 0.739   4.261     0.000   3.148
## y5~~y5        2.351 0.480   4.895     0.000   2.351
## y6~~y6        4.954 0.914   5.419     0.000   4.954
## y7~~y7        3.431 0.713   4.814     0.000   3.431
## y8~~y8        3.254 0.695   4.685     0.000   3.254
## ind60~~ind60  0.448 0.087   5.173     0.000   0.448
## dem60~~dem60  3.956 0.921   4.295     0.000   3.956
## dem65~~dem65  0.172 0.215   0.803     0.422   0.172
##             MLR.se   MLR.z MLR.pvalue
## ind60=~x1    0.000      NA         NA
## ind60=~x2    0.145  15.044      0.000
## ind60=~x3    0.140  12.950      0.000
## dem60=~y1    0.000      NA         NA
## dem60=~y2    0.150   8.392      0.000
## dem60=~y3    0.130   8.107      0.000
## dem60=~y4    0.146   8.661      0.000
## dem65=~y5    0.000      NA         NA
## dem65=~y6    0.181   6.541      0.000
```

```
## dem65=~y7      0.173   7.415      0.000
## dem65=~y8      0.189   6.685      0.000
## dem60~ind60    0.342   4.332      0.000
## dem65~ind60    0.225   2.546      0.011
## dem65~dem60    0.087   9.595      0.000
## y1~~y5         0.454   1.373      0.170
## y2~~y4         0.727   1.807      0.071
## y2~~y6         0.862   2.497      0.013
## y3~~y7         0.615   1.292      0.196
## y4~~y8         0.434   0.803      0.422
## y6~~y8         0.738   1.837      0.066
## x1~~x1         0.019   4.406      0.000
## x2~~x2         0.073   1.652      0.099
## x3~~x3         0.083   5.632      0.000
## y1~~y1         0.477   3.965      0.000
## y2~~y2         1.297   5.686      0.000
## y3~~y3         1.086   4.665      0.000
## y4~~y4         0.793   3.968      0.000
## y5~~y5         0.605   3.885      0.000
## y6~~y6         0.902   5.492      0.000
## y7~~y7         0.622   5.520      0.000
## y8~~y8         0.900   3.615      0.000
## ind60~~ind60   0.073   6.149      0.000
## dem60~~dem60   0.916   4.320      0.000
## dem65~~dem65   0.227   0.759      0.448
```

As seen form this output, the parameter estimates from both model fittings are the same since they are both fitted by the maximum likelihood estimation. However, the standard errors are different since in the *robust* estimation (i.e., *estimator="MLR"*) are estimated by the robust Huber-White estimation to account for non-normally distributed data, which should be used for statistical inference to get the Wald's *z*-values and the associated *p*-values.

4.4   Data Analysis Using SAS

The CALIS procedure of *SAS* currently supports only one *robust* version of maximum likelihood estimation—the *MLM* or *MLSB*. This section illustrates the steps of using the *MLSB* method of the CALIS procedure to analyze the *PoliticalDemocracy* data set. First, you input an external data set by the following statements:

```
data PoliticalDemocracy;
  set "c:\PoliticalDemocracy";
  y1 = PRESS60; y2 = FREOP60; y3 = FREEL60; y4 = LEGIS60;
```

```
    y5 = PRESS65; y6 = FREOP65; y7 = FREEL65; y8 = LEGIS65;
    x1 = GNPPC60; x2 = ENPC60;  x3 = INDLF60;
run;
```

Then you specify the same model as in Chapter 3, but with the METHOD=MLSB option added on to the CALIS statement, as shown in the following:

```
proc calis data=PoliticalDemocracy method=MLSB;
    path
        ind60 ==> x1-x3 = 1,
        dem60 ==> y1-y4 = 1,
        dem65 ==> y5-y8 = 1,
        ind60 ==> dem60 dem65,
        dem60 ==> dem65;
    pcov y1 y5, y2 y6, y3 y7, y4 y8,
         y2 y4, y6 y8;
    fitindex on(only)=[chisq df probchi SBchisq probSBchi
            RMSEA RMSEA_LL RMSEA_UL PROBCLFIT SRMR CFI BENTLERNNFI]
            noindextype;
run;
```

In the FITINDEX statement, you request the Satorra-Bentler scaled model chi-square (SBchisq) and its p-value (probSBchi) be shown, in addition to other fit statistics that have been mentioned in previous chapters.

Fit Summary	
Chi-Square	37.6169
Chi-Square DF	35
Pr > Chi-Square	0.3503
SB-Scaled Model Chi-Square	40.5115
Pr > SB-Scaled Model Chi-Square	0.2401
Standardized RMR (SRMR)	0.0444
RMSEA Estimate	0.0461
RMSEA Lower 90% Confidence Limit	0.0000
RMSEA Upper 90% Confidence Limit	0.0993
Probability of Close Fit	0.5086
Bentler Comparative Fit Index	0.9922
Bentler-Bonett Non-normed Index	0.9877

The uncorrected chi-square value is 37.62. With the adjustment of non-normality, the chi-square value is 40.51 ($p = 0.24$). These values are very similar to the *MLR* method that was illustrated by using *lavaan*. Because the METHOD=MLSB is used, all the remaining fit indices shown in the above table are based on the SB-scaled chi-square. All fit indices, but the confidence interval for RMSEA, indicate good model fit. The 90% confidence interval for

RMSEA is (0, 0.09), which is not entirely in the region that is below 0.05. Together with the point estimate of RMSEA at 0.044, you conclude that this is a good model fit, but your conclusion might subject to sampling error.

The next table shows the estimates of path coefficients and their standard error estimates. As expected, the ML estimates here match closely with those produced by the *MLR* method of *lavaan*—due to the fact that both are ML estimates. The standard error estimates estimated by *MLSB* (or *MLM*) are also quite close to that of the *MLR*. An overall impression is that the corresponding estimates would not differ by more than 0.02.

PATH List							
Path			Parameter	Estimate	Standard Error	t Value	Pr > |t|
ind60	===>	x1		1.00000			
ind60	===>	x2	_Parm01	2.18037	0.14263	15.2872	<.0001
ind60	===>	x3	_Parm02	1.81851	0.13821	13.1575	<.0001
dem60	===>	y1		1.00000			
dem60	===>	y2	_Parm03	1.25676	0.13927	9.0240	<.0001
dem60	===>	y3	_Parm04	1.05772	0.13329	7.9355	<.0001
dem60	===>	y4	_Parm05	1.26480	0.12599	10.0389	<.0001
dem65	===>	y5		1.00000			
dem65	===>	y6	_Parm06	1.18571	0.17005	6.9726	<.0001
dem65	===>	y7	_Parm07	1.27952	0.16519	7.7456	<.0001
dem65	===>	y8	_Parm08	1.26596	0.16996	7.4484	<.0001
ind60	===>	dem60	_Parm09	1.48301	0.33561	4.4189	<.0001
ind60	===>	dem65	_Parm10	0.57231	0.20478	2.7947	0.0052
dem60	===>	dem65	_Parm11	0.83735	0.08585	9.7537	<.0001

Other estimates from *MLSB* are not shown. See Exercise 3 for running this example with the asymptotically distribution-free method (ADF).

4.5 Discussions

In this chapter, we illustrated the structural equation modeling for non-normally distributed data. Non-normally distributed data should be a norm in real SEM data analysis and therefore the *robust* maximum likelihood should be adopted whenever there is any doubt that the assumption of normally distributed data is violated. In particular, we demonstrated the uses of two robust ML methods: MLR and MLSB (or MLM).

Before closing, some important notes are in order. First, as discussed throughout this chapter, although all these robust versions of ML produce *robust* standard errors and a *robust* (or *scaled*) test statistic, the parameter estimates produced are still that of the original *ML* estimation. Second, the *robustness* of these adjustments is with respect to the violation of the normality assumption, but not to the existence of outliers, which is a separate topic that has been dealt with in SEM. Third, the *robustness* of these adjustments is also not to the violation of model specifications, which is yet another topic that has been researched in the SEM literature. We made no attempt to cover all robustness topics other than the one in this chapter.

4.6   Exercises

1. In Chapter 2, we used *lavaan* built-in data set *HolzingerSwineford1939* to build a CFA model assuming that the nine observed items $x1$ to $x9$ are jointly distributed as a multivariate normally distributed data. In this exercise, examine the distribution assumption following the approach in this chapter graphically as well as statistically test this assumption.

2. Examine the distributions for the 22 observed items in *Maslach Burnout Inventory (MBI)* discussed in Chapter 2 on whether they are normally distributed. If there are not, re-fit the CFA model using *MLR* and compare the differences in model fittings.

   Hint: Use the following *R* code chunk:

   ```
   shapiro.test(HolzingerSwineford1939$x1)
   shapiro.test(HolzingerSwineford1939$x2)
   shapiro.test(HolzingerSwineford1939$x3)
   shapiro.test(HolzingerSwineford1939$x4)
   shapiro.test(HolzingerSwineford1939$x5)
   shapiro.test(HolzingerSwineford1939$x6)
   shapiro.test(HolzingerSwineford1939$x7)
   shapiro.test(HolzingerSwineford1939$x8)
   shapiro.test(HolzingerSwineford1939$x9)
   ```

3. The asymptotically distribution free (ADF) method is one of the methods you can use to analyze nonnormal data in SEM. Fit the same model that is described in this chapter to the *PoliticalDemocracy* data by using the ADF method. To do this with *SAS*, you can use the METHOD = ADF option in place of the METHOD = MLSB option in the PROC CALIS code of this chapter. Compare and contrast the estimation results between that of the ADF and MLSB methods. Do both methods support a good model? Does one method provide larger standard error estimates consistently than the other? Which method should you trust more and why?

5

Structural Equation Modeling with Categorical Data

In previous chapters, we discussed the structural equation modeling assuming that the data are continuous with a joint multivariate normal distribution. In Chapter 4, we relaxed this assumption of joint multivariate normal distribution and discussed the robust estimation when the data are not normally distributed.

However, most of the data collected for CFA/SEM are categorical (such as binary, ordinal, and nominal) from questionnaires using Likert-type scales. For example, the *Maslach Burnout Inventory (MBI)* data illustrated in Chapter 1 used the *MBI* measurement tool with a 22-item instrument structured on a 7-point Likert-type scale that ranges from 1 (*feeling has never been experienced*) to 7 (*feeling experienced daily*). As such, there is an ordinality in this type of data, which are not continuous. Although they are categorical, they have been treated as continuous data in previous chapters and the structural equation modeling was conducted assuming that they were continuous with a multivariate normal distribution. The treatments and assumptions in previous chapters are further investigated in this chapter to see if they are appropriate. If not, SEM methodology that addresses the ordinal data type should be used.

Thus, in this chapter, we discuss the structural equation modeling for categorical data and to use this methodology to further analyze the *MBI* data.

5.1 Exploratory Data Analysis

The *Maslach Burnout Inventory (MBI)* data set was detailed in Chapter 1. The data can be loaded into *R* as follows:

```
# Read the data into R session using 'read.csv'
dMBI = read.csv("data/ELEMmbi.csv", header=F)
colnames(dMBI) = paste("I",1:22,sep="")
# Check the data dimension
dim(dMBI)
```

```
## [1] 372   22
```

DOI: 10.1201/9781003365860-5

As discussed previously, the items in the data were collected as ordinal responses, which range from 1 (*feeling has never been experienced*) to 7 (*feeling experienced daily*). We can show the ordinal data distributions for the first six items in bivariate tables as follows:

```
# Counts for I1 X I2
xtabs(~I1+I2,dMBI)
```

```
##     I2
## I1   1  2  3  4  5  6  7
##  1   1  1  0  2  0  0  0
##  2   3 27 18 12  7  6  0
##  3   0  4  9 10  7  4  2
##  4   0  2  4 36 17 24  2
##  5   0  1  2 11 18 18  2
##  6   0  0  0  4 10 62 12
##  7   0  0  0  0  0  5 29
```

```
# Counts for I3 X I4
xtabs(~I3+I4,dMBI)
```

```
##     I4
## I3   2  3  4  5  6  7
##  1   1  0  2  3  6 24
##  2   2  1  5 12 26 60
##  3   0  0  3  3 17 32
##  4   0  1  1  6 25 28
##  5   0  0  4  4 13 21
##  6   1  2  2  2 21 31
##  7   0  0  0  1  4  8
```

```
# Counts for I5 X I6
xtabs(~I5+I6,dMBI)
```

```
##     I6
## I5   1  2  3  4  5  6  7
##  1  60 53 19 18  4  6  2
##  2  19 46 16 11  5  5  1
##  3   5 12 11  6  4  1  0
##  4   3  9  6  8  2  4  2
##  5   1  2  0  2  2  2  0
##  6   1  2  5  3  2  8  2
##  7   1  0  0  0  0  1  0
```

In addition to the bivariate ordinal distributions in these tables, we can observe that *I1* and *I2* are highly correlated since most of the data are in

the diagonal. However, most of *I4* were observed at the higher Likert scales
(i.e., 6 and 7) and most of the *I5* at the lower Likert scales(i.e., 1 and 2).
Moreover, the middle table shows that there were no responses to category 1
for *I4*. Together these tables show that the bivariate distributions or univariate
distributions of these items could be quite different.

We can also graphically examine the ordinality and distributions of items
under the associated three subscales: the *Emotional Exhaustion* subscale (*EE*)
(nine items) in Figure 5.1, the *Depersonalization* subscale (*DP*) (five items)
in Figure 5.2, and the *Personal Accomplishment* (*PA*) subscale (eight items)
in Figure 5.3. To produce these figures, we make use of *R* package *lattice* to
make these *histograms*, as follows:

```
library(lattice)
histogram(~ I1+I2+I3+I6+I8+I13+I14+I16+I20,dMBI,
xlab="Seven-Point Likert Scale", ylab="",
        main="Items Associated with EE")


histogram(~ I5+I10+I11+I15+I22,dMBI,
xlab="Seven-Point Likert Scale", ylab="",
        main="Items Associated with DP")
```

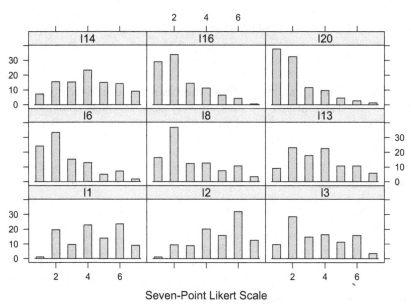

FIGURE 5.1
The Ordinality and Distribution of Latent Factor-Emotional Exhaustion.

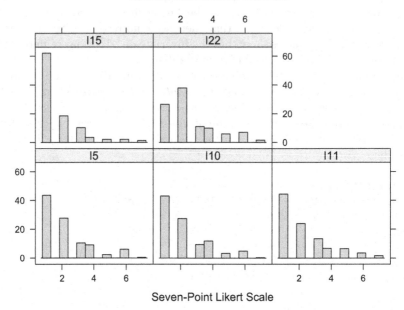

FIGURE 5.2
The Ordinality and Distribution of Latent Factor-Personal Accomplishment.

```
histogram(~I4+I7+I9+I12+I17+I18+I19+I21, dMBI,
xlab="Seven-Point Likert Scale", ylab="",
       main="Items Associated with PA")
```

Again, the non-normality of these data can be seen clearly from these figures. Some distributions of items are very skewed. Such nonnormality in the data leads us to explore better methods for parameter estimation and the associated statistical inference.

5.2   Model and Estimation

So far, all analyses have been based on maximum likelihood estimation or robust ML estimation. As we have known, a common assumption for both estimation methods is that the observed variables are continuous. This is not the case for the *MBI* data where we have 22 items observed with 7-point Likert scales—they are not continuous and some are far from normally distributed. Fortunately, new methods have been developed to take the categorical nature of the data into account and we will describe these methods for a high level of understanding.

Items Associated with PA

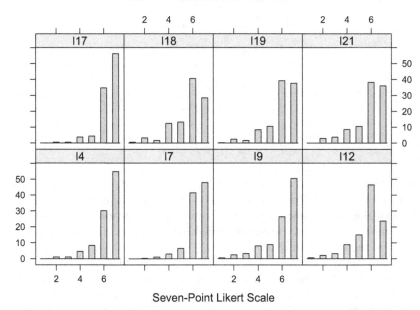

FIGURE 5.3
The Ordinality and Distribution of Latent Factor-DReduced Personal Accomplishment.

5.2.1 The Threshold Model for Ordered Categorical Responses

The basic theory is built on the concept that the responses of a categorical variable can be thought of as being generated from a corresponding underlying continuous variable that is normally distributed. This underlying continuous variable is a latent and unobserved variable that represents the propensity of response in a certain direction. The categorical responses are regarded as only crude measurements of these unobserved latent continuous variables. The *thresholds* (or scale points), which partition the whole range of the latent continuous scale into a finite number of categories, are used to define the mapping of the latent propensity responses to the observed categorical responses.

For the purpose of illustration, let us use the measurement instrument for *MBI* in this chapter. As we know now, each item is structured on a 7-point Likert scale, thereby representing a categorical variable with seven ordinal classes. Let I denote generically any one of these 22 items and represent an observed categorical variable (i.e., or *item*, which is the *dependent indicator variable* in this *CFA/SEM* model), and I^* represent the unobserved underlying continuous variable corresponding to I. The relation between I and

I^* is described by the following set of equations:

$$
\begin{aligned}
I &= 1 \quad \text{if} \quad -\infty < I^* \le t_1 \\
I &= 2 \quad \text{if} \quad t_1 < I^* \le t_2 \\
I &= 3 \quad \text{if} \quad t_2 < I^* \le t_3 \\
I &= 4 \quad \text{if} \quad t_3 < I^* \le t_4 \\
I &= 5 \quad \text{if} \quad t_4 < I^* \le t_5 \\
I &= 6 \quad \text{if} \quad t_5 < I^* \le t_6 \\
I &= 7 \quad \text{if} \quad t_6 < I^* < \infty
\end{aligned}
\tag{5.1}
$$

where t_i $(i = 1, \ldots, 6;\ -\infty < t_1 < t_2 < t_3 < t_4 < t_5 < t_6 < \infty)$ are thresholds that define regions of I^* for yielding the categorical responses I. The number of *thresholds* is always one less than the number of categories, which is similar to the number of dummy variables needed for coding a classification variable in classical regression analysis.

The response categorization of I^* is also graphically illustrated in Figure 5.4. In this figure, the underlying latent continuous variable I^* is a standard normal distribution represented by the *black* curve. The six vertical lines are the hypothetical *thresholds* with t_1 to t_6 marked on the top of the lines. The possible response categories are from $I = 1$ to $I = 7$. The value of I thus depends on which regions (which are demarcated by the thresholds) that I^* falls into.

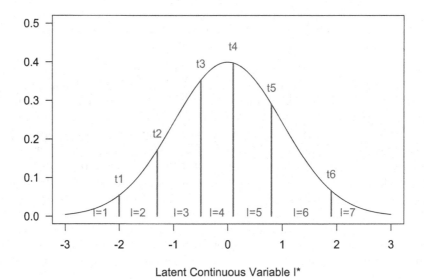

FIGURE 5.4
Categorization of Latent Continuous Variable to Categorical Variable.

The categorization process depicted in Figure 5.4 to produce Likert-scaled data is certainly subject to measurement and sampling errors. As pointed out by O'Brien (1985), this process can produce (1) categorization error resulting from the splitting of the continuous scale into the categorical scale, and (2) transformation error resulting from categories of unequal widths. Therefore, the *thresholds* would have to be estimated along with all other parameters in structural equation modeling.

5.2.2 Modeling Endogenous and Exogenous Categorical Variables in SEM

The idea in the preceding section applies mostly to endogenous categorical responses in CFA or SEM. For categorical variables that are exogenous in a model (i.e., exogenous categorical variables), dummy coding of the categorical variables might be a more convenient modeling strategy. Hence, the exposition here focuses on endogenous ordered categorical responses.

Now that each ordinal categorical variable I links to a latent propensity response I^*, modeling of the categorical variable is essentially through the modeling of the latent response in model equations. For example, the following equation predicts the latent propensity response I^* by p predictors $x_1, \ldots, x_p$s.

$$I_i^* = \beta_0 + \beta_1 x_{1i} + \cdots + \beta_P x_{Pi} + e_i \tag{5.2}$$

If I is a measurement indicator of a factor ξ in a confirmatory factor model, the corresponding model equation is written as:

$$I_i^* = \beta_0 + \beta_1 \xi_i + e_i \tag{5.3}$$

In this case, β_1 represents the factor loading. This equation is exactly the same form as the confirmatory factor model for continuous measurement indicators, which has been discussed in Chapter 2. Therefore, from this similarity we can expect that some covariance structure modeling techniques that have been discussed previously should be applicable to the current case with categorical indicators. This expectation is certainly reasonable. As we will see, the weighted least squares (WLS) estimation is such a method that can be adapted to the current modeling. However, there are also additional details that must be supplemented for modeling ordered categorical responses. The next few sections discuss these details.

5.2.3 The Covariance Structures

Throughout Chapters 1–4, we have emphasized the test of the theoretical model by setting up a statistical null hypothesis. To simplify presentations in this section, let us focus on the covariance structure hypothesis, which is stated as:

$$H_0 : \Sigma = \Sigma(\theta) \tag{5.4}$$

The meaning of this hypothesis is that if our hypothesized model is true, then the unstructured population covariance matrix Σ can be reproduced exactly as a function $\Sigma(\cdot)$ by plugging in the parameter vector θ. Accordingly, our task in practical model fitting is to define the hypothesized model (hence the implied covariance structure $\Sigma(\cdot)$) and to estimate the θ value (denoted as $\hat{\theta}$) that best describes (or reproduces) the sample counterpart of Σ—which is the sample covariance matrix S. Evaluation of model fit is then essentially a measure of discrepancy between $\Sigma(\hat{\theta})$ and S.

So far the computation of S is based on the Pearson product-moment correlation coefficients, which implicitly assume that the observed variables are all continuous. With categorical responses in a model, the null hypothesis can be stated essentially in the same form, but the variables involved in null hypothesis must be clarified.

Suppose now $y_1, y_2, \ldots, y_p$ are the analysis variables in the model. If at least one variable in this model is ordinal and we use the threshold model (Equation 5.1) to represent its categorical responses, then the unstructured population matrix in the null hypothesis H_0 would have to be denoted as Σ^* for variables $y_1^*, y_2^*, \ldots, y_p^*$, in which y_j^* is defined similarly to the threshold model (Equation 5.1) if it is as an ordinal response and is defined as the original y_j otherwise. That is, the null hypothesis is now stated as:

$$H_0 : \Sigma^* = \Sigma(\theta) \tag{5.5}$$

The hypothesis (Equation 5.5) is not simply a notational difference from the original hypothesis (Equation 5.4). First, although the hypothesized covariance structure $\Sigma(\theta)$ is the same form as the original one, the target variables are different. For example, if the original model is a CFA model for continuous measurement variables in hypothesis (Equation 5.4), the numerical or matrix *structure* $\Sigma(\cdot)$ is still that of the original CFA model in hypothesis (Equation 5.5). However, the set of measurement variables described by the covariance structure in Equation (5.5) is now that of y_i^*s, not y_is. Second, by the same logic, Σ^* is now the unstructured population covariance matrix for y_i^*s in hypothesis (Equation 5.5), but not for y_is as in hypothesis (Equation 5.4).

5.2.4   The Fitting of Covariance Structures in Samples

Following the arguments in the preceding section, now that the null hypothesis (Equation 5.5) is about the structure of Σ^*, what would be the sample consistent estimate of Σ^*? It is certainly not the usual sample covariance matrix S, which contains the sample Pearson product-moment correlation coefficients. The correct answer is denoted as S^*, which is a consistent estimate of Σ^*. But how is S^* computed?

First, each of the covariance elements of S^* can be computed one-at-a-time by considering the corresponding pair of observed variables. If both variables are numeric (continuous), the usual formula based on Pearson correlations is used. If both variables are ordinal categorical, then the polychoric correlation is used. If one variable is numeric and the other is ordinal, then the polyserial correlation is used. Finally, binary variables can be treated as a special case of ordinal categorical.

The computations of covariance elements involving ordinal variables are generally not by using closed-form formulas. The related correlation must be estimated along with the thresholds in the threshold model (Equation 5.1), either simultaneously or separately in multistage estimation. See Olsson (1979) and Olsson et al. (1982) for details. Once all elements of S^* are estimated, some mathematical procedures might be applied to "smooth" the estimation of S^* so that it would have those proper mathematical properties of a covariance matrix. Details for this are beyond the scope of our discussion.

With S^* and the specified covariance structure $\Sigma(\theta)$, estimation of the model parameters as if we have measured all continuous variables in the data—which might mean that the estimation methods that are described in Chapters 1 and 4 can now be used. However, although S^* now appears like a sample covariance matrix for continuous data, the target set of variables y_j^* are not normally distributed. This limits the choice of appropriate discrepancy functions for analyzing models with categorical responses. For example, the maximum likelihood estimation (based on minimizing the $F_{\mathrm{ML}}(\cdot)$ function with the input S^*) might still yield consistent estimates, but due to the mis-specified distribution, its model fit χ^2 test statistic and standard error estimates might be incorrect. We therefore would like to use estimator that does not require the normal distribution assumption.

5.2.5 The Diagonally Weighted Least Squares (DWLS) Discrepancy Function for Parameter Estimation

Recall in Chapters 1 and 4 that an estimator that is suitable for any arbitrary distribution is the *Asymptotically Distribution-Free (ADF)* estimator (Browne, 1984), or more commonly called the *Weighted Least Squares (WLS)* estimator, whose discrepancy function is formulated as follows:

$$F_{\mathrm{WLS}}\left(S, \Sigma(\theta)\right) = \mathrm{vecs}(S - \Sigma(\theta))'W\mathrm{vecs}(S - \Sigma(\theta)) \qquad (5.6)$$

where vecs(.) extracts the lower triangular elements of the symmetric matrix in the argument and puts the elements into a vector, and W is a symmetric weight matrix. If S is of dimension p, vecs(S) is a vector of order $p^* = p(p + 1)/2$ and W is a $p^* \times p^*$ symmetric matrix. For optimality reasons, W is usually computed as the inverse of a consistent estimate of the asymptotic covariance matrix of $\mathrm{vecs}(\sqrt{n}S)$ in applications. When the mean structures are also modeled, vecs(S) and vecs($\Sigma(\theta)$) can be extended to include the samples means and the hypothesized mean structures. For simplicity, only the sample covariance matrix and the covariance structures are included in the notation.

With S^* as the sample input information, the adaptation of the WLS estimator is straightforward. The corresponding discrepancy function that should be minimized for estimating θ is:

$$F_{\text{WLS}}\left(S^*, \Sigma(\theta)\right) = \text{vecs}(S^* - \Sigma(\theta))'W\text{vecs}(S^* - \Sigma(\theta)) \qquad (5.7)$$

An obvious advantage of this WLS estimator is that we no longer need to worry about the mis-specification of the distribution because WLS is distribution free (although it requires a large sample size). But now the computation of the weight matrix W might become demanding due to its size and the matrix inversion needed. One simplification is to substitute W with an identity matrix so that it becomes the unweighted least squares (ULS) estimator. However, the ULS estimator might not be as optimal as desired, and so a more popular choice is to use a diagonal weight matrix whose diagonal elements are the inverse of asymptotic variance of parameter estimates in the model. This yields the so-called diagonally weighted least squares (DWLS) estimator. With the DWLS estimation, the diagonal weight matrix still needs to be estimated but the inversion of a large matrix is avoided. The DWLS estimates would be more optimal than that of the ULS estimator.

Additional methods have been proposed to improve the ULS or the DWLS estimator. Two aspects of improvement are distinguished. One is about the improvement of the model fit χ^2 test statistic and the other is about a better estimation of the standard errors. The improvement of the chi-square statistic can be mean-adjusted or both mean and variance-adjusted. The improved standard errors are oftentimes called *robust* standard errors, which are based on a *sandwich* approach that corrects for certain mis-specifications. Computer software attaches *new* estimator names when the produced results use one or more of these improvements. For example, the following Mplus terminology has been used:

- ULSMV: mean and variance-adjusted χ^2 test statistic with the ULS estimator

- WLSM: mean-adjusted χ^2 test statistic with the DWLS estimates

- WLSMV: mean- and variance-adjusted χ^2 test statistic with DWLS estimates

All these estimation are accompanied with *robust* standard error estimates.

Among these corrections, the *WLSMV* estimator performed the best in CFA/SEM for modeling of categorical data as demonstrated in Brown (2015). The WLSMV estimator was developed by Muthen et al. (1997) and designed specifically to fit categorical data with small and moderate sample sizes. Further research and development based on extensive Monte-Carlo simulation by Flora and Curran (2004) on *WLSMV* estimator confirmed that it can yield accurate test statistics, parameter estimates, and standard errors under both normal and non-normal latent response distributions across sample sizes ranging from 100 to 1,000, as well as across different CFA models. Beauducel and Herzberg (2006) also reported satisfactory results for the *WLSMV* estimator with findings of superior model fit and more precise factor loadings when the number of categories is low (e.g., two or three categories, compared with four, five, or six). This *WLSMV* estimator has become the **gold standard** in the SEM ever since. Starting in version 6, the WLSMV estimation uses a new type of χ^2 correction called *scaled and shifted test statistic*. The *lavaan* package also mimics this new WLSMV behavior for its default estimation.

5.2.6 A Note about the Parameterization Related to the Latent Propensity Responses

Unlike the CFA model discussed in Chapter 2, I^*s are not observed and therefore their scales are arbitrary. Some parameterization that fixes the scales of these latent propensity responses is needed. Recall the general model equation (5.2) for I^* is:

$$I_i^* = \beta_0 + \beta_1 x_{1i} + \cdots + \beta_P x_{Pi} + e_i \qquad (5.8)$$

Two different *parametrizations* are available in *mplus* and *R/lavaan*:

1. *Delta* Parameterization: *Delta* parametrization is the default parameterization when using a WLS estimator. In this *Delta* parameterization, the scale of I^* is fixed. That is, $I^* \sim N(0, \sigma^*)$ with $\sigma^* = \frac{1}{\Delta^2}$ and Δ is referred to as the *scale factor*. The *scale factors* are set to 1 by default, even if it can be set to other values to scale I^* in other ways, and can only be estimated under certain conditions (not recommended).

2. *Theta* Parameterization: *Theta* parameterization is an alternative parameterization when using a WLS estimator. In the *theta* parameterization, the scale of the residual is fixed. That is, $e_i \sim N(0, \theta)$, where θ is the residual variance (this θ is certainly not the same as the model parameter θ; Unfortunately, because of the name of this parameterization method, the use of θ symbol is unavoidable). This residual variance is set to 1 by default, even if it can be set to other values. And, it can only be estimated under certain conditions (not recommended). The *theta* parameterization is the standard parameterization of the probit model in regression analysis with the ML estimation.

We will use this default *delta* parameterization in this chapter without specifying *parameterization= "delta"*, and use the *theta* parameterization in Chapter 7.

5.2.7   Some Practical Assumptions

Applications of categorical data in structural equation modeling are based on three critical assumptions:

- The latent propensity response variables that generate the observed categorical responses are assumed to be normally distributed. This assumption is essential but not verifiable.

- Sample size should be sufficiently large to enable reliable estimation of the related covariance/correlation matrix S^*.

- The number of observed ordinal variables is kept to a minimum to avoid identification problems due to empty cells for observing specific response patterns.

5.3   Data Analysis Using *R*

We will fit the CFA model to this data using three estimators to compare their performance. The first is to use the default MLE assuming the data are normally distributed as what we discussed in Chapter 2 and the second is to use the robust MLE discussed in Chapter 4, which assumes that the data are continuous with adjustments on robust standard errors and a Satorra-Bentler scaled test statistic. The third is to use the *WLSMV* estimator to account for the categorical and ordinal nature from the data, which would be the most appropriate estimator for this categorical data set.

5.3.1   The CFA Model

We start with the final CFA model in Chapter 2 as follows:

```
# The final CFA model
CFA4MBI3 <- '
# Latent variables
EE =~ I1+I2+I3+I6+I8+I13+I14+I16+I20
DP =~ I5+I10+I11+I15+I22
PA =~ I4+I7+I9+I12+I17+I18+I19+I21

# Additional item correlations based on MIs
I6   ~~ I16
I1   ~~ I2
I10  ~~ I11
# Cross-loading from EE to Item12 based on MI
EE =~ I12
'
```

5.3.2 Model Fitting with MLE Assuming Normal Distribution - MLE Estimation

As done in Chapter 2, *link not working* the default *cfa* is based on the jointly multivariate normal distribution with the maximum likelihood estimation implemented as follows:

```
# fit the CFA model
fitMBI.Default = cfa(CFA4MBI3, data=dMBI)
# Print the model summary with the fitting measures
summary(fitMBI.Default, fit.measures=TRUE)
```

```
## lavaan 0.6-12 ended normally after 52 iterations
##
##   Estimator                                      ML
##   Optimization method                        NLMINB
##   Number of model parameters                      51
##
##   Number of observations                        372
##
## Model Test User Model:
##
##   Test statistic                            446.419
##   Degrees of freedom                            202
##   P-value (Chi-square)                        0.000
##
## Model Test Baseline Model:
##
##   Test statistic                           3452.269
##   Degrees of freedom                            231
##   P-value                                     0.000
##
## User Model versus Baseline Model:
##
##   Comparative Fit Index (CFI)                 0.924
##   Tucker-Lewis Index (TLI)                    0.913
##
## Loglikelihood and Information Criteria:
##
##   Loglikelihood user model (H0)          -12686.394
##   Loglikelihood unrestricted model (H1)  -12463.184
##
##   Akaike (AIC)                            25474.787
##   Bayesian (BIC)                          25674.651
##   Sample-size adjusted Bayesian (BIC)     25512.844
```

```
##
## Root Mean Square Error of Approximation:
##
##    RMSEA                                               0.057
##    90 Percent confidence interval - lower              0.050
##    90 Percent confidence interval - upper              0.064
##    P-value RMSEA <= 0.05                               0.052
##
## Standardized Root Mean Square Residual:
##
##    SRMR                                                0.057
##
## Parameter Estimates:
##
##    Standard errors                              Standard
##    Information                                  Expected
##    Information saturated (h1) model             Structured
##
## Latent Variables:
##                       Estimate  Std.Err  z-value  P(>|z|)
##    EE =~
##      I1                 1.000
##      I2                 0.878    0.049    17.969    0.000
##      I3                 1.073    0.075    14.303    0.000
##      I6                 0.764    0.069    11.011    0.000
##      I8                 1.215    0.075    16.299    0.000
##      I13                1.072    0.073    14.739    0.000
##      I14                0.880    0.075    11.673    0.000
##      I16                0.727    0.063    11.556    0.000
##      I20                0.806    0.061    13.113    0.000
##    DP =~
##      I5                 1.000
##      I10                0.889    0.114     7.829    0.000
##      I11                1.105    0.125     8.819    0.000
##      I15                0.921    0.105     8.791    0.000
##      I22                0.776    0.115     6.730    0.000
##    PA =~
##      I4                 1.000
##      I7                 0.973    0.147     6.611    0.000
##      I9                 1.763    0.248     7.105    0.000
##      I12                1.131    0.188     6.022    0.000
##      I17                1.327    0.176     7.555    0.000
##      I18                1.890    0.255     7.421    0.000
##      I19                1.695    0.232     7.300    0.000
##      I21                1.342    0.213     6.290    0.000
```

```
##   EE =~
##     I12                 -0.316    0.050    -6.366    0.000
##
## Covariances:
##                        Estimate  Std.Err  z-value  P(>|z|)
##   .I6 ~~
##     .I16                 0.706    0.090     7.869    0.000
##   .I1 ~~
##     .I2                  0.588    0.082     7.131    0.000
##   .I10 ~~
##     .I11                 0.517    0.101     5.106    0.000
##   EE ~~
##     DP                   0.747    0.104     7.197    0.000
##     PA                  -0.167    0.040    -4.168    0.000
##   DP ~~
##     PA                  -0.181    0.038    -4.768    0.000
##
## Variances:
##                        Estimate  Std.Err  z-value  P(>|z|)
##     .I1                  1.268    0.106    11.992    0.000
##     .I2                  1.238    0.101    12.310    0.000
##     .I3                  1.285    0.109    11.814    0.000
##     .I6                  1.636    0.127    12.902    0.000
##     .I8                  0.783    0.080     9.788    0.000
##     .I13                 1.115    0.097    11.536    0.000
##     .I14                 1.822    0.143    12.781    0.000
##     .I16                 1.281    0.100    12.793    0.000
##     .I20                 1.031    0.083    12.361    0.000
##     .I5                  1.407    0.126    11.211    0.000
##     .I10                 1.455    0.126    11.510    0.000
##     .I11                 1.355    0.130    10.397    0.000
##     .I15                 1.004    0.094    10.735    0.000
##     .I22                 2.008    0.159    12.631    0.000
##     .I4                  0.795    0.062    12.846    0.000
##     .I7                  0.515    0.041    12.477    0.000
##     .I9                  1.108    0.093    11.860    0.000
##     .I12                 0.898    0.072    12.545    0.000
##     .I17                 0.374    0.035    10.642    0.000
##     .I18                 0.906    0.081    11.130    0.000
##     .I19                 0.838    0.073    11.459    0.000
##     .I21                 1.240    0.097    12.723    0.000
##     EE                   1.486    0.186     7.982    0.000
##     DP                   0.800    0.144     5.567    0.000
##     PA                   0.199    0.048     4.112    0.000
```

5.3.3   Model Fitting with Robust MLE

As seen from Figures 5.1–5.3, the data are not normal distributed. We can make use the robust MLE to obtain a robust standard errors and a Satorra-Bentler scaled test statistic. This can be implemented using option *estimator = "MLR"* as follows:

```
# fit the CFA model
fitMBI.Robust = cfa(CFA4MBI3, data=dMBI, estimator="MLR")
# Print the model summary with the fitting measures
summary(fitMBI.Robust, fit.measures=TRUE)
```

```
## lavaan 0.6-12 ended normally after 52 iterations
##
##    Estimator                                        ML
##    Optimization method                          NLMINB
##    Number of model parameters                       51
##
##    Number of observations                          372
##
## Model Test User Model:
##                                           Standard       Robust
##    Test Statistic                          446.419      385.882
##    Degrees of freedom                          202          202
##    P-value (Chi-square)                      0.000        0.000
##    Scaling correction factor                              1.157
##       Yuan-Bentler correction (Mplus variant)
##
## Model Test Baseline Model:
##
##    Test statistic                         3452.269     2832.328
##    Degrees of freedom                          231          231
##    P-value                                   0.000        0.000
##    Scaling correction factor                              1.219
##
## User Model versus Baseline Model:
##
##    Comparative Fit Index (CFI)               0.924        0.929
##    Tucker-Lewis Index (TLI)                  0.913        0.919
##
##    Robust Comparative Fit Index (CFI)                     0.933
##    Robust Tucker-Lewis Index (TLI)                        0.923
##
## Loglikelihood and Information Criteria:
##
##    Loglikelihood user model (H0)        -12686.394   -12686.394
```

```
## Scaling correction factor                                    1.606
##      for the MLR correction
## Loglikelihood unrestricted model (H1)     -12463.184  -12463.184
## Scaling correction factor                                    1.247
##      for the MLR correction
##
## Akaike (AIC)                               25474.787   25474.787
## Bayesian (BIC)                             25674.651   25674.651
## Sample-size adjusted Bayesian (BIC)        25512.844   25512.844
##
## Root Mean Square Error of Approximation:
##
## RMSEA                                          0.057       0.049
## 90 Percent confidence interval - lower         0.050       0.042
## 90 Percent confidence interval - upper         0.064       0.056
## P-value RMSEA <= 0.05                          0.052       0.541
##
## Robust RMSEA                                               0.053
## 90 Percent confidence interval - lower                     0.045
## 90 Percent confidence interval - upper                     0.061
##
## Standardized Root Mean Square Residual:
##
## SRMR                                           0.057       0.057
##
## Parameter Estimates:
##
## Standard errors                               Sandwich
## Information bread                             Observed
## Observed information based on                  Hessian
##
## Latent Variables:
##                    Estimate  Std.Err  z-value  P(>|z|)
## EE =~
##    I1                 1.000
##    I2                 0.878    0.041   21.187    0.000
##    I3                 1.073    0.056   19.144    0.000
##    I6                 0.764    0.079    9.654    0.000
##    I8                 1.215    0.069   17.672    0.000
##    I13                1.072    0.074   14.401    0.000
##    I14                0.880    0.063   13.985    0.000
##    I16                0.727    0.076    9.625    0.000
##    I20                0.806    0.070   11.516    0.000
## DP =~
##    I5                 1.000
##    I10                0.889    0.129    6.884    0.000
```

```
##      I11            1.105    0.143    7.720    0.000
##      I15            0.921    0.114    8.058    0.000
##      I22            0.776    0.120    6.443    0.000
##   PA =~
##      I4             1.000
##      I7             0.973    0.124    7.852    0.000
##      I9             1.763    0.347    5.077    0.000
##      I12            1.131    0.217    5.221    0.000
##      I17            1.327    0.213    6.239    0.000
##      I18            1.890    0.355    5.329    0.000
##      I19            1.695    0.342    4.956    0.000
##      I21            1.342    0.215    6.249    0.000
##   EE =~
##      I12           -0.316    0.055   -5.770    0.000
##
## Covariances:
##                  Estimate  Std.Err  z-value  P(>|z|)
##   .I6 ~~
##     .I16           0.706    0.124    5.710    0.000
##   .I1 ~~
##     .I2            0.588    0.089    6.629    0.000
##   .I10 ~~
##     .I11           0.517    0.108    4.770    0.000
##   EE ~~
##      DP            0.747    0.104    7.218    0.000
##      PA           -0.167    0.039   -4.272    0.000
##   DP ~~
##      PA           -0.181    0.039   -4.588    0.000
##
## Variances:
##                  Estimate  Std.Err  z-value  P(>|z|)
##     .I1            1.268    0.108   11.738    0.000
##     .I2            1.238    0.100   12.385    0.000
##     .I3            1.285    0.110   11.675    0.000
##     .I6            1.636    0.143   11.402    0.000
##     .I8            0.783    0.080    9.779    0.000
##     .I13           1.115    0.132    8.455    0.000
##     .I14           1.822    0.148   12.351    0.000
##     .I16           1.281    0.117   10.908    0.000
##     .I20           1.031    0.138    7.456    0.000
##     .I5            1.407    0.189    7.435    0.000
##     .I10           1.455    0.151    9.653    0.000
##     .I11           1.355    0.158    8.605    0.000
##     .I15           1.004    0.150    6.710    0.000
##     .I22           2.008    0.182   11.008    0.000
##     .I4            0.795    0.116    6.867    0.000
```

##	.I7	0.515	0.077	6.730	0.000
##	.I9	1.108	0.151	7.333	0.000
##	.I12	0.898	0.105	8.574	0.000
##	.I17	0.374	0.057	6.618	0.000
##	.I18	0.906	0.151	6.005	0.000
##	.I19	0.838	0.117	7.140	0.000
##	.I21	1.240	0.137	9.058	0.000
##	EE	1.486	0.156	9.529	0.000
##	DP	0.800	0.179	4.467	0.000
##	PA	0.199	0.057	3.479	0.001

5.3.4   Model Fitting Accounting for Categorization - WLSMV Estimator

With the observed ordinality of these 22 items, we fit the *cfa* model to this data assuming *categorical* observed variables. The implementation in *lavaan* is very straight-forward to use the option *ordered* to specify the ordinality of these observed items as follows:

```
# fit the CFA model with "ordinality*
fitMBI.Cat = cfa(CFA4MBI3, data=dMBI,
    ordered=c("I1","I2","I3","I4","I5","I6","I7","I8",
              "I9","I10","I11","I12","I13","I14","I15",
              "I16","I17","I18","I19","I20","I21","I22"))
# Print the model summary with the fitting measures
summary(fitMBI.Cat, fit.measures=TRUE)
```

```
## lavaan 0.6-12 ended normally after 47 iterations
##
##   Estimator                                     DWLS
##   Optimization method                         NLMINB
##   Number of model parameters                     157
##
##   Number of observations                         372
##
## Model Test User Model:
##                                          Standard      Robust
##   Test Statistic                          454.653     527.628
##   Degrees of freedom                          202         202
##   P-value (Chi-square)                      0.000       0.000
##   Scaling correction factor                               1.022
##   Shift parameter                                        82.961
##     simple second-order correction
##
## Model Test Baseline Model:
##
```

```
## Test statistic                                       21660.267    8323.550
## Degrees of freedom                                         231         231
## P-value                                                  0.000       0.000
## Scaling correction factor                                             2.648
##
## User Model versus Baseline Model:
##
## Comparative Fit Index (CFI)                              0.988       0.960
## Tucker-Lewis Index (TLI)                                 0.987       0.954
##
## Robust Comparative Fit Index (CFI)                                      NA
## Robust Tucker-Lewis Index (TLI)                                         NA
##
## Root Mean Square Error of Approximation:
##
## RMSEA                                                    0.058       0.066
## 90 Percent confidence interval - lower                   0.051       0.059
## 90 Percent confidence interval - upper                   0.065       0.073
## P-value RMSEA <= 0.05                                    0.031       0.000
##
## Robust RMSEA                                                            NA
## 90 Percent confidence interval - lower                                  NA
## 90 Percent confidence interval - upper                                  NA
##
## Standardized Root Mean Square Residual:
##
## SRMR                                                     0.065       0.065
##
## Parameter Estimates:
##
## Standard errors                                    Robust.sem
## Information                                          Expected
## Information saturated (h1) model                 Unstructured
##
## Latent Variables:
##                    Estimate  Std.Err  z-value  P(>|z|)
## EE =~
##   I1                  1.000
##   I2                  0.954    0.031   31.211    0.000
##   I3                  1.065    0.039   27.410    0.000
##   I6                  0.872    0.048   18.045    0.000
##   I8                  1.210    0.042   28.953    0.000
##   I13                 1.092    0.040   27.619    0.000
##   I14                 0.868    0.043   20.030    0.000
##   I16                 0.915    0.048   19.044    0.000
##   I20                 1.009    0.045   22.440    0.000
```

```
##    DP =~
##     I5                  1.000
##     I10                 1.017    0.090   11.315    0.000
##     I11                 1.144    0.092   12.421    0.000
##     I15                 1.108    0.090   12.247    0.000
##     I22                 0.735    0.077    9.602    0.000
##    PA =~
##     I4                  1.000
##     I7                  1.097    0.112    9.824    0.000
##     I9                  1.216    0.136    8.927    0.000
##     I12                 0.861    0.101    8.513    0.000
##     I17                 1.491    0.145   10.283    0.000
##     I18                 1.319    0.138    9.583    0.000
##     I19                 1.325    0.136    9.769    0.000
##     I21                 1.071    0.124    8.642    0.000
##    EE =~
##     I12                -0.436    0.057   -7.636    0.000
##
## Covariances:
##                     Estimate  Std.Err  z-value  P(>|z|)
##   .I6 ~~
##     .I16               0.276    0.032    8.558    0.000
##   .I1 ~~
##     .I2                0.271    0.030    8.982    0.000
##   .I10 ~~
##     .I11               0.182    0.041    4.441    0.000
##    EE ~~
##     DP                 0.331    0.033   10.083    0.000
##     PA                -0.145    0.025   -5.891    0.000
##    DP ~~
##     PA                -0.194    0.028   -6.971    0.000
##
## Intercepts:
##                     Estimate  Std.Err  z-value  P(>|z|)
##    .I1                0.000
##    .I2                0.000
##    .I3                0.000
##    .I6                0.000
##    .I8                0.000
##    .I13               0.000
##    .I14               0.000
##    .I16               0.000
##    .I20               0.000
##    .I5                0.000
##    .I10               0.000
##    .I11               0.000
```

```
##      .I15                0.000
##      .I22                0.000
##      .I4                 0.000
##      .I7                 0.000
##      .I9                 0.000
##      .I12                0.000
##      .I17                0.000
##      .I18                0.000
##      .I19                0.000
##      .I21                0.000
##      EE                  0.000
##      DP                  0.000
##      PA                  0.000
##
## Thresholds:
##                     Estimate  Std.Err  z-value  P(>|z|)
##      I1|t1            -2.299    0.189  -12.190    0.000
##      I1|t2            -0.817    0.074  -11.098    0.000
##      I1|t3            -0.514    0.068   -7.521    0.000
##      I1|t4             0.081    0.065    1.243    0.214
##      I1|t5             0.446    0.067    6.604    0.000
##      I1|t6             1.332    0.091   14.626    0.000
##      I2|t1            -2.299    0.189  -12.190    0.000
##      I2|t2            -1.254    0.088  -14.326    0.000
##      I2|t3            -0.865    0.075  -11.573    0.000
##      I2|t4            -0.266    0.066   -4.034    0.000
##      I2|t5             0.135    0.065    2.071    0.038
##      I2|t6             1.144    0.083   13.753    0.000
##      I3|t1            -1.300    0.090  -14.513    0.000
##      I3|t2            -0.301    0.066   -4.550    0.000
##      I3|t3             0.074    0.065    1.139    0.255
##      I3|t4             0.506    0.068    7.420    0.000
##      I3|t5             0.865    0.075   11.573    0.000
##      I3|t6             1.813    0.124   14.672    0.000
##      I6|t1            -0.700    0.071   -9.831    0.000
##      I6|t2             0.190    0.065    2.898    0.004
##      I6|t3             0.608    0.070    8.735    0.000
##      I6|t4             1.069    0.081   13.272    0.000
##      I6|t5             1.332    0.091   14.626    0.000
##      I6|t6             2.079    0.153   13.548    0.000
##      I8|t1            -0.978    0.078  -12.582    0.000
##      I8|t2             0.081    0.065    1.243    0.214
##      I8|t3             0.401    0.067    5.989    0.000
##      I8|t4             0.780    0.073   10.713    0.000
##      I8|t5             1.069    0.081   13.272    0.000
##      I8|t6             1.813    0.124   14.672    0.000
```

##	I13|t1	-1.332	0.091	-14.626	0.000
##	I13|t2	-0.460	0.068	-6.808	0.000
##	I13|t3	0.000	0.065	0.000	1.000
##	I13|t4	0.600	0.070	8.634	0.000
##	I13|t5	0.967	0.077	12.493	0.000
##	I13|t6	1.562	0.104	15.023	0.000
##	I14|t1	-1.457	0.098	-14.931	0.000
##	I14|t2	-0.744	0.072	-10.323	0.000
##	I14|t3	-0.301	0.066	-4.550	0.000
##	I14|t4	0.294	0.066	4.447	0.000
##	I14|t5	0.726	0.072	10.127	0.000
##	I14|t6	1.332	0.091	14.626	0.000
##	I16|t1	-0.552	0.069	-8.029	0.000
##	I16|t2	0.329	0.066	4.962	0.000
##	I16|t3	0.753	0.072	10.421	0.000
##	I16|t4	1.211	0.086	14.123	0.000
##	I16|t5	1.661	0.111	14.977	0.000
##	I16|t6	2.551	0.246	10.363	0.000
##	I20|t1	-0.315	0.066	-4.756	0.000
##	I20|t2	0.529	0.068	7.725	0.000
##	I20|t3	0.905	0.076	11.946	0.000
##	I20|t4	1.366	0.093	14.729	0.000
##	I20|t5	1.747	0.118	14.837	0.000
##	I20|t6	2.213	0.174	12.755	0.000
##	I5|t1	-0.162	0.065	-2.484	0.013
##	I5|t2	0.560	0.069	8.130	0.000
##	I5|t3	0.905	0.076	11.946	0.000
##	I5|t4	1.332	0.091	14.626	0.000
##	I5|t5	1.497	0.100	14.984	0.000
##	I5|t6	2.551	0.246	10.363	0.000
##	I10|t1	-0.176	0.065	-2.691	0.007
##	I10|t2	0.537	0.069	7.826	0.000
##	I10|t3	0.836	0.074	11.289	0.000
##	I10|t4	1.383	0.094	14.776	0.000
##	I10|t5	1.635	0.109	15.001	0.000
##	I10|t6	2.784	0.324	8.581	0.000
##	I11|t1	-0.142	0.065	-2.174	0.030
##	I11|t2	0.476	0.068	7.012	0.000
##	I11|t3	0.905	0.076	11.946	0.000
##	I11|t4	1.197	0.085	14.051	0.000
##	I11|t5	1.635	0.109	15.001	0.000
##	I11|t6	2.141	0.162	13.195	0.000
##	I15|t1	0.308	0.066	4.653	0.000
##	I15|t2	0.865	0.075	11.573	0.000
##	I15|t3	1.332	0.091	14.626	0.000
##	I15|t4	1.585	0.106	15.023	0.000

##	I15|t5	1.813	0.124	14.672	0.000
##	I15|t6	2.213	0.174	12.755	0.000
##	I22|t1	-0.625	0.070	-8.935	0.000
##	I22|t2	0.372	0.067	5.579	0.000
##	I22|t3	0.692	0.071	9.733	0.000
##	I22|t4	1.057	0.080	13.188	0.000
##	I22|t5	1.366	0.093	14.729	0.000
##	I22|t6	2.141	0.162	13.195	0.000
##	I4|t1	-2.299	0.189	-12.190	0.000
##	I4|t2	-2.024	0.146	-13.834	0.000
##	I4|t3	-1.497	0.100	-14.984	0.000
##	I4|t4	-1.034	0.079	-13.018	0.000
##	I4|t5	-0.122	0.065	-1.864	0.062
##	I7|t1	-2.784	0.324	-8.581	0.000
##	I7|t2	-2.213	0.174	-12.755	0.000
##	I7|t3	-1.717	0.115	-14.896	0.000
##	I7|t4	-1.240	0.087	-14.260	0.000
##	I7|t5	0.054	0.065	0.828	0.407
##	I9|t1	-2.551	0.246	-10.363	0.000
##	I9|t2	-1.887	0.131	-14.426	0.000
##	I9|t3	-1.540	0.103	-15.016	0.000
##	I9|t4	-1.069	0.081	-13.272	0.000
##	I9|t5	-0.735	0.072	-10.225	0.000
##	I9|t6	-0.013	0.065	-0.207	0.836
##	I12|t1	-2.551	0.246	-10.363	0.000
##	I12|t2	-1.929	0.135	-14.265	0.000
##	I12|t3	-1.562	0.104	-15.023	0.000
##	I12|t4	-1.046	0.080	-13.104	0.000
##	I12|t5	-0.529	0.068	-7.725	0.000
##	I12|t6	0.717	0.072	10.029	0.000
##	I17|t1	-2.551	0.246	-10.363	0.000
##	I17|t2	-2.299	0.189	-12.190	0.000
##	I17|t3	-1.661	0.111	-14.977	0.000
##	I17|t4	-1.332	0.091	-14.626	0.000
##	I17|t5	-0.156	0.065	-2.381	0.017
##	I18|t1	-2.551	0.246	-10.363	0.000
##	I18|t2	-1.779	0.120	-14.763	0.000
##	I18|t3	-1.609	0.107	-15.016	0.000
##	I18|t4	-0.925	0.076	-12.130	0.000
##	I18|t5	-0.498	0.068	-7.318	0.000
##	I18|t6	0.568	0.069	8.231	0.000
##	I19|t1	-2.784	0.324	-8.581	0.000
##	I19|t2	-1.929	0.135	-14.265	0.000
##	I19|t3	-1.717	0.115	-14.896	0.000
##	I19|t4	-1.144	0.083	-13.753	0.000
##	I19|t5	-0.735	0.072	-10.225	0.000

##					
##	I19|t6	0.315	0.066	4.756	0.000
##	I21|t1	-1.887	0.131	-14.426	0.000
##	I21|t2	-1.497	0.100	-14.984	0.000
##	I21|t3	-1.023	0.079	-12.932	0.000
##	I21|t4	-0.649	0.070	-9.235	0.000
##	I21|t5	0.358	0.067	5.373	0.000

```
## 
## Variances:
##                    Estimate   Std.Err   z-value   P(>|z|)
##     .I1              0.452
##     .I2              0.501
##     .I3              0.379
##     .I6              0.584
##     .I8              0.198
##     .I13             0.347
##     .I14             0.587
##     .I16             0.541
##     .I20             0.442
##     .I5              0.588
##     .I10             0.574
##     .I11             0.460
##     .I15             0.494
##     .I22             0.777
##     .I4              0.695
##     .I7              0.633
##     .I9              0.549
##     .I12             0.561
##     .I17             0.323
##     .I18             0.470
##     .I19             0.465
##     .I21             0.651
##      EE              0.548      0.036    15.330    0.000
##      DP              0.412      0.051     8.121    0.000
##      PA              0.305      0.055     5.499    0.000
## 
## Scales y*:
##                    Estimate   Std.Err   z-value   P(>|z|)
##      I1              1.000
##      I2              1.000
##      I3              1.000
##      I6              1.000
##      I8              1.000
##      I13             1.000
##      I14             1.000
##      I16             1.000
##      I20             1.000
```

##	I5	1.000
##	I10	1.000
##	I11	1.000
##	I15	1.000
##	I22	1.000
##	I4	1.000
##	I7	1.000
##	I9	1.000
##	I12	1.000
##	I17	1.000
##	I18	1.000
##	I19	1.000
##	I21	1.000

For fitting categorical data, *lavaan* will automatically using *WLSMV* estimator, where the parameters are estimated with the *diagonally weighted least squares(DWLS)*. The *robust* standard errors are computed by the full weight matrix and the associated test statistics are mean- and variance-adjusted. All these changes can be seen from the above output.

Different from the default fitting using MLE, the model fitting with categorical data has an extra component denoted by *Thresholds* for all the 22 items, each with 6 thresholds. This is similar to the classical regression analysis we have known before where when a categorical variable with K categories, we need to construct a set of $K - 1$ dummy variables to account for these K categories. In this data, we have $K = 7$ categories, which is why *lavaan* estimated 6 (i.e., $K - 1$) thresholds.

5.3.5   Comparison among MLE, MLR and WLSMV

Examine the outputs from these three estimators, we can see that there are visible differences among the fitting measures and the parameter estimations.

We examine the parameter estimation more closely here. We first extract all the information associated with parameter estimation using *lavaan* function *parameterEstimates* as follows:

```
# Extract all the tables related to the parameter estimates
Est.Default   = parameterEstimates(fitMBI.Default);
Est.Robust    = parameterEstimates(fitMBI.Robust);
Est.Cat       = parameterEstimates(fitMBI.Cat);
```

We do not intent to print any of them here since there are printed out in the previous model fittings and very lengthy for all of them to be printed again.

To compare the three model fitting, we only examine the estimated factor loadings, which are associated with the latent variables and related to the *lavaan* operator (*op*) of "=˜". Therefore, the following *R* code chunk will extract the parameter estimates and their associated standard errors:

```
# Extract from the default MLE
Load.Default = Est.Default[Est.Default[,"op"]==c("=~"),
                           c("lhs","op","rhs","est","se")]
colnames(Load.Default) = c("lhs","op","rhs","est.Default",
                           "se.Default")
# Extract from the MLR
Load.Robust  = Est.Robust[Est.Robust[,"op"]==c("=~"),
                          c("lhs","op","rhs","est","se")]
colnames(Load.Robust) = c("lhs","op","rhs","est.Robust",
                          "se.Robust")
# Extract from WLSMV
Load.Cat = Est.Cat[Est.Cat[,"op"]==c("=~"),
                   c("lhs","op","rhs","est","se")]
colnames(Load.Cat) = c("lhs","op","rhs","est.Cat","se.Cat")
# merge all three together
Load = merge(merge(Load.Default, Load.Robust), Load.Cat)
# Print the estimates from three methods
Load
```

```
##     lhs op rhs est.Default se.Default est.Robust
## 1   DP =~ I10       0.889      0.1136      0.889
## 2   DP =~ I11       1.105      0.1253      1.105
## 3   DP =~ I15       0.921      0.1048      0.921
## 4   DP =~ I22       0.776      0.1154      0.776
## 5   DP =~  I5       1.000      0.0000      1.000
## 6   EE =~  I1       1.000      0.0000      1.000
## 7   EE =~ I12      -0.316      0.0496     -0.316
## 8   EE =~ I13       1.072      0.0727      1.072
## 9   EE =~ I14       0.880      0.0754      0.880
## 10  EE =~ I16       0.727      0.0629      0.727
## 11  EE =~  I2       0.878      0.0488      0.878
## 12  EE =~ I20       0.806      0.0615      0.806
## 13  EE =~  I3       1.073      0.0750      1.073
## 14  EE =~  I6       0.764      0.0694      0.764
## 15  EE =~  I8       1.215      0.0746      1.215
## 16  PA =~ I12       1.131      0.1878      1.131
## 17  PA =~ I17       1.327      0.1756      1.327
## 18  PA =~ I18       1.890      0.2547      1.890
## 19  PA =~ I19       1.695      0.2322      1.695
## 20  PA =~ I21       1.342      0.2133      1.342
## 21  PA =~  I4       1.000      0.0000      1.000
## 22  PA =~  I7       0.973      0.1472      0.973
## 23  PA =~  I9       1.763      0.2481      1.763
```

```
##      se.Robust est.Cat se.Cat
## 1      0.1292    1.017 0.0899
## 2      0.1431    1.144 0.0921
## 3      0.1143    1.108 0.0905
## 4      0.1205    0.735 0.0766
## 5      0.0000    1.000 0.0000
## 6      0.0000    1.000 0.0000
## 7      0.0547   -0.436 0.0571
## 8      0.0745    1.092 0.0395
## 9      0.0629    0.868 0.0433
## 10     0.0755    0.915 0.0481
## 11     0.0414    0.954 0.0306
## 12     0.0700    1.009 0.0450
## 13     0.0561    1.065 0.0388
## 14     0.0791    0.872 0.0483
## 15     0.0688    1.210 0.0418
## 16     0.2166    0.861 0.1011
## 17     0.2127    1.491 0.1450
## 18     0.3546    1.319 0.1377
## 19     0.3421    1.325 0.1356
## 20     0.2147    1.071 0.1239
## 21     0.0000    1.000 0.0000
## 22     0.1240    1.097 0.1117
## 23     0.3473    1.216 0.1363
```

It can be seen that the parameter estimates for both the default MLE and
MLR are the same, but all the SEs are different among the three estimators.
To make this clearer, we first graphically show the parameter estimates and
SEs as seen in Figure 5.5. This figure is produced with the *R* function *plot* to
plot the default MLE *est* and *se* with points. Then we add the *MLR est/se*
on the top of the plots using *R* function *points* with option *pch-2* as *triangles*.
The *abline* is to add a 1-to-1 line to the figure to show whether they are equal.

```
par(mfrow=c(1,2))
plot(est.Cat~est.Default, Load,
     xlim=c(-0.5, 1.5), ylim=c(-0.5, 1.5),las=1,
     xlab="Estimated Loadings from MLE",
     ylab="Estimated Loadings from MLR/DWLS");
points(est.Robust~est.Default, Load, pch=2)
abline(0,1, lwd=3, col="red")
plot(se.Cat~se.Default, Load,
     ylim=c(0, 0.35), xlim=c(0, 0.35),las=1,
     xlab="SEs for Loadings from MLE",
     ylab="SEs for Loadings from MLR/DWLS");
```

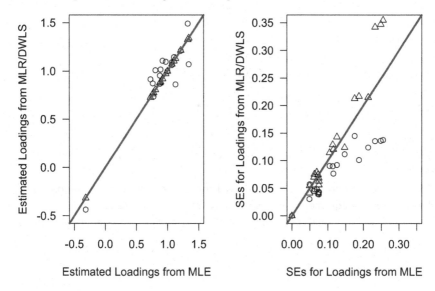

FIGURE 5.5
Parameter Estimates and Standard Errors from Three Estimates.

```
points(se.Robust~se.Default, Load, pch=2)
abline(0,1, lwd=3, col="red")
```

Closely examine Figure 5.5, the parameter estimates are generally close to the 1-to-1 line as seen in the left-side of the figure. For SEs on the right-side of the figure, we can seen that the SEs for the default MLE are close to those from the *MLR* (i.e., the *triangles*), but the SEs from the *WLSMV* are less than those from MLE/MLR, which should be intuitively true since the data are not normally distributed anymore and they are categorical with ordinality. The estimated standard errors should take the categorical nature into account, resulting better estimation of standard errors.

We can further examine on average how less the SEs from the *WLSMV* using a regression through origin as follows:

```
summary(lm(se.Cat~-1+se.Default, Load))
```

```
##
## Call:
## lm(formula = se.Cat ~ -1 + se.Default, data = Load)
##
## Residuals:
##      Min       1Q   Median       3Q      Max
## -0.02123 -0.00692  0.00000  0.01137  0.03540
```

```
##
## Coefficients:
##            Estimate Std. Error t value
## se.Default   0.6240     0.0238    26.3
##                        Pr(>|t|)
## se.Default <0.0000000000000002 ***
## ---
## Signif. codes:
## 0 '***' 0.001 '**' 0.01 '*' 0.05 '.' 0.1 ' ' 1
##
## Residual standard error: 0.0151 on 22 degrees of freedom
## Multiple R-squared:  0.969,  Adjusted R-squared:  0.968
## F-statistic:  690 on 1 and 22 DF,  p-value: <0.0000000000000002
```

This shows that on average, the SEs from *WLSMV* are only 0.624 times of those SEs from the default MLE.

5.4   Data Analysis Using SAS

At present, *PROC CALIS* does not have estimation methods implemented specifically for analyzing categorical responses.

5.5   Discussions

In this chapter, we introduced *WLSMV* estimator for categorical data with application to the *MBI* data and we demonstrated that the *WLSMV* estimator is more efficient with smaller standard error estimation than the default *MLE* and the robust *MLE*.

To be emphasized, the *WLSMV* estimator is a general method for categorical data in SEM which has also been implemented in most of the software, such as, *MPlus*. *WLSMV* is a general keyword that simultaneously requests the *DWLS* estimator and a *mean- and variance-adjusted (MV)* chi-squared test statistic. Therefore in *lavaan*, we can just call *lavaan(..., estimator = "WLSMV")*, which is equivalent to *lavaan(..., estimator = "DWLS", se = "robust.sem", test = "scaled.shifted")*.

Besides the *WLSMV*, *lavaan* also has an experimental *marginal maximum likelihood* estimator (*estimator = "MML"*). The *MML* may be slow in estimation due to the numerical integration associated with this estimator and more likely to have convergence problems.

Another estimator is the *ADF* (*estimator* = *"WLS"*), which does not make any specific distributional assumptions about the data. However, this *WLS* estimator does require a very large sample size ($N > 5000$) for the parameter estimation (i.e., parameter estimates, SEs, and the test statistics) to be reasonably trusted.

5.6   Exercises

To further investigate the model fitting on categorical data,

1. Dichotomize the *dMBI* data using the following *R* code:

```
# Dichotomized the MBI
dMBIBinary = as.data.frame(lapply(dMBI, cut, 2,
                           labels=FALSE) )
# Print the first six observations
head(dMBIBinary)
```

2. Re-fit the dichotomized binary data *dMBIBinary* with the *MLE*, *MLR* and *WLSMV*.

3. Compare the parameter estimates and SEs associated with these three estimators.

6

Multi-Group Data Analysis: Continuous Data

So far we have discussed structural equation modeling based on *single* samples. In this chapter, we will discuss SEM for multiple samples, i.e., multi-group (or multiple-group) analysis, for continuous data, and further consider multi-group analysis for categorical data in Chapter 7.

In multi-group SEM, we focus on whether or not the SEM components, such as the measurement model and/or the structural model, are *invariant* (i.e., equivalent) across the multi-group of interest, such as gender, age, culture, intervention treatment groups, etc. For example, in Chapter 2, we used a data set (i.e., *elemmbi.csv*) on *Maslach Burnout Inventory (MBI)* from 372 elementary school male teachers to build a CFA model to examine the three dimensions of burnout, including *Emotional Exhaustion (EE)*, *Depersonalization (DP)*, and *Reduced Personal Accomplishment (PA)*. The *MBI* was further applied to 580 elementary school teachers and 692 secondary school teachers. We would be interested in confirming whether this *MBI* instrument is invariant (or equivalent) for both groups of teachers.

Invariance in a multi-group SEM analysis can refer to different kinds of model aspects. A very convenient way is to separate the study of the invariance into two components in the model:

1. Invariance in the measurement models: this is to test whether the measurement instruments are invariant or equivalent across multiple groups. In studying this type of invariance, we test the equality of the parameters associated with the measurement model across groups. The set of parameters includes the factor loadings, intercepts of measurement variables, and residual variances and covariances (of measurement variables).

2. Invariance in structural models: this is to test whether the latent structures are invariant or equivalent across multiple groups. In studying this type of invariance, we test for the equality of the parameters associated with the latent structure model across groups. The set of parameters includes the path coefficients for describing the functional relationships among factors, means and variances of factors, and covariances between factors.

DOI: 10.1201/9781003365860-6

In general, model parsimony (simplicity) is sought after in multi-group analysis so that measurement and structural invariance of the highest possible degree is desirable. However, invariance models are restrictive and they trade model fit for parsimony. In practice, therefore, a researcher must strike a balance between model parsimony and model fit. As a result, when fitting a multi-group model, a researcher usually settle on models with a certain degree of invariance with an adequate model fit—it could be measurement invariance only, structural invariance only, or invariance of a subset of parameters in the measurement and structural components.

This chapter demonstrates how a partial invariance model can be built with satisfactory model fit by fitting a multi-group CFA model to two independent samples (groups). We will use the two extra data sets on *Maslach Burnout Inventory (MBI)* with one from 580 elementary school male teachers and the other from 692 secondary school teachers to illustrate the test of invariance of the CFA models between these two groups of teachers. The data set and the associated analysis can be found in Chapter 7 of Byrne (2012).

6.1   Data Descriptions

As discussed in Chapter 1, the *MBI* measures three dimensions of burnout which include *Emotional Exhaustion (EE)*, *Depersonalization (DP)*, and *Reduced Personal Accomplishment (PA)*. As a measurement tool, the *MBI* is a 22-item instrument structured on a 7-point Likert-type scale that ranges from 0 (*feeling has never been experienced*) to 6 (*feeling experienced daily*). Note that here the 7-point Likert-type scale is coded from 0 to 6 instead of being coded from 1 to 7 as in Chapter 2.

The data set for elementary school teachers is named *mbielm1.dat*, which contains the 22 item-level data for the 580 elementary male teachers. Since this data was saved in a *tab delimited text* file, We can read this data into *R* using *R* function *read.table* as follows:

```
# Read the elementary school teachers' data into R using
  'read.table'
dElem = read.table("data/mbielm1.dat", header=F)
# Name the variables
colnames(dElem) = c("I01","I02","I03","I04","I05","I06","I07",
                    "I08","I09","I10","I11","I12","I13","I14",
                    "I15","I16","I17","I18","I19","I20","I21",
                    "I22")
# Check the data dimension
dim(dElem)

## [1] 580   22
```

```
# Print the data summary
#summary(dElem)
```

Similarly, we can read the data set for the 692 secondary school teachers into *R* as follows:

```
# Read the secondary school teachers data into R using
  'read.table'
dSec = read.table("data/mbisec1.dat", header=F)
colnames(dSec) = c("I01","I02","I03","I04","I05","I06","I07",
                   "I08","I09","I10","I11","I12","I13","I14",
                   "I15","I16","I17","I18","I19","I20","I21",
                   "I22")
# Check the data dimension
dim(dSec)
```

```
## [1] 692  22
```

```
# Print the data summary
#summary(dSec)
```

For multi-group analysis, we need to combine these two data sets together to create one data set for further analysis. This can be done using *R* function *rbind* to *row-bind* these two data sets as follows:

```
# Stack these two data together
dMBI = rbind(data.frame(School="Elem", dElem),
             data.frame(School="Sec", dSec))
# dimension of the new data
dim(dMBI)
```

```
## [1] 1272    23
```

```
# Print the data summary
#summary(dMBI)
```

As seen from the data summary, the new data has 1272 observations (i.e., 580 + 692) and 23 variables (the newly created *School* along with the 22 items). The data distribution by the two *School* groups can be seen in Figure 6.1.

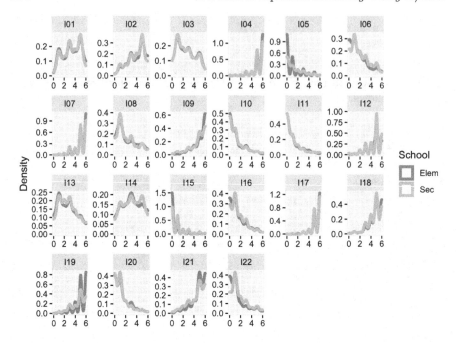

FIGURE 6.1
The Density Plot of all 22 Items by the Two School Groups.

```
# Call tidyverse to reshape the data
library(tidyverse)
dMBIlong <- dMBI %>%
  gather("I01","I02","I03","I04","I05","I06","I07","I08","I09",
         "I10","I11","I12","I13","I14","I15","I16","I17","I18",
         "I19","I20","I21","I22",
         key="Item", value="Value")
# The dimension of the reshaped data
dim(dMBIlong)
```

```
## [1] 27984      3
```

```
# Call ggplot2 to make the density plot
library(ggplot2)
ggplot(dMBIlong, aes(Value, col=School))+
  geom_density(size=1.5)+
  facet_wrap(~Item, nrow=4, scales="free")+
  labs(x=" ", y="Density")
```

6.2   Model and Estimation

In Chapter 2, we have discussed the mean and covariance structures of CFA models in the context of a single-group analysis. Estimation of model parameters given a sample can be done by selecting or finding the best set of estimates that minimizes a discrepancy function. The maximum likelihood (ML) discrepancy function of the following form is the most popular one applied in practice:

$$F_{ML}\left((S,\bar{x}),(\Sigma(\theta),\mu(\theta))\right) = \log|\Sigma(\theta)| - \log|S| + \text{trace}\left(S\Sigma(\theta)^{-1}\right) - p$$
$$+(\bar{x}-\mu(\theta))'\Sigma(\theta)^{-1}(\bar{x}-\mu(\theta)) \tag{6.1}$$

where $\mu(\cdot)$ and $\Sigma(\cdot)$ represent the hypothesized mean and covariance structures respectively, $\bar{x}$ and S are the sample mean vector and covariance matrix respectively, and θ is the parameter vector that includes intercepts, factor loadings, factor means, variances and covariances, and residual variances and covariances in the CFA model. See Chapter 2 for a detailed explanation of the mean and covariance structures of the CFA model.

Extension of this single-sample estimation method to the multi-group CFA analysis is straightforward. First, an individual discrepancy function for each group or sample is defined and is indexed by a group indicator, say, g. That is, for the g-th group, the component ML discrepancy function is:

$$F_{\text{ML}}^{(g)}\left((S_g,\bar{x}_g),(\Sigma(\theta_g),\mu(\theta_g))\right) = \log|\Sigma(\theta_g)| - \log|S_g| + \text{trace}\left(S_g\Sigma(\theta_g)^{-1}\right) - p$$
$$+(\bar{x}_g-\mu(\theta_g))'\Sigma(\theta_g)^{-1}(\bar{x}_g-\mu(\theta_g)) \tag{6.2}$$

In particular, the discrepancy function $F_{\text{ML}}^{(1)}\left((S_1,\bar{x}_1),(\Sigma(\theta_1),\mu(\theta_1))\right)$ is for the first group, $F_{\text{ML}}^{(2)}\left((S_2,\bar{x}_2),(\Sigma(\theta_2),\mu(\theta_2))\right)$ is for the second group, and so on. To simplify notation for the purpose of defining multi-group maximum likelihood estimation, we use $F_{\text{ML}}^{(1)}(\theta_1)$, $F_{\text{ML}}^{(2)}(\theta_2)$, ..., and so on to denote these individual discrepancy functions.

Notice that in the current context of multi-group analysis, the mean and covariance structures $\mu(\cdot)$ and $\Sigma(\cdot)$ remain the same for all groups, even though the parameter values θ_g could be different across groups. In most situations of practical multi-group SEM modeling, interests are placed on testing partial or complete constraints on the values of θ_g's across groups. Depending on the particular sets of parameters that are constrained across groups, different kinds of invariance models are defined. The next few sections describe different types of invariance models in more detail.

In order to estimate all parameters in independent groups simultaneously, an overall discrepancy function is defined as a weighted sum of the group-specific discrepancy functions. That is, the following objective function is minimized for the multi-group maximum likelihood estimation of the structural equation model:

$$F_{\mathrm{ML}}(\theta_1, \theta_2, ..., \theta_G) = \sum_{g=1}^{G} \frac{N_g}{N} F_{\mathrm{ML}}^{(g)}(\theta_g) \tag{6.3}$$

where $N = \sum_{g=1}^{G} N_g$ is the sum of sample sizes N_g's for indepdenet groups or samples. By definition, the maximum likelihood estimates of θ_g's would minimize the discrepancy function (6.3). In some software, the weight $\frac{N_g-1}{N-G}$, instead $\frac{N_g}{N}$, is being used in defining the multi-group maximum likelihood discrepancy function (6.3). The two types of weighting are asymptotically equivalent, although they might produce slightly different sets of estimates in practical applications when the sample size is not large.

We now explain different kinds of invariance models in the context of confirmatory factor analysis.

6.2.1   Unconstrained θ_g's Across Groups: Configural Invariance Model

In this case, the hypothesized model for the multi-group analysis does **not** place any constraints on the parameters θ_g's for groups. That is, it does not assume that $\theta_1 = \theta_2 = \cdots = \theta_g = \cdots = \theta_G$ in the hypothesized model. Each group can therefore have distinct estimates of the CFA model parameters. However, the form of the mean and covariance structures, as represented by $\mu(\cdot)$ and $\Sigma(\cdot)$, of the CFA model remains invariant across groups.

Put in another way, in a configural invariance model, the reproduced or implied mean and covariance structures for the groups are using exactly the same mean and covariance functions, $\mu(\cdot)$ and $\Sigma(\cdot)$, but with different parameter estimates being plugged in. Notationally, the implied mean and covariance structures for group 1 are $\mu(\theta_1)$ and $\Sigma(\theta_1)$ respectively, that for group 2 are $\mu(\theta_2)$ and $\Sigma(\theta_2)$ respectively, and so on. Therefore, even with unconstrained θ_g's across groups, this least restrictive model is still being labeled as a configural invariance model. It is important to note that configural invariance does not require that the CFA model for groups must be a "standard' one (e.g., no error covariances among observed variables), it only requires that the model *structure*, as reflected by the mean and covariance functions $\mu(\cdot)$ and $\Sigma(\cdot)$, remains the same across groups.

From a software application perspective, a configural invariance model can be implemented simply by specifying the *same* model for each group in a multi-group analysis. Usually, the parameters across models for groups are not constrained by default and so this accomplishes the specification of a configural invariance model for the groups. Note that because no restrictions are actually placed on the parameters across groups, another way to obtain the estimates of $\theta_1, \theta_2, ..., \theta_G$ is to fit the same model to the groups separately.

6.2.2 Completely Constrained θ_g's Across Groups: *Ideal* Model

In this case, $\theta_1 = \theta_2 = \cdots . = \theta_g = \cdots = \theta_G = \theta$ is assumed in the hypothesized CFA model for all groups and θ is the common parameter vector that is applied to each group to yield the reproduced or implied mean and covariance structures for all groups. This model is highly constrained and is certainly quite parsimonious. However, it is also highly restrictive and rarely to fit the data well in applications. In this sense, this model is labeled here as an *ideal* model.

6.2.3 Partial Invariance Models

In terms of model restrictions, partial invariance models are between the configural invariance model and the *ideal* model. Although it does not impose the strict equality $\theta_1 = \theta_2 = \cdots = \theta_g = \cdots = \theta_G$ in the model for multiple groups, it hypothesizes that a non-empty subset of parameters in each of the parameter vectors $\theta_1, \theta_2, \ldots, \theta_G$ is invariant across groups. That is, some (but not all) parameters in $\theta_1, \theta_2, \theta_g, \ldots, \theta_G$ have the same values across groups.

In the beginning of this chapter, we mentioned that parameters in an SEM or CFA can be classified into either the measurement component or structural component of the model. The measurement component contains the factor loadings, intercepts for observed variables, and residual variances and covariances (for observed variables) as parameters. The structural component contains the path coefficients for describing the functional relationships among factors (which are not present in a CFA model but can be present in a general structural equation model), means and variances of factors, and covariances between factors.

In multi-group analyses, (strict) measurement invariance refers to the fact that *all* parameters in the measurement component are constrained to be equal in all groups. In addition, (strict) structural invariance refers to the fact that *all* parameters in the structural component are constrained to be equal in all groups. An *ideal* model is being fitted when you require both measurement and structural invariance in a multi-group analysis. As mentioned, an *ideal* model is usually unrealistic in practice. Sometimes, we cannot even expect a strict measurement invariance or strict structural invariance to fit the data well. Partial invariance models are usually more reasonable. However, partial invariance can be specified with any combination of parameters in the measurement and structural components. Thus, searching for a suitable partial invariance model for the data involves a lot of possibilities of selecting the sets of parameters to be constrained for invariance.

Fortunately, there are some workable strategies one can follow to establish a partial invariance model. For example, you might like to establish configural invariance, measurement invariance, and structural invariance in sequence. To be more specific, the next section describes a step-by-step strategy to find a good multi-group model with partial invariance.

Note that the partial invariance property described in this section (and generally this chapter) refers to the invariance of specific types of parameters in the multi-group model. Implicitly, the invariance property applies to *all* variables in question. For example, if the factor loadings are invariant across groups, we implicitly refer to the invariance of all factor loadings for *all* variables in the model. With a broader definition, such a partial invariance might also refer to the invariance of factor loadings in only *some* of the variables across groups. However, due to the complex interpretations incurred by such a broader definition, we would not consider it further.

6.2.4   Strategy of Multi-group Analysis

A typical strategy for multi-group analysis includes four main steps. The first step is to establish the group-specific *baseline* models for all groups. In this step, a satisfactory model should be built *separately* with the best model fitting for each group based on the principles of model parsimony and the substantive knowledge.

Once the group-specific *baseline* models are established, the second step is to fit the multi-group *baseline* model with all the available data from all groups where all the path parameters are group-specific. This multi-group *baseline* model is sometimes called *configural* model, which is the first model to be tested. As discussed previously, this model is also the *configural invariance* model (Horn and McArdle, 1992), which emphasizes the invariance of the pattern of the model. At this *configural* step, the *configuration* of the group-specific *baseline* model is re-established for all groups as the multi-group *baseline* model with no invariance requirement for all the model parameters. This is the least restrictive model to be established and this *configural* model will be *group-variant*, which is obviously not the *best* model we are aiming for—more parsimonious models might be sought after.

Opposite to the *configural* model, the third step is to fit the most restrictive model where all parameters (including all factor loading, residual variances, regression paths, etc.) are constraint to be equal, which is *group-invariant*. In other words, both the measurement and structural components are strictly invariant across models for groups. This *ideal* model is easy to fit, but rare to be established in real data analysis due to many different unique aspects from the groups. But if established with suitable model fit, an *ideal* model is found for all groups, which would be the *best* model for the data and we are done with the analysis.

More realistically, the *best* model for any data analysis would be a more *practical* model between the *configural* model and the *ideal* model. Therefore, the fourth step is to search for *partial invariance* within the measurement model and structural model. This will be an iterative process involving not only model fitting and selection, but mostly the substantive knowledge of the data. Therefore, this step is the hardest step in multi-group model search.

In this fourth step, we usually start with the test for *measurement model invariance* or simply *measurement invariance*. In measurement model invariance, we test the parameters associated with the observed variables and their links to the latent variables. Specifically, these parameters include the factor loadings (i.e., the links between the observed variables and the latent variables), the intercepts on the observed variables (i.e., the means of these observed variables), and the residual variances of the observed variables. Three types of measurement invariance are distinguished:

- Weak or metric invariance: only factor loadings are invariant

- Strong or scalar invariance: factor loadings and measurement intercepts are invariant

- Strict (measurement) invariance: all parameters in the measurement components are invariant

A general consensus is that we must at least establish the weak or metric invariance before seeking structural invariance in the multi-group model. The reason is that weak invariance would support the notion that latent factors in multiple groups are measuring the *same* constructs and thus it would make sense to compare the factors across groups to see if they have the same latent structures. A model with structure invariance but lacking measurement invariance, even if it fits the data well, is not interpretable.

After we establish the *measurement model invariance*, we can test for the *structural model invariance* next, focusing on the structural parameters— that is, the means and variances/covariances of the latent variables, and the regression paths among latent factors (if present in the model).

6.3 Data Analysis Using *R*

We now demonstrate our step-by-step strategy to fit a multi-group model to the MBI data for elementary and secondary school teachers.

6.3.1 Establish Group-Specific Baseline Models

As the first step, we fit the CFA model for each group (i.e., elementary teachers, v.s., secondary teachers) to identify satisfactory model fitting. We do not intend to repeat all the steps in the illustration described in Chapter 2. Instead, we show the results in model fitting of these two groups.

Following the results in Chapter 2, we first specify the CFA model as follows:

```
#
# CFA model for MBI
#
CFA4MBI <- '
# Latent variables
EE =~ I01+I02+I03+I06+I08+I13+I14+I16+I20
DP =~ I05+I10+I11+I15+I22
PA =~ I04+I07+I09+I12+I17+I18+I19+I21

# Additional item correlations based on MIs
I01 ~~ I02
I06 ~~ I16
I10 ~~ I11
I04 ~~ I07
I09 ~~ I19
# Additional cross-loading from EE based on MI
EE =~ I12
EE =~ I11

# Latent variable covariance
EE ~~ DP
EE ~~ PA
DP ~~ PA

# Intercepts for items
I01 ~ 1; I02 ~ 1; I03 ~ 1; I04 ~ 1; I05 ~ 1
I06 ~ 1; I07 ~ 1; I08 ~ 1; I09 ~ 1; I10 ~ 1
I11 ~ 1; I12 ~ 1; I13 ~ 1; I14 ~ 1; I15 ~ 1
I16 ~ 1; I17 ~ 1; I18 ~ 1; I19 ~ 1; I20 ~ 1
I21 ~ 1; I22 ~ 1
' # end of "CFA4MBI" model
```

With this model, we can call *sem* to fit the aforementioned model for *elementary school teachers* as follows:

```
# Fit the CFA model for elementary teachers
fitMBI.Elem = cfa(CFA4MBI, data=dElem)
# Print the model summary with the fitting measures
summary(fitMBI.Elem, fit.measures=TRUE)


## lavaan 0.6-12 ended normally after 55 iterations
##
##   Estimator                                         ML
##   Optimization method                           NLMINB
##   Number of model parameters                        76
```

```
##
##    Number of observations                          580
##
## Model Test User Model:
##
##    Test statistic                              527.608
##    Degrees of freedom                              199
##    P-value (Chi-square)                          0.000
##
## Model Test Baseline Model:
##
##    Test statistic                             5608.130
##    Degrees of freedom                              231
##    P-value                                       0.000
##
## User Model versus Baseline Model:
##
##    Comparative Fit Index (CFI)                   0.939
##    Tucker-Lewis Index (TLI)                      0.929
##
## Loglikelihood and Information Criteria:
##
##    Loglikelihood user model (H0)            -19651.508
##    Loglikelihood unrestricted model (H1)    -19387.704
##
##    Akaike (AIC)                              39455.016
##    Bayesian (BIC)                            39786.606
##    Sample-size adjusted Bayesian (BIC)       39545.336
##
## Root Mean Square Error of Approximation:
##
##    RMSEA                                         0.053
##    90 Percent confidence interval - lower        0.048
##    90 Percent confidence interval - upper        0.059
##    P-value RMSEA <= 0.05                         0.155
##
## Standardized Root Mean Square Residual:
##
##    SRMR                                          0.048
##
## Parameter Estimates:
##
##    Standard errors                            Standard
##    Information                                Expected
##    Information saturated (h1) model         Structured
```

```
##
## Latent Variables:
##                     Estimate  Std.Err  z-value  P(>|z|)
##    EE =~
##      I01             1.000
##      I02             0.916    0.040    23.030   0.000
##      I03             1.032    0.057    18.120   0.000
##      I06             0.807    0.056    14.352   0.000
##      I08             1.264    0.060    21.184   0.000
##      I13             1.050    0.057    18.302   0.000
##      I14             0.956    0.061    15.570   0.000
##      I16             0.758    0.051    14.740   0.000
##      I20             0.868    0.048    18.120   0.000
##    DP =~
##      I05             1.000
##      I10             0.977    0.083    11.772   0.000
##      I11             0.750    0.100    7.536    0.000
##      I15             0.653    0.060    10.870   0.000
##      I22             0.612    0.078    7.868    0.000
##    PA =~
##      I04             1.000
##      I07             1.097    0.121    9.078    0.000
##      I09             1.886    0.233    8.105    0.000
##      I12             1.342    0.190    7.071    0.000
##      I17             1.521    0.172    8.846    0.000
##      I18             1.978    0.228    8.679    0.000
##      I19             1.856    0.215    8.643    0.000
##      I21             1.350    0.193    7.001    0.000
##    EE =~
##      I12            -0.380    0.043    -8.753   0.000
##      I11             0.222    0.058    3.835    0.000
##
## Covariances:
##                     Estimate  Std.Err  z-value  P(>|z|)
##   .I01 ~~
##     .I02             0.506    0.063    8.082    0.000
##   .I06 ~~
##     .I16             0.860    0.080    10.763   0.000
##   .I10 ~~
##     .I11             0.617    0.081    7.605    0.000
##   .I04 ~~
##     .I07             0.170    0.030    5.676    0.000
##   .I09 ~~
##     .I19             0.099    0.049    2.019    0.043
##    EE ~~
```

```
##      DP              0.782    0.086     9.137    0.000
##      PA             -0.208    0.034    -6.154    0.000
##   DP ~~
##      PA             -0.215    0.033    -6.452    0.000
##
## Intercepts:
##                   Estimate  Std.Err   z-value  P(>|z|)
##      .I01            3.409    0.069    49.484    0.000
##      .I02            3.976    0.065    60.903    0.000
##      .I03            2.572    0.071    36.229    0.000
##      .I04            5.412    0.039   139.751    0.000
##      .I05            1.053    0.061    17.177    0.000
##      .I06            1.676    0.069    24.353    0.000
##      .I07            5.338    0.036   146.968    0.000
##      .I08            2.184    0.075    29.012    0.000
##      .I09            5.031    0.055    90.946    0.000
##      .I10            1.164    0.061    18.990    0.000
##      .I11            1.122    0.063    17.931    0.000
##      .I12            4.693    0.053    87.789    0.000
##      .I13            2.548    0.072    35.621    0.000
##      .I14            3.122    0.076    41.263    0.000
##      .I15            0.545    0.047    11.713    0.000
##      .I16            1.433    0.063    22.715    0.000
##      .I17            5.416    0.037   147.600    0.000
##      .I18            4.883    0.050    96.794    0.000
##      .I19            5.007    0.047   106.468    0.000
##      .I20            1.281    0.060    21.456    0.000
##      .I21            4.841    0.054    90.467    0.000
##      .I22            1.328    0.064    20.729    0.000
##      EE              0.000
##      DP              0.000
##      PA              0.000
##
## Variances:
##                   Estimate  Std.Err   z-value  P(>|z|)
##      .I01            1.213    0.081    14.931    0.000
##      .I02            1.182    0.078    15.195    0.000
##      .I03            1.284    0.086    14.996    0.000
##      .I06            1.745    0.108    16.105    0.000
##      .I08            0.829    0.067    12.281    0.000
##      .I13            1.273    0.085    14.907    0.000
##      .I14            1.913    0.121    15.864    0.000
##      .I16            1.424    0.089    16.033    0.000
##      .I20            0.908    0.061    14.996    0.000
##      .I05            1.197    0.097    12.381    0.000
```

```
##    .I10              1.237    0.098   12.662    0.000
##    .I11              1.381    0.099   13.894    0.000
##    .I15              0.835    0.059   14.236    0.000
##    .I22              2.009    0.125   16.030    0.000
##    .I04              0.711    0.044   16.017    0.000
##    .I07              0.573    0.037   15.555    0.000
##    .I09              1.208    0.083   14.629    0.000
##    .I12              0.938    0.061   15.362    0.000
##    .I17              0.412    0.031   13.133    0.000
##    .I18              0.853    0.062   13.858    0.000
##    .I19              0.734    0.054   13.474    0.000
##    .I21              1.371    0.085   16.127    0.000
##    EE                1.539    0.150   10.244    0.000
##    DP                0.985    0.126    7.818    0.000
##    PA                0.159    0.033    4.813    0.000
```

As seen from the model fitting for the group of *elementary teachers*, even though the Chi-square test is significant (p-value < 0.0001), the other model fitting measures are very satisfactory as indicated by the *Comparative Fit Index (CFI)* = 0.939, the *Tucker-Lewis Index (TLI)* = 0.929, the *Root Mean Square Error of Approximation (RMSEA)* = 0.053, and the *Standardized Root Mean Square Residual (SRMR)* = 0.048. All the factor loadings, intercepts, and variances/covariances are highly statistically significant.

Similarly, we can call *sem* to fit above model for *secondary school teachers* as follows:

```
# Fit the CFA model for secondary teachers
fitMBI.Sec = cfa(CFA4MBI, data=dSec)
# Print the model summary with the fitting measures
summary(fitMBI.Sec, fit.measures=TRUE)
```

```
## lavaan 0.6-12 ended normally after 52 iterations
##
##    Estimator                                     ML
##    Optimization method                       NLMINB
##    Number of model parameters                    76
##
##    Number of observations                       692
##
## Model Test User Model:
##
##    Test statistic                           639.803
##    Degrees of freedom                           199
##    P-value (Chi-square)                       0.000
##
```

```
## Model Test Baseline Model:
##
##    Test statistic                              6432.706
##    Degrees of freedom                               231
##    P-value                                        0.000
##
## User Model versus Baseline Model:
##
##    Comparative Fit Index (CFI)                    0.929
##    Tucker-Lewis Index (TLI)                       0.917
##
## Loglikelihood and Information Criteria:
##
##    Loglikelihood user model (H0)             -24279.510
##    Loglikelihood unrestricted model (H1)     -23959.608
##
##    Akaike (AIC)                               48711.019
##    Bayesian (BIC)                             49056.028
##    Sample-size adjusted Bayesian (BIC)        48814.715
##
## Root Mean Square Error of Approximation:
##
##    RMSEA                                          0.057
##    90 Percent confidence interval - lower         0.052
##    90 Percent confidence interval - upper         0.062
##    P-value RMSEA <= 0.05                          0.014
##
## Standardized Root Mean Square Residual:
##
##    SRMR                                           0.052
##
## Parameter Estimates:
##
##    Standard errors                             Standard
##    Information                                 Expected
##    Information saturated (h1) model          Structured
##
## Latent Variables:
##                    Estimate  Std.Err  z-value  P(>|z|)
##    EE =~
##      I01              1.000
##      I02              0.933    0.038   24.269    0.000
##      I03              1.092    0.061   17.925    0.000
##      I06              0.881    0.059   14.941    0.000
##      I08              1.293    0.062   20.935    0.000
```

```
## I13                1.138    0.060   18.984   0.000
## I14                0.981    0.063   15.474   0.000
## I16                0.745    0.052   14.400   0.000
## I20                0.915    0.050   18.185   0.000
## DP =~
## I05                1.000
## I10                1.036    0.090   11.496   0.000
## I11                0.579    0.086    6.701   0.000
## I15                0.991    0.084   11.853   0.000
## I22                0.912    0.088   10.379   0.000
## PA =~
## I04                1.000
## I07                1.628    0.205    7.921   0.000
## I09                2.390    0.307    7.784   0.000
## I12                1.430    0.206    6.950   0.000
## I17                1.435    0.187    7.672   0.000
## I18                2.167    0.268    8.097   0.000
## I19                2.181    0.276    7.891   0.000
## I21                1.628    0.240    6.772   0.000
## EE =~
## I12               -0.472    0.045  -10.533   0.000
## I11                0.417    0.051    8.092   0.000
##
## Covariances:
##                 Estimate  Std.Err  z-value  P(>|z|)
## .I01 ~~
##    .I02            0.624    0.060   10.388   0.000
## .I06 ~~
##    .I16            0.663    0.069    9.605   0.000
## .I10 ~~
##    .I11            0.952    0.085   11.170   0.000
## .I04 ~~
##    .I07            0.083    0.041    2.038   0.042
## .I09 ~~
##    .I19            0.332    0.059    5.600   0.000
## EE ~~
##    DP              0.559    0.065    8.649   0.000
##    PA             -0.150    0.027   -5.491   0.000
## DP ~~
##    PA             -0.172    0.029   -5.963   0.000
##
## Intercepts:
##                 Estimate  Std.Err  z-value  P(>|z|)
## .I01             3.371    0.060   55.937   0.000
## .I02             3.890    0.058   66.973   0.000
```

```
##    .I03              2.526    0.065    39.101     0.000
##    .I04              5.168    0.042   123.343     0.000
##    .I05              1.217    0.056    21.817     0.000
##    .I06              1.999    0.063    31.871     0.000
##    .I07              5.014    0.043   116.064     0.000
##    .I08              2.143    0.065    33.039     0.000
##    .I09              4.702    0.057    82.848     0.000
##    .I10              1.275    0.059    21.431     0.000
##    .I11              1.166    0.059    19.748     0.000
##    .I12              4.527    0.051    88.954     0.000
##    .I13              2.653    0.063    41.827     0.000
##    .I14              3.147    0.067    46.680     0.000
##    .I15              1.078    0.054    20.011     0.000
##    .I16              1.548    0.055    28.112     0.000
##    .I17              5.303    0.037   145.039     0.000
##    .I18              4.705    0.045   105.646     0.000
##    .I19              4.600    0.049    93.051     0.000
##    .I20              1.211    0.053    22.713     0.000
##    .I21              4.462    0.057    77.884     0.000
##    .I22              1.790    0.061    29.412     0.000
##     EE               0.000
##     DP               0.000
##     PA               0.000
##
## Variances:
##                   Estimate  Std.Err   z-value   P(>|z|)
##    .I01              1.249    0.075    16.598     0.000
##    .I02              1.234    0.073    16.818     0.000
##    .I03              1.379    0.084    16.493     0.000
##    .I06              1.739    0.099    17.499     0.000
##    .I08              0.798    0.059    13.477     0.000
##    .I13              1.146    0.072    15.837     0.000
##    .I14              1.929    0.111    17.392     0.000
##    .I16              1.395    0.079    17.612     0.000
##    .I20              0.908    0.056    16.353     0.000
##    .I05              1.306    0.092    14.256     0.000
##    .I10              1.540    0.105    14.652     0.000
##    .I11              1.640    0.097    16.870     0.000
##    .I15              1.178    0.085    13.863     0.000
##    .I22              1.861    0.116    16.095     0.000
##    .I04              1.063    0.060    17.853     0.000
##    .I07              0.889    0.054    16.396     0.000
##    .I09              1.362    0.090    15.180     0.000
##    .I12              0.998    0.060    16.655     0.000
##    .I17              0.612    0.038    16.172     0.000
```

##	.I18	0.659	0.049	13.382	0.000
##	.I19	0.969	0.066	14.705	0.000
##	.I21	1.869	0.106	17.590	0.000
##	EE	1.265	0.122	10.359	0.000
##	DP	0.846	0.109	7.748	0.000
##	PA	0.152	0.036	4.246	0.000

Similarly, for the group of *elementary teachers*, even though the Chi-square test is significant (p-value < 0.0001), the other model fitting measures are very satisfactory as indicated by the *Comparative Fit Index (CFI)* = 0.929, the *Tucker-Lewis Index (TLI)* = 0.917, the *Root Mean Square Error of Approximation (RMSEA)* = 0.057, and the *Standardized Root Mean Square Residual (SRMR)* = 0.052. All the factor loadings, intercepts, and variances/covariances are highly statistically significant.

We therefore established the group-specific *baseline* models for these two groups.

The group-specific model can be graphically seen in Figure 6.2 with the following *R* code:

```
semPaths(fitMBI.Sec)
```

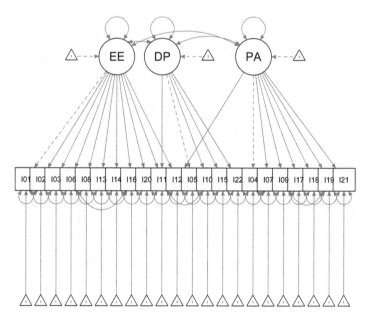

FIGURE 6.2
Model Path Diagram for MBI.

6.3.2 Establish the Configural Invariance Model

With the same CFA model *CFA4MBI*, we can call *sem* to fit the *configural* model for both groups of *elementary school teachers* and *secondary school teachers* together using the *lavaan* function *cfa* with option *group = "School"* as described in Rosseel (2012). The *R* code chunk is as follows:

```
# Fit the CFA model for elementary teachers
fitMBI.Configural = cfa(CFA4MBI, data=dMBI, group="School")
# Print the model summary with the fitting measures
summary(fitMBI.Configural, fit.measures=TRUE)
```

```
## lavaan 0.6-12 ended normally after 91 iterations
##
##   Estimator                                      ML
##   Optimization method                        NLMINB
##   Number of model parameters                    152
##
##   Number of observations per group:
##     Elem                                        580
##     Sec                                         692
##
## Model Test User Model:
##
##   Test statistic                           1167.411
##   Degrees of freedom                            398
##   P-value (Chi-square)                        0.000
##   Test statistic for each group:
##     Elem                                    527.608
##     Sec                                     639.803
##
## Model Test Baseline Model:
##
##   Test statistic                          12040.836
##   Degrees of freedom                            462
##   P-value                                     0.000
##
## User Model versus Baseline Model:
##
##   Comparative Fit Index (CFI)                 0.934
##   Tucker-Lewis Index (TLI)                    0.923
##
## Loglikelihood and Information Criteria:
##
##   Loglikelihood user model (H0)          -43931.018
##   Loglikelihood unrestricted model (H1)  -43347.312
```

```
##
##    Akaike (AIC)                                      88166.035
##    Bayesian (BIC)                                    88948.584
##    Sample-size adjusted Bayesian (BIC)               88465.759
##
## Root Mean Square Error of Approximation:
##
##    RMSEA                                                 0.055
##    90 Percent confidence interval - lower                0.051
##    90 Percent confidence interval - upper                0.059
##    P-value RMSEA <= 0.05                                 0.011
##
## Standardized Root Mean Square Residual:
##
##    SRMR                                                  0.050
##
## Parameter Estimates:
##
##    Standard errors                                    Standard
##    Information                                        Expected
##    Information saturated (h1) model                 Structured
##
##
## Group 1 [Elem]:
##
## Latent Variables:
##                      Estimate  Std.Err  z-value  P(>|z|)
##    EE =~
##      I01               1.000
##      I02               0.916    0.040   23.030    0.000
##      I03               1.032    0.057   18.120    0.000
##      I06               0.807    0.056   14.352    0.000
##      I08               1.264    0.060   21.184    0.000
##      I13               1.050    0.057   18.302    0.000
##      I14               0.956    0.061   15.569    0.000
##      I16               0.758    0.051   14.740    0.000
##      I20               0.868    0.048   18.120    0.000
##    DP =~
##      I05               1.000
##      I10               0.977    0.083   11.772    0.000
##      I11               0.750    0.100    7.536    0.000
##      I15               0.653    0.060   10.870    0.000
##      I22               0.612    0.078    7.868    0.000
##    PA =~
##      I04               1.000
```

```
##       I07          1.097    0.121    9.078    0.000
##       I09          1.886    0.233    8.105    0.000
##       I12          1.342    0.190    7.071    0.000
##       I17          1.521    0.172    8.846    0.000
##       I18          1.978    0.228    8.679    0.000
##       I19          1.856    0.215    8.643    0.000
##       I21          1.350    0.193    7.001    0.000
##    EE =~
##       I12         -0.380    0.043   -8.753    0.000
##       I11          0.222    0.058    3.835    0.000
##
## Covariances:
##                  Estimate  Std.Err  z-value  P(>|z|)
##    .I01 ~~
##       .I02         0.506    0.063    8.082    0.000
##    .I06 ~~
##       .I16         0.860    0.080   10.763    0.000
##    .I10 ~~
##       .I11         0.617    0.081    7.605    0.000
##    .I04 ~~
##       .I07         0.170    0.030    5.676    0.000
##    .I09 ~~
##       .I19         0.099    0.049    2.019    0.043
##    EE ~~
##       DP           0.782    0.086    9.137    0.000
##       PA          -0.208    0.034   -6.154    0.000
##    DP ~~
##       PA          -0.215    0.033   -6.452    0.000
##
## Intercepts:
##                  Estimate  Std.Err  z-value  P(>|z|)
##    .I01           3.409    0.069   49.485    0.000
##    .I02           3.976    0.065   60.903    0.000
##    .I03           2.572    0.071   36.229    0.000
##    .I04           5.412    0.039  139.751    0.000
##    .I05           1.053    0.061   17.177    0.000
##    .I06           1.676    0.069   24.353    0.000
##    .I07           5.338    0.036  146.968    0.000
##    .I08           2.184    0.075   29.012    0.000
##    .I09           5.031    0.055   90.946    0.000
##    .I10           1.164    0.061   18.990    0.000
##    .I11           1.122    0.063   17.931    0.000
##    .I12           4.693    0.053   87.789    0.000
##    .I13           2.548    0.072   35.621    0.000
##    .I14           3.122    0.076   41.263    0.000
```

```
## .I15                    0.545   0.047   11.713    0.000
## .I16                    1.433   0.063   22.715    0.000
## .I17                    5.416   0.037  147.600    0.000
## .I18                    4.883   0.050   96.794    0.000
## .I19                    5.007   0.047  106.468    0.000
## .I20                    1.281   0.060   21.456    0.000
## .I21                    4.841   0.054   90.467    0.000
## .I22                    1.328   0.064   20.730    0.000
## EE                      0.000
## DP                      0.000
## PA                      0.000
##
## Variances:
##                       Estimate  Std.Err  z-value   P(>|z|)
## .I01                    1.213   0.081    14.931    0.000
## .I02                    1.182   0.078    15.195    0.000
## .I03                    1.284   0.086    14.996    0.000
## .I06                    1.745   0.108    16.105    0.000
## .I08                    0.829   0.067    12.281    0.000
## .I13                    1.273   0.085    14.907    0.000
## .I14                    1.913   0.121    15.864    0.000
## .I16                    1.424   0.089    16.033    0.000
## .I20                    0.908   0.061    14.996    0.000
## .I05                    1.197   0.097    12.381    0.000
## .I10                    1.237   0.098    12.662    0.000
## .I11                    1.381   0.099    13.894    0.000
## .I15                    0.835   0.059    14.236    0.000
## .I22                    2.009   0.125    16.030    0.000
## .I04                    0.711   0.044    16.017    0.000
## .I07                    0.573   0.037    15.555    0.000
## .I09                    1.208   0.083    14.629    0.000
## .I12                    0.938   0.061    15.362    0.000
## .I17                    0.412   0.031    13.133    0.000
## .I18                    0.853   0.062    13.858    0.000
## .I19                    0.734   0.054    13.474    0.000
## .I21                    1.371   0.085    16.127    0.000
## EE                      1.539   0.150    10.244    0.000
## DP                      0.985   0.126     7.818    0.000
## PA                      0.159   0.033     4.813    0.000
##
##
## Group 2 [Sec]:
##
## Latent Variables:
##                       Estimate  Std.Err  z-value   P(>|z|)
```

```
## EE =~
## I01                1.000
## I02                0.933    0.038    24.269    0.000
## I03                1.092    0.061    17.925    0.000
## I06                0.881    0.059    14.941    0.000
## I08                1.293    0.062    20.935    0.000
## I13                1.138    0.060    18.984    0.000
## I14                0.981    0.063    15.474    0.000
## I16                0.745    0.052    14.400    0.000
## I20                0.915    0.050    18.185    0.000
## DP =~
## I05                1.000
## I10                1.036    0.090    11.496    0.000
## I11                0.579    0.086     6.701    0.000
## I15                0.991    0.084    11.853    0.000
## I22                0.912    0.088    10.379    0.000
## PA =~
## I04                1.000
## I07                1.628    0.205     7.921    0.000
## I09                2.390    0.307     7.784    0.000
## I12                1.430    0.206     6.950    0.000
## I17                1.435    0.187     7.672    0.000
## I18                2.167    0.268     8.097    0.000
## I19                2.181    0.276     7.891    0.000
## I21                1.628    0.240     6.772    0.000
## EE =~
## I12               -0.472    0.045   -10.533    0.000
## I11                0.417    0.051     8.092    0.000
##
## Covariances:
##                 Estimate  Std.Err  z-value  P(>|z|)
## .I01 ~~
##   .I02             0.624    0.060    10.388    0.000
## .I06 ~~
##   .I16             0.663    0.069     9.605    0.000
## .I10 ~~
##   .I11             0.952    0.085    11.170    0.000
## .I04 ~~
##   .I07             0.083    0.041     2.038    0.042
## .I09 ~~
##   .I19             0.332    0.059     5.600    0.000
## EE ~~
##   DP               0.559    0.065     8.649    0.000
##   PA              -0.150    0.027    -5.491    0.000
## DP ~~
```

```
## PA              -0.172   0.029  -5.963   0.000
##
## Intercepts:
##                Estimate  Std.Err  z-value  P(>|z|)
## .I01              3.371    0.060   55.937    0.000
## .I02              3.890    0.058   66.973    0.000
## .I03              2.526    0.065   39.101    0.000
## .I04              5.168    0.042  123.343    0.000
## .I05              1.217    0.056   21.817    0.000
## .I06              1.999    0.063   31.871    0.000
## .I07              5.014    0.043  116.064    0.000
## .I08              2.143    0.065   33.039    0.000
## .I09              4.702    0.057   82.848    0.000
## .I10              1.275    0.059   21.431    0.000
## .I11              1.166    0.059   19.748    0.000
## .I12              4.527    0.051   88.954    0.000
## .I13              2.653    0.063   41.827    0.000
## .I14              3.147    0.067   46.680    0.000
## .I15              1.078    0.054   20.011    0.000
## .I16              1.548    0.055   28.112    0.000
## .I17              5.303    0.037  145.039    0.000
## .I18              4.705    0.045  105.646    0.000
## .I19              4.600    0.049   93.051    0.000
## .I20              1.211    0.053   22.713    0.000
## .I21              4.462    0.057   77.884    0.000
## .I22              1.790    0.061   29.412    0.000
## EE                0.000
## DP                0.000
## PA                0.000
##
## Variances:
##                Estimate  Std.Err  z-value  P(>|z|)
## .I01              1.249    0.075   16.598    0.000
## .I02              1.234    0.073   16.818    0.000
## .I03              1.379    0.084   16.493    0.000
## .I06              1.739    0.099   17.499    0.000
## .I08              0.798    0.059   13.477    0.000
## .I13              1.146    0.072   15.837    0.000
## .I14              1.929    0.111   17.392    0.000
## .I16              1.395    0.079   17.612    0.000
## .I20              0.908    0.056   16.353    0.000
## .I05              1.306    0.092   14.256    0.000
## .I10              1.540    0.105   14.652    0.000
## .I11              1.640    0.097   16.870    0.000
## .I15              1.178    0.085   13.863    0.000
```

##	.I22	1.861	0.116	16.095	0.000
##	.I04	1.063	0.060	17.853	0.000
##	.I07	0.889	0.054	16.396	0.000
##	.I09	1.362	0.090	15.180	0.000
##	.I12	0.998	0.060	16.655	0.000
##	.I17	0.612	0.038	16.172	0.000
##	.I18	0.659	0.049	13.382	0.000
##	.I19	0.969	0.066	14.705	0.000
##	.I21	1.869	0.106	17.590	0.000
##	EE	1.265	0.122	10.359	0.000
##	DP	0.846	0.109	7.748	0.000
##	PA	0.152	0.036	4.246	0.000

Observed from the output, we can see the number of observations per group is 580 for *Elem* and 692 for *Sec*. The Chi-square test statistic is 1167.411, which is the sum from each group (i.e., 527.608 from *Elem* + 639.803 from *Sec*). With the degrees of freedom of 398, the associated p-value is <0.00001, which is highly significant. However, the other model fitting measures are very satisfactory as indicated by the *Comparative Fit Index (CFI)* = 0.934, the *Tucker-Lewis Index (TLI)* = 0.923, the *Root Mean Square Error of Approximation (RMSEA)* = 0.055, and the *Standardized Root Mean Square Residual (SRMR)* = 0.050.

Further observed from the output, we can see that all the factor loadings, intercepts, and variances/covariances are highly statistically significant for both groups, but there are freely estimated without any constraints and they are *group-variant*. This established the *configural* model for these two groups.

6.3.3 Fit the *Ideal* Model

As the most restricted CFA model, we need to restrict all the factor loadings from these 22 observed variables to the 3 latent variables, intercepts and variances/covariances from these 22 observed variables and 3 latent variables. This can be accomplished by using *lavaan* option *group.equal* in *cfa* function as follows:

```
# Fit the CFA model for all teachers
fitMBI.Ideal = cfa(CFA4MBI, data=dMBI, group="School",
   group.equal=c(
     #factor loadings
     "loadings",
     #intercepts for 22 items
     "intercepts",
     # residual variance for the 22 items
     "residuals",
     # residual covariances for the 22 items
```

```
    "residual.covariances",
    # means for 3 latent variables
    "means",
    # variance for the 3 latent variables
    "lv.variances",
    # covariances for the 3 latent variables
    "lv.covariances"))
# Print the model summary with the fitting measures
summary(fitMBI.Ideal, fit.measures=TRUE)
```

```
## lavaan 0.6-12 ended normally after 82 iterations
##
##   Estimator                                         ML
##   Optimization method                           NLMINB
##   Number of model parameters                       152
##   Number of equality constraints                    76
##
##   Number of observations per group:
##      Elem                                          580
##      Sec                                           692
##
## Model Test User Model:
##
##   Test statistic                              1560.385
##   Degrees of freedom                               474
##   P-value (Chi-square)                           0.000
##   Test statistic for each group:
##      Elem                                      772.336
##      Sec                                       788.050
##
## Model Test Baseline Model:
##
##   Test statistic                             12040.836
##   Degrees of freedom                               462
##   P-value                                        0.000
##
## User Model versus Baseline Model:
##
##   Comparative Fit Index (CFI)                    0.906
##   Tucker-Lewis Index (TLI)                       0.909
##
## Loglikelihood and Information Criteria:
##
##   Loglikelihood user model (H0)             -44127.505
##   Loglikelihood unrestricted model (H1)     -43347.312
```

```
##
##   Akaike (AIC)                                      88407.010
##   Bayesian (BIC)                                    88798.285
##   Sample-size adjusted Bayesian (BIC)               88556.872
##
## Root Mean Square Error of Approximation:
##
##   RMSEA                                                 0.060
##   90 Percent confidence interval - lower                0.057
##   90 Percent confidence interval - upper                0.063
##   P-value RMSEA <= 0.05                                 0.000
##
## Standardized Root Mean Square Residual:
##
##   SRMR                                                  0.071
##
## Parameter Estimates:
##
##   Standard errors                                    Standard
##   Information                                        Expected
##   Information saturated (h1) model                 Structured
##
##
## Group 1 [Elem]:
##
## Latent Variables:
##                     Estimate  Std.Err  z-value  P(>|z|)
##   EE =~
##     I01                1.000
##     I02     (.p2.)     0.924    0.028   33.475    0.000
##     I03     (.p3.)     1.063    0.042   25.473    0.000
##     I06     (.p4.)     0.844    0.041   20.618    0.000
##     I08     (.p5.)     1.277    0.043   29.735    0.000
##     I13     (.p6.)     1.094    0.041   26.393    0.000
##     I14     (.p7.)     0.970    0.044   21.974    0.000
##     I16     (.p8.)     0.750    0.037   20.544    0.000
##     I20     (.p9.)     0.890    0.035   25.628    0.000
##   DP =~
##     I05                1.000
##     I10     (.11.)     0.999    0.061   16.269    0.000
##     I11     (.12.)     0.636    0.063   10.015    0.000
##     I15     (.13.)     0.853    0.053   16.171    0.000
##     I22     (.14.)     0.801    0.060   13.442    0.000
##   PA =~
##     I04                1.000
```

```
## 	   I07     (.16.)    1.400    0.115    12.172    0.000
## 	   I09     (.17.)    2.119    0.184    11.513    0.000
## 	   I12     (.18.)    1.350    0.133    10.146    0.000
## 	   I17     (.19.)    1.407    0.120    11.711    0.000
## 	   I18     (.20.)    2.013    0.166    12.097    0.000
## 	   I19     (.21.)    2.027    0.170    11.920    0.000
## 	   I21     (.22.)    1.523    0.151    10.078    0.000
## 	 EE =~
## 	   I12     (.28.)   -0.431    0.031   -13.926    0.000
## 	   I11     (.29.)    0.336    0.037     9.013    0.000
##
## Covariances:
## 	                   Estimate  Std.Err   z-value   P(>|z|)
## 	 .I01 ~~
## 	   .I02    (.23.)    0.570    0.043    13.125    0.000
## 	 .I06 ~~
## 	   .I16    (.24.)    0.763    0.053    14.455    0.000
## 	 .I10 ~~
## 	   .I11    (.25.)    0.815    0.059    13.758    0.000
## 	 .I04 ~~
## 	   .I07    (.26.)    0.123    0.026     4.774    0.000
## 	 .I09 ~~
## 	   .I19    (.27.)    0.228    0.039     5.780    0.000
## 	 EE ~~
## 	   DP      (.30.)    0.644    0.052    12.397    0.000
## 	   PA      (.31.)   -0.175    0.022    -8.106    0.000
## 	 DP ~~
## 	   PA      (.32.)   -0.203    0.023    -8.912    0.000
##
## Intercepts:
## 	                   Estimate  Std.Err   z-value   P(>|z|)
## 	   .I01    (.33.)    3.388    0.045    74.620    0.000
## 	   .I02    (.34.)    3.929    0.043    90.475    0.000
## 	   .I03    (.35.)    2.547    0.048    53.300    0.000
## 	   .I04    (.36.)    5.279    0.029   181.824    0.000
## 	   .I05    (.37.)    1.142    0.041    27.641    0.000
## 	   .I06    (.38.)    1.851    0.047    39.756    0.000
## 	   .I07    (.39.)    5.162    0.029   177.353    0.000
## 	   .I08    (.40.)    2.162    0.049    43.908    0.000
## 	   .I09    (.41.)    4.852    0.040   120.898    0.000
## 	   .I10    (.42.)    1.224    0.043    28.613    0.000
## 	   .I11    (.43.)    1.146    0.043    26.690    0.000
## 	   .I12    (.44.)    4.603    0.037   124.532    0.000
## 	   .I13    (.45.)    2.605    0.048    54.840    0.000
## 	   .I14    (.46.)    3.136    0.050    62.271    0.000
```

```
##    .I15    (.47.)    0.835    0.037    22.604    0.000
##    .I16    (.48.)    1.495    0.042    35.983    0.000
##    .I17    (.49.)    5.355    0.026   205.631    0.000
##    .I18    (.50.)    4.786    0.034   142.868    0.000
##    .I19    (.51.)    4.785    0.035   137.267    0.000
##    .I20    (.52.)    1.243    0.040    31.235    0.000
##    .I21    (.53.)    4.635    0.040   116.061    0.000
##    .I22    (.54.)    1.579    0.045    35.393    0.000
##    EE                0.000
##    DP                0.000
##    PA                0.000
##
## Variances:
##                    Estimate  Std.Err  z-value  P(>|z|)
##    .I01    (.55.)    1.232    0.055    22.299    0.000
##    .I02    (.56.)    1.211    0.053    22.645    0.000
##    .I03    (.57.)    1.334    0.060    22.263    0.000
##    .I06    (.58.)    1.769    0.074    23.798    0.000
##    .I08    (.59.)    0.816    0.045    18.220    0.000
##    .I13    (.60.)    1.207    0.055    21.752    0.000
##    .I14    (.61.)    1.917    0.082    23.517    0.000
##    .I16    (.62.)    1.414    0.059    23.813    0.000
##    .I20    (.63.)    0.912    0.041    22.184    0.000
##    .I05    (.64.)    1.261    0.067    18.811    0.000
##    .I10    (.65.)    1.418    0.073    19.529    0.000
##    .I11    (.66.)    1.545    0.069    22.243    0.000
##    .I15    (.67.)    1.071    0.054    19.742    0.000
##    .I22    (.68.)    1.948    0.086    22.609    0.000
##    .I04    (.69.)    0.905    0.038    23.942    0.000
##    .I07    (.70.)    0.751    0.034    22.369    0.000
##    .I09    (.71.)    1.300    0.062    20.950    0.000
##    .I12    (.72.)    0.972    0.043    22.650    0.000
##    .I17    (.73.)    0.532    0.025    21.252    0.000
##    .I18    (.74.)    0.752    0.039    19.427    0.000
##    .I19    (.75.)    0.861    0.044    19.652    0.000
##    .I21    (.76.)    1.642    0.069    23.734    0.000
##    EE      (.77.)    1.390    0.095    14.580    0.000
##    DP      (.78.)    0.912    0.083    10.964    0.000
##    PA      (.79.)    0.167    0.026     6.493    0.000
##
##
## Group 2 [Sec]:
##
## Latent Variables:
##                    Estimate  Std.Err  z-value  P(>|z|)
```

```
##    EE =~
##      I01                        1.000
##      I02       (.p2.)           0.924     0.028    33.475    0.000
##      I03       (.p3.)           1.063     0.042    25.473    0.000
##      I06       (.p4.)           0.844     0.041    20.618    0.000
##      I08       (.p5.)           1.277     0.043    29.735    0.000
##      I13       (.p6.)           1.094     0.041    26.393    0.000
##      I14       (.p7.)           0.970     0.044    21.974    0.000
##      I16       (.p8.)           0.750     0.037    20.544    0.000
##      I20       (.p9.)           0.890     0.035    25.628    0.000
##    DP =~
##      I05                        1.000
##      I10       (.11.)           0.999     0.061    16.269    0.000
##      I11       (.12.)           0.636     0.063    10.015    0.000
##      I15       (.13.)           0.853     0.053    16.171    0.000
##      I22       (.14.)           0.801     0.060    13.442    0.000
##    PA =~
##      I04                        1.000
##      I07       (.16.)           1.400     0.115    12.172    0.000
##      I09       (.17.)           2.119     0.184    11.513    0.000
##      I12       (.18.)           1.350     0.133    10.146    0.000
##      I17       (.19.)           1.407     0.120    11.711    0.000
##      I18       (.20.)           2.013     0.166    12.097    0.000
##      I19       (.21.)           2.027     0.170    11.920    0.000
##      I21       (.22.)           1.523     0.151    10.078    0.000
##    EE =~
##      I12       (.28.)          -0.431     0.031   -13.926    0.000
##      I11       (.29.)           0.336     0.037     9.013    0.000
##
## Covariances:
##                            Estimate  Std.Err  z-value  P(>|z|)
##   .I01 ~~
##      .I02      (.23.)           0.570     0.043    13.125    0.000
##   .I06 ~~
##      .I16      (.24.)           0.763     0.053    14.455    0.000
##   .I10 ~~
##      .I11      (.25.)           0.815     0.059    13.758    0.000
##   .I04 ~~
##      .I07      (.26.)           0.123     0.026     4.774    0.000
##   .I09 ~~
##      .I19      (.27.)           0.228     0.039     5.780    0.000
##    EE ~~
##       DP       (.30.)           0.644     0.052    12.397    0.000
##       PA       (.31.)          -0.175     0.022    -8.106    0.000
##    DP ~~
```

```
##      PA        (.32.)   -0.203   0.023   -8.912   0.000
##
## Intercepts:
##                         Estimate Std.Err  z-value  P(>|z|)
##      .I01       (.33.)   3.388   0.045    74.620   0.000
##      .I02       (.34.)   3.929   0.043    90.475   0.000
##      .I03       (.35.)   2.547   0.048    53.300   0.000
##      .I04       (.36.)   5.279   0.029   181.824   0.000
##      .I05       (.37.)   1.142   0.041    27.641   0.000
##      .I06       (.38.)   1.851   0.047    39.756   0.000
##      .I07       (.39.)   5.162   0.029   177.353   0.000
##      .I08       (.40.)   2.162   0.049    43.908   0.000
##      .I09       (.41.)   4.852   0.040   120.898   0.000
##      .I10       (.42.)   1.224   0.043    28.613   0.000
##      .I11       (.43.)   1.146   0.043    26.690   0.000
##      .I12       (.44.)   4.603   0.037   124.532   0.000
##      .I13       (.45.)   2.605   0.048    54.840   0.000
##      .I14       (.46.)   3.136   0.050    62.271   0.000
##      .I15       (.47.)   0.835   0.037    22.604   0.000
##      .I16       (.48.)   1.495   0.042    35.983   0.000
##      .I17       (.49.)   5.355   0.026   205.631   0.000
##      .I18       (.50.)   4.786   0.034   142.868   0.000
##      .I19       (.51.)   4.785   0.035   137.267   0.000
##      .I20       (.52.)   1.243   0.040    31.235   0.000
##      .I21       (.53.)   4.635   0.040   116.061   0.000
##      .I22       (.54.)   1.579   0.045    35.393   0.000
##      EE                  0.000
##      DP                  0.000
##      PA                  0.000
##
## Variances:
##                         Estimate Std.Err  z-value  P(>|z|)
##      .I01       (.55.)   1.232   0.055    22.299   0.000
##      .I02       (.56.)   1.211   0.053    22.645   0.000
##      .I03       (.57.)   1.334   0.060    22.263   0.000
##      .I06       (.58.)   1.769   0.074    23.798   0.000
##      .I08       (.59.)   0.816   0.045    18.220   0.000
##      .I13       (.60.)   1.207   0.055    21.752   0.000
##      .I14       (.61.)   1.917   0.082    23.517   0.000
##      .I16       (.62.)   1.414   0.059    23.813   0.000
##      .I20       (.63.)   0.912   0.041    22.184   0.000
##      .I05       (.64.)   1.261   0.067    18.811   0.000
##      .I10       (.65.)   1.418   0.073    19.529   0.000
##      .I11       (.66.)   1.545   0.069    22.243   0.000
##      .I15       (.67.)   1.071   0.054    19.742   0.000
```

##	.I22	(.68.)	1.948	0.086	22.609	0.000
##	.I04	(.69.)	0.905	0.038	23.942	0.000
##	.I07	(.70.)	0.751	0.034	22.369	0.000
##	.I09	(.71.)	1.300	0.062	20.950	0.000
##	.I12	(.72.)	0.972	0.043	22.650	0.000
##	.I17	(.73.)	0.532	0.025	21.252	0.000
##	.I18	(.74.)	0.752	0.039	19.427	0.000
##	.I19	(.75.)	0.861	0.044	19.652	0.000
##	.I21	(.76.)	1.642	0.069	23.734	0.000
##	EE	(.77.)	1.390	0.095	14.580	0.000
##	DP	(.78.)	0.912	0.083	10.964	0.000
##	PA	(.79.)	0.167	0.026	6.493	0.000

Note that, the option *group.equal* can further include "regressions" for regression paths and "thresholds" for categorical data. These options are not used for this CFA model since there are no regression paths in this CFA model and no thresholds since the data are treated as continuous.

Examining the output, we can see that all the factor loadings to the 3 latent variables, intercepts, variances/covariances are restricted and estimated to be equal for both groups. These parameters are all statistically significant.

However, in this *Ideal* CFA model, the Chi-square test is highly significant, The other model fitting measures are not satisfactory as indicated by the *Comparative Fit Index (CFI)* = 0.906, the *Tucker-Lewis Index (TLI)* = 0.909, the *Root Mean Square Error of Approximation (RMSEA)* = 0.060, and the *Standardized Root Mean Square Residual (SRMR)* = 0.071, which indicates that this *Ideal* model is too restrictive and needs to be relaxed with some parameters freely estimated.

We can also call *R* function *anova* to perform a likelihood ratio test between the *configural* and the *Ideal* models to see whether these two models are equivalent as follows:

```
anova(fitMBI.Ideal, fitMBI.Configural)
```

```
## Chi-Squared Difference Test
##
##                       Df    AIC    BIC  Chisq Chisq diff
## fitMBI.Configural 398 88166 88949  1167
## fitMBI.Ideal      474 88407 88798  1560            393
##                       Df diff          Pr(>Chisq)
## fitMBI.Configural
## fitMBI.Ideal              76 <0.0000000000000002 ***
## ---
## Signif. codes:
## 0 '***' 0.001 '**' 0.01 '*' 0.05 '.' 0.1 ' ' 1
```

It can be seen that these two models are statistically highly significantly different. Because the *configural* invariance fit the data well but the *Ideal* model does not, a partial invariance model that is in-between these extremes might strike a good balance of model fit and parsimony.

6.3.4   Partial Invariance

There are a vast number of *partial invariance* models that can be tested when the *Ideal* model is not the final model. The *partial invariance* can be tested for all or some factor loadings, means, residual variance and covariance, as well as any of these combinations.

The general guideline is to test invariances for factor loadings first, to intercepts and to regression parameters (which are present in this example with CFA models). We illustrate how easy to implement this in *R lavaan* without printing out all the model fitting details (to save printing pages).

We fit three invariance models with (1) Metric invariance: Only factor loadings invariance using *group.equal=c("loadings")*), (2) Only intercepts invariance using *group.equal=c("intercepts")*), and (3) Scalar invariance: Both factor loadings invariance and intercepts invariance using *group.equal= c("loadings", "intercept")*) as follows:

```
# Only factor Loadings invariance
fitMBI.Loadings = cfa(CFA4MBI, data=dMBI, group="School",
    group.equal=c("loadings"))
# Only intercepts invariance
fitMBI.Intercepts = cfa(CFA4MBI, data=dMBI, group="School",
    group.equal=c("intercepts"))
# Both factor loadings and intercepts invariance
fitMBI.LoadingsIntercepts = cfa(CFA4MBI, data=dMBI,
            group="School",
            group.equal=c("loadings","intercepts"))
```

Then we can call the *R* function *anova* to test these three models by the likelihood ratio Chi-square test to the *configural* and the *Ideal* models due to their nested structures. From the model complexity, the three invariance models (i.e., *fitMBI.LoadingsIntercepts*, *fitMBI.Loadings*, *fitMBI.Intercepts*) are nested under the *configural* model (i.e., *fitMBI.Configural*), and the *Ideal* model (i.e., *fitMBI.Ideal*) is nested under all of these three invariance models. Among the three invariance models, the invariance model for both factor loadings and intercepts (i.e., *fitMBI.LoadingsIntercepts*) is nested under the other two models(i.e., *fitMBI.Loadings*, *fitMBI.Intercepts*). This can be implemented in the following *R* code chunk:

```
# Test the three invariance models to the Ideal Model
anova(fitMBI.Ideal,fitMBI.LoadingsIntercepts)
```

```
## Chi-Squared Difference Test
##
##                                Df   AIC    BIC  Chisq
## fitMBI.LoadingsIntercepts 438 88232 88809   1314
## fitMBI.Ideal             474 88407 88798   1560
##                                Chisq diff Df diff
## fitMBI.LoadingsIntercepts
## fitMBI.Ideal                           247       36
##                                     Pr(>Chisq)
## fitMBI.LoadingsIntercepts
## fitMBI.Ideal                  <0.0000000000000002 ***
## ---
## Signif. codes:
## 0 '***' 0.001 '**' 0.01 '*' 0.05 '.' 0.1 ' ' 1
```

```
anova(fitMBI.Ideal,fitMBI.Loadings)
```

```
## Chi-Squared Difference Test
##
##                    Df   AIC    BIC  Chisq Chisq diff
## fitMBI.Loadings 419 88167 88842   1210
## fitMBI.Ideal    474 88407 88798   1560            350
##                    Df diff        Pr(>Chisq)
## fitMBI.Loadings
## fitMBI.Ideal            55 <0.0000000000000002 ***
## ---
## Signif. codes:
## 0 '***' 0.001 '**' 0.01 '*' 0.05 '.' 0.1 ' ' 1
```

```
anova(fitMBI.Ideal,fitMBI.Intercepts)
```

```
## Chi-Squared Difference Test
##
##                     Df   AIC    BIC  Chisq Chisq diff
## fitMBI.Intercepts 417 88209 88894   1249
## fitMBI.Ideal      474 88407 88798   1560            312
##                     Df diff        Pr(>Chisq)
## fitMBI.Intercepts
## fitMBI.Ideal             57 <0.0000000000000002 ***
## ---
## Signif. codes:
## 0 '***' 0.001 '**' 0.01 '*' 0.05 '.' 0.1 ' ' 1
```

```
# Test the three invariance model to configural invariance model
anova(fitMBI.Configural,fitMBI.Loadings)
```

```
## Chi-Squared Difference Test
##
##                     Df    AIC   BIC Chisq Chisq diff
## fitMBI.Configural 398 88166 88949  1167
## fitMBI.Loadings   419 88167 88842  1210         43.1
##                     Df diff Pr(>Chisq)
## fitMBI.Configural
## fitMBI.Loadings          21     0.0031 **
## ---
## Signif. codes:
## 0 '***' 0.001 '**' 0.01 '*' 0.05 '.' 0.1 ' ' 1
```

```
anova(fitMBI.Configural,fitMBI.Intercepts)
```

```
## Chi-Squared Difference Test
##
##                     Df    AIC   BIC Chisq Chisq diff
## fitMBI.Configural 398 88166 88949  1167
## fitMBI.Intercepts 417 88209 88894  1249         81.4
##                     Df diff  Pr(>Chisq)
## fitMBI.Configural
## fitMBI.Intercepts       19 0.000000001 ***
## ---
## Signif. codes:
## 0 '***' 0.001 '**' 0.01 '*' 0.05 '.' 0.1 ' ' 1
```

```
anova(fitMBI.Configural,fitMBI.LoadingsIntercepts)
```

```
## Chi-Squared Difference Test
##
##                            Df    AIC   BIC Chisq
## fitMBI.Configural         398 88166 88949  1167
## fitMBI.LoadingsIntercepts 438 88232 88809  1314
##                            Chisq diff Df diff
## fitMBI.Configural
## fitMBI.LoadingsIntercepts         146      40
##                                         Pr(>Chisq)
## fitMBI.Configural
## fitMBI.LoadingsIntercepts 0.000000000000051 ***
## ---
## Signif. codes:
## 0 '***' 0.001 '**' 0.01 '*' 0.05 '.' 0.1 ' ' 1
```

```
# Test the both invariance to only one invariance
anova(fitMBI.LoadingsIntercepts,fitMBI.Loadings)
```

```
## Chi-Squared Difference Test
##
##                               Df    AIC   BIC Chisq
## fitMBI.Loadings              419  88167 88842  1210
## fitMBI.LoadingsIntercepts    438  88232 88809  1314
##                               Chisq diff Df diff
## fitMBI.Loadings
## fitMBI.LoadingsIntercepts          103      19
##                               Pr(>Chisq)
## fitMBI.Loadings
## fitMBI.LoadingsIntercepts 0.00000000000015 ***
## ---
## Signif. codes:
## 0 '***' 0.001 '**' 0.01 '*' 0.05 '.' 0.1 ' ' 1
```

```
anova(fitMBI.LoadingsIntercepts,fitMBI.Intercepts)
```

```
## Chi-Squared Difference Test
##
##                               Df    AIC   BIC Chisq
## fitMBI.Intercepts            417  88209 88894  1249
## fitMBI.LoadingsIntercepts    438  88232 88809  1314
##                               Chisq diff Df diff
## fitMBI.Intercepts
## fitMBI.LoadingsIntercepts          64.8     21
##                               Pr(>Chisq)
## fitMBI.Intercepts
## fitMBI.LoadingsIntercepts   0.0000024 ***
## ---
## Signif. codes:
## 0 '***' 0.001 '**' 0.01 '*' 0.05 '.' 0.1 ' ' 1
```

As seen from the tests, the three invariance models are less restrictive
and are statistically significantly better than the *Ideal* model with smaller
Chi-square statistics. In turn, the three invariance models are more restrictive
than the *configural* model and they are statistically significantly different.

Among the three invariance models, the invariance for both factor loadings
and intercepts (i.e., *fitMBI.LoadingsIntercepts*) are nested within the other two
invariance models (i.e., only factor loadings invariance model-*fitMBI.Loadings*
and only intercepts invariance-*fitMBI.Intercepts*). By the likelihood ratio Chi-
square test, these two invariance models are statistically significantly better

then the invariance model for both factor loadings and intercepts with smaller Chi-square test statistics.

We can further include other model fitting measures to select the best model among the three invariance models as follows:

```
# Both factor loadings and intercepts
fitMeasures(fitMBI.LoadingsIntercepts, c("cfi","tli","rmsea",
          "srmr", "aic","bic","chisq"))
```

```
##       cfi       tli      rmsea      srmr       aic
##     0.924     0.920     0.056     0.056 88232.250
##       bic     chisq
## 88808.865  1313.625
```

```
# Factor loadings only
fitMeasures(fitMBI.Loadings, c("cfi","tli","rmsea","srmr",
          "aic","bic","chisq"))
```

```
##       cfi       tli      rmsea      srmr       aic
##     0.932     0.925     0.054     0.054 88167.127
##       bic     chisq
## 88841.560  1210.502
```

```
# Intercepts only
fitMeasures(fitMBI.Intercepts, c("cfi","tli","rmsea","srmr",
          "aic","bic","chisq"))
```

```
##       cfi       tli      rmsea      srmr       aic
##     0.928     0.920     0.056     0.053 88209.476
##       bic     chisq
## 88894.206  1248.851
```

From all the model fitting measures, the invariance model on factor loadings only is the best among the three invariance models. We can print this model fitting to examine all the parameter estimates:

```
summary(fitMBI.Loadings)
```

```
## lavaan 0.6-12 ended normally after 69 iterations
##
##   Estimator                                         ML
##   Optimization method                           NLMINB
##   Number of model parameters                       152
##   Number of equality constraints                    21
##
```

```
##    Number of observations per group:
##       Elem                                            580
##       Sec                                             692
##
## Model Test User Model:
##
##       Test statistic                             1210.502
##       Degrees of freedom                              419
##       P-value (Chi-square)                          0.000
##       Test statistic for each group:
##          Elem                                      550.173
##          Sec                                       660.329
##
## Parameter Estimates:
##
##       Standard errors                            Standard
##       Information                                Expected
##       Information saturated (h1) model         Structured
##
##
## Group 1 [Elem]:
##
## Latent Variables:
##                       Estimate  Std.Err  z-value  P(>|z|)
##    EE =~
##       I01                1.000
##       I02      (.p2.)    0.925    0.028   33.504    0.000
##       I03      (.p3.)    1.063    0.042   25.513    0.000
##       I06      (.p4.)    0.841    0.041   20.730    0.000
##       I08      (.p5.)    1.277    0.043   29.772    0.000
##       I13      (.p6.)    1.095    0.041   26.430    0.000
##       I14      (.p7.)    0.969    0.044   21.967    0.000
##       I16      (.p8.)    0.753    0.036   20.653    0.000
##       I20      (.p9.)    0.891    0.035   25.676    0.000
##    DP =~
##       I05                1.000
##       I10      (.11.)    1.013    0.062   16.222    0.000
##       I11      (.12.)    0.648    0.065    9.965    0.000
##       I15      (.13.)    0.815    0.051   15.876    0.000
##       I22      (.14.)    0.780    0.059   13.143    0.000
##    PA =~
##       I04                1.000
##       I07      (.16.)    1.304    0.107   12.173    0.000
##       I09      (.17.)    2.096    0.183   11.481    0.000
##       I12      (.18.)    1.356    0.134   10.083    0.000
```

```
##    I17      (.19.)    1.451    0.123    11.803    0.000
##    I18      (.20.)    2.042    0.168    12.137    0.000
##    I19      (.21.)    1.978    0.167    11.875    0.000
##    I21      (.22.)    1.456    0.148     9.854    0.000
## EE =~
##    I12      (.28.)   -0.426    0.031   -13.726    0.000
##    I11      (.29.)    0.335    0.038     8.868    0.000
##
## Covariances:
##                    Estimate  Std.Err  z-value  P(>|z|)
## .I01 ~~
##    .I02              0.515    0.062     8.265    0.000
## .I06 ~~
##    .I16              0.858    0.080    10.794    0.000
## .I10 ~~
##    .I11              0.673    0.076     8.861    0.000
## .I04 ~~
##    .I07              0.167    0.030     5.587    0.000
## .I09 ~~
##    .I19              0.080    0.048     1.664    0.096
##  EE ~~
##    DP                0.694    0.071     9.756    0.000
##    PA               -0.191    0.028    -6.755    0.000
##  DP ~~
##    PA               -0.184    0.026    -7.183    0.000
##
## Intercepts:
##                    Estimate  Std.Err  z-value  P(>|z|)
##    .I01              3.409    0.068    50.016    0.000
##    .I02              3.976    0.065    61.220    0.000
##    .I03              2.572    0.071    36.131    0.000
##    .I04              5.412    0.039   140.515    0.000
##    .I05              1.053    0.060    17.678    0.000
##    .I06              1.676    0.069    24.201    0.000
##    .I07              5.338    0.037   142.940    0.000
##    .I08              2.184    0.075    29.242    0.000
##    .I09              5.031    0.056    89.818    0.000
##    .I10              1.164    0.060    19.288    0.000
##    .I11              1.122    0.062    18.039    0.000
##    .I12              4.693    0.054    86.924    0.000
##    .I13              2.548    0.072    35.261    0.000
##    .I14              3.122    0.075    41.406    0.000
##    .I15              0.545    0.048    11.349    0.000
##    .I16              1.433    0.062    22.953    0.000
##    .I17              5.416    0.036   151.601    0.000
```

```
##     .I18              4.883    0.050   97.198   0.000
##     .I19              5.007    0.047  106.298   0.000
##     .I20              1.281    0.060   21.423   0.000
##     .I21              4.841    0.054   90.241   0.000
##     .I22              1.328    0.065   20.312   0.000
##      EE               0.000
##      DP               0.000
##      PA               0.000
##
## Variances:
##                     Estimate  Std.Err  z-value  P(>|z|)
##     .I01              1.224    0.081   15.136   0.000
##     .I02              1.189    0.078   15.341   0.000
##     .I03              1.280    0.085   15.051   0.000
##     .I06              1.741    0.108   16.111   0.000
##     .I08              0.840    0.066   12.662   0.000
##     .I13              1.267    0.085   14.914   0.000
##     .I14              1.918    0.120   15.939   0.000
##     .I16              1.427    0.088   16.132   0.000
##     .I20              0.907    0.060   15.070   0.000
##     .I05              1.255    0.091   13.752   0.000
##     .I10              1.285    0.094   13.716   0.000
##     .I11              1.440    0.094   15.374   0.000
##     .I15              0.803    0.059   13.585   0.000
##     .I22              1.988    0.126   15.827   0.000
##     .I04              0.716    0.044   16.181   0.000
##     .I07              0.563    0.037   15.275   0.000
##     .I09              1.184    0.081   14.561   0.000
##     .I12              0.937    0.060   15.524   0.000
##     .I17              0.435    0.031   14.269   0.000
##     .I18              0.860    0.060   14.312   0.000
##     .I19              0.720    0.053   13.598   0.000
##     .I21              1.362    0.084   16.156   0.000
##      EE               1.470    0.122   12.027   0.000
##      DP               0.805    0.092    8.787   0.000
##      PA               0.145    0.024    6.115   0.000
##
##
## Group 2 [Sec]:
##
## Latent Variables:
##                     Estimate  Std.Err  z-value  P(>|z|)
##   EE =~
##     I01               1.000
##     I02      (.p2.)   0.925    0.028   33.504   0.000
```

```
##     I03      (.p3.)     1.063     0.042    25.513     0.000
##     I06      (.p4.)     0.841     0.041    20.730     0.000
##     I08      (.p5.)     1.277     0.043    29.772     0.000
##     I13      (.p6.)     1.095     0.041    26.430     0.000
##     I14      (.p7.)     0.969     0.044    21.967     0.000
##     I16      (.p8.)     0.753     0.036    20.653     0.000
##     I20      (.p9.)     0.891     0.035    25.676     0.000
##   DP =~
##     I05                 1.000
##     I10      (.11.)     1.013     0.062    16.222     0.000
##     I11      (.12.)     0.648     0.065     9.965     0.000
##     I15      (.13.)     0.815     0.051    15.876     0.000
##     I22      (.14.)     0.780     0.059    13.143     0.000
##   PA =~
##     I04                 1.000
##     I07      (.16.)     1.304     0.107    12.173     0.000
##     I09      (.17.)     2.096     0.183    11.481     0.000
##     I12      (.18.)     1.356     0.134    10.083     0.000
##     I17      (.19.)     1.451     0.123    11.803     0.000
##     I18      (.20.)     2.042     0.168    12.137     0.000
##     I19      (.21.)     1.978     0.167    11.875     0.000
##     I21      (.22.)     1.456     0.148     9.854     0.000
##   EE =~
##     I12      (.28.)    -0.426     0.031   -13.726     0.000
##     I11      (.29.)     0.335     0.038     8.868     0.000
##
## Covariances:
##                       Estimate  Std.Err  z-value  P(>|z|)
##  .I01 ~~
##    .I02                 0.616     0.060    10.336     0.000
##  .I06 ~~
##    .I16                 0.663     0.069     9.623     0.000
##  .I10 ~~
##    .I11                 0.916     0.085    10.775     0.000
##  .I04 ~~
##    .I07                 0.086     0.040     2.141     0.032
##  .I09 ~~
##    .I19                 0.359     0.058     6.220     0.000
##   EE ~~
##     DP                  0.616     0.065     9.444     0.000
##     PA                 -0.169     0.026    -6.409     0.000
##   DP ~~
##     PA                 -0.204     0.028    -7.257     0.000
##
```

```
## Intercepts:
##                      Estimate  Std.Err   z-value  P(>|z|)
##    .I01               3.371     0.061    55.393    0.000
##    .I02               3.890     0.058    66.616    0.000
##    .I03               2.526     0.064    39.199    0.000
##    .I04               5.168     0.042   122.296    0.000
##    .I05               1.217     0.057    21.310    0.000
##    .I06               1.999     0.062    32.111    0.000
##    .I07               5.014     0.042   119.153    0.000
##    .I08               2.143     0.065    32.809    0.000
##    .I09               4.702     0.056    83.697    0.000
##    .I10               1.275     0.060    21.139    0.000
##    .I11               1.166     0.059    19.605    0.000
##    .I12               4.527     0.050    89.700    0.000
##    .I13               2.653     0.063    42.180    0.000
##    .I14               3.147     0.068    46.538    0.000
##    .I15               1.078     0.052    20.565    0.000
##    .I16               1.548     0.056    27.808    0.000
##    .I17               5.303     0.037   141.503    0.000
##    .I18               4.705     0.045   105.354    0.000
##    .I19               4.600     0.049    93.388    0.000
##    .I20               1.211     0.053    22.744    0.000
##    .I21               4.462     0.057    78.050    0.000
##    .I22               1.790     0.060    29.873    0.000
##     EE                0.000
##     DP                0.000
##     PA                0.000
##
## Variances:
##                      Estimate  Std.Err   z-value  P(>|z|)
##    .I01               1.240     0.075    16.582    0.000
##    .I02               1.228     0.073    16.834    0.000
##    .I03               1.379     0.083    16.606    0.000
##    .I06               1.744     0.099    17.596    0.000
##    .I08               0.794     0.059    13.557    0.000
##    .I13               1.151     0.072    16.049    0.000
##    .I14               1.922     0.110    17.415    0.000
##    .I16               1.393     0.079    17.593    0.000
##    .I20               0.911     0.055    16.474    0.000
##    .I05               1.272     0.092    13.863    0.000
##    .I10               1.505     0.104    14.518    0.000
##    .I11               1.619     0.098    16.484    0.000
##    .I15               1.249     0.081    15.488    0.000
```

```
##     .I22          1.888    0.113   16.700   0.000
##     .I04          1.058    0.059   17.841   0.000
##     .I07          0.923    0.054   17.174   0.000
##     .I09          1.403    0.088   15.914   0.000
##     .I12          1.001    0.060   16.778   0.000
##     .I17          0.598    0.038   15.891   0.000
##     .I18          0.639    0.048   13.424   0.000
##     .I19          0.983    0.064   15.313   0.000
##     .I21          1.885    0.106   17.753   0.000
##     EE            1.324    0.105   12.547   0.000
##     DP            0.984    0.103    9.527   0.000
##     PA            0.178    0.029    6.232   0.000
```

We can see that all the factor loadings are estimated to be the same for both *elementary school teachers* and *secondary school teachers*. However, all other parameters associated with intercepts, residual variances and covariances are different. Therefore, we have established the metric invariance model.

6.4   Data Analysis Using SAS

The CALIS procedure of *SAS* supports flexible specifications of multi-group analysis. You can specify any kind of invariance among groups. Although the procedure does not support options that specify the invariance of a particular group of parameters (e.g., invariance of loadings) directly, you can use model referencing syntax (e.g., REFMODEL statement and options) to simplify the specifications of different kinds of invariance models.

To illustrate the applications to the *MBI* data, two SAS data sets, *mbielm1* and *mbisec1*, are created and they are read into the CALIS procedure in the following code for analyzing a configural invariance model for two groups:

```
proc calis meanstr;
    group 1/ name='Elementary' data='c:\mbielm1';
    group 2/ name='Secondary' data='c:\mbisec1';
    model 1/ group = 1;
        path
            EE ==> I01 I02 I03 I06 I08 I13 I14 I16 I20 = 1,
            PA ==> I04 I07 I09 I12 I17 I18 I19 I21 = 1,
            DP ==> I05 I10 I11 I15 I22 = 1,
            EE ==> I11 I12;
        pcov
            I01 I02, I06 I16, I10 I11, I04 I07, I09 I19;
```

```
     model 2/ group = 2;
        refmodel 1/ allnewparms;
     fitindex on(only)=[chisq df probchi RMSEA RMSEA_LL RMSEA_UL
             PROBCLFIT SRMR CFI BENTLERNNFI] noindextype;
run;
```

In the GROUP statements, the two independent samples for elementary and secondary school teachers are defined by using two distinct data sets. The first MODEL statement specifies a model for the elementary school teachers, which is designated as Group 1 in the first GROUP statement. Within the scope of Model 1, a PATH statement specifies the required measurement models, including the hypothesized loadings for measuring the three latent variables and the two additional loadings from *EE* to *I11* and *I12*. The PCOV statement specifies five error covariances for observed variables. Because you specify the MEANSTR option in the PROC CALIS statement, the model includes all default mean and intercept parameters, in addition to other default parameters such as error variances of the observed variables and variances and covariances of the latent factors.

Next, the second MODEL statement specifies the MODEL for the secondary school teachers, which is designated as Group 2 in the second GROUP statement. The REFMODEL statement makes reference to the specifications of Model 1. This means that all explicit specifications in Model 1 are now duplicated in the second model. The ALLNEWPARMS option ensures that all free parameters that have been specified in Model 1 are now copied to Model 2 as new and unconstrained free parameters. These specifications realize the configural invariance model for the two groups.

Finally, the FITINDEX statement simplifies the fit summary table to display only those fit indices of interests, as shown in the following table:

Fit Summary	
Chi-Square	1165.5763
Chi-Square DF	398
Pr > Chi-Square	<.0001
Standardized RMR (SRMR)	0.0501
RMSEA Estimate	0.0551
RMSEA Lower 90% Confidence Limit	0.0515
RMSEA Upper 90% Confidence Limit	0.0588
Probability of Close Fit	0.0111
Bentler Comparative Fit Index	0.9336
Bentler-Bonett Non-normed Index	0.9229

In addition to the overall fit summary of the configural invariance model, which replicates the *lavaan* results, PROC CALIS displays a fit comparison table between groups, as shown in the following table:

	Fit Comparison Among Groups	Overall	Elementary	Secondary
Modeling Info	Number of Observations	1272	580	692
	Number of Variables	22	22	22
	Number of Moments	550	275	275
	Number of Parameters	152	76	76
	Number of Active Constraints	0	0	0
	Baseline Model Function Value	9.4660	9.6692	9.2958
	Baseline Model Chi-Square	12021.8710	5598.4609	6423.4100
	Baseline Model Chi-Square DF	462	231	231
Fit Index	Fit Function	0.9178	0.9097	0.9246
	Percent Contribution to Chi-Square	100	45	55
	Root Mean Square Residual (RMR)	0.1059	0.0969	0.1128
	Standardized RMR (SRMR)	0.0501	0.0481	0.0518
	Goodness of Fit Index (GFI)	0.9838	0.9858	0.9821
	Bentler-Bonett NFI	0.9030	0.9059	0.9005

Note that the fit comparison table shown here was produced with the FITINDEX statement deleted from the previous code. Had you used the previous CALIS code with the FITINDEX statement specification, the comparison would have shown the SRMR statistics only. The comparison table shows that the quality of fit of these groups are pretty much the same in the context of the configural invariance model. The parameter estimates and other statistics are not shown here.

Next, the *ideal* model with both configural and measurement invariance is fitted with the following CALIS code:

```
proc calis;
    group 1/ name='Elementary' data='c:\mbielm1';
    group 2/ name='Secondary' data='c:\mbisec1';
    model 1/ group = 1;
       path
          1  ==> I01-I22,
          EE ==> I01 I02 I03 I06 I08 I13 I14 I16 I20 = 1,
          PA ==> I04 I07 I09 I12 I17 I18 I19 I21 = 1,
          DP ==> I05 I10 I11 I15 I22 = 1,
          EE ==> I11 I12;
```

```
      pvar
            EE PA DP I01-I22;
      pcov
            EE PA DP, I01 I02, I06 I16, I10 I11, I04 I07, I09 I19;
      Model 2/ group = 2;
         refmodel 1;
      fitindex on(only)=[chisq df probchi RMSEA RMSEA_LL RMSEA_UL
                  PROBCLFIT SRMR CFI BENTLERNNFI] noindextype;
run;
```

One new specification here is the use of **1-paths** in the PATH statement
for the 22 observed items. This specifies the intercept parameters of these
items. The need for these explicit intercept parameter specification (instead
of using the MEANSTR option in the PROC CALIS statement) is due to
the fact that model referencing by using the REFMODEL statement in the
specification of the model for the secondary school teachers (Group 2) applies
only to the parameters that have been *explicitly* specified. In the current case,
Model 2 makes reference to all parameter specifications in Model 1 and thus
effectively constrain all intercepts, loadings, variances and error variances, and
error covariances to be the same in the two models. Together with the fact that
the factor means for the two groups are zeros by default, the *ideal* model is
thus specified as required. Had the **1-paths** not been specified, the intercepts
would have been unconstrained for the two groups.

The following fit summary table shows the fit of the *ideal* model:

Fit Summary	
Chi-Square	1557.4793
Chi-Square DF	474
Pr > Chi-Square	<.0001
Standardized RMR (SRMR)	0.0721
RMSEA Estimate	0.0600
RMSEA Lower 90% Confidence Limit	0.0567
RMSEA Upper 90% Confidence Limit	0.0633
Probability of Close Fit	<.0001
Bentler Comparative Fit Index	0.9063
Bentler-Bonett Non-normed Index	0.9086

The following table shows the estimates of intercepts as coefficients of the
specified 1-paths. These estimates are invariant across the groups, although
only Model 1 results are shown here.

	Model 1. PATH List					
Path		Parameter	Estimate	Standard Error	t Value	Pr > |t|
1 ===> I01		_Parm01	3.38836	0.04548	74.5025	<.0001
1 ===> I02		_Parm02	3.92924	0.04350	90.3330	<.0001
1 ===> I03		_Parm03	2.54717	0.04786	53.2162	<.0001
1 ===> I04		_Parm04	5.27907	0.02908	181.5	<.0001
1 ===> I05		_Parm05	1.14231	0.04139	27.5980	<.0001
1 ===> I06		_Parm06	1.85144	0.04664	39.6944	<.0001
1 ===> I07		_Parm07	5.16193	0.02915	177.1	<.0001
1 ===> I08		_Parm08	2.16195	0.04932	43.8389	<.0001
1 ===> I09		_Parm09	4.85218	0.04020	120.7	<.0001
1 ===> I10		_Parm10	1.22406	0.04285	28.5679	<.0001
1 ===> I11		_Parm11	1.14623	0.04301	26.6481	<.0001
1 ===> I12		_Parm12	4.60298	0.03702	124.3	<.0001
1 ===> I13		_Parm13	2.60535	0.04758	54.7535	<.0001
1 ===> I14		_Parm14	3.13601	0.05044	62.1735	<.0001
1 ===> I15		_Parm15	0.83494	0.03699	22.5705	<.0001
1 ===> I16		_Parm16	1.49529	0.04162	35.9268	<.0001
1 ===> I17		_Parm17	5.35455	0.02608	205.3	<.0001
1 ===> I18		_Parm18	4.78615	0.03355	142.6	<.0001
1 ===> I19		_Parm19	4.78535	0.03492	137.1	<.0001
1 ===> I20		_Parm20	1.24292	0.03985	31.1862	<.0001
1 ===> I21		_Parm21	4.63519	0.04000	115.9	<.0001
1 ===> I22		_Parm22	1.57943	0.04469	35.3388	<.0001

Other estimation results should now be familiar and are not shown here. To fit the *intercept-invariance* model, you can use the following code:

```
proc calis;
    group 1/ name='Elementary' data='c:\mbielm1';
    group 2/ name='Secondary' data='c:\mbisec1';
    model 1/ group = 1;
        path
            1   ==> I01-I22  = Intercept01-Intercept22,
            1   ==> EE PA DP = 3*0.,
            EE ==> I01 I02 I03 I06 I08 I13 I14 I16 I20 = 1,
            PA ==> I04 I07 I09 I12 I17 I18 I19 I21 = 1,
            DP ==> I05 I10 I11 I15 I22 = 1,
            EE ==> I11 I12;
        pcov
            I01 I02, I06 I16, I10 I11, I04 I07, I09 I19;
```

```
model 2/ group = 2;
   path
      1  ==> I01-I22   = Intercept01-Intercept22,
      1  ==> EE PA DP = FactMeans01-FactMeans03,
      EE ==> I01 I02 I03 I06 I08 I13 I14 I16 I20 = 1,
      PA ==> I04 I07 I09 I12 I17 I18 I19 I21 = 1,
      DP ==> I05 I10 I11 I15 I22 = 1,
      EE ==> I11 I12;
   pcov
      I01 I02, I06 I16, I10 I11, I04 I07, I09 I19;
   fitindex on(only)=[chisq df probchi RMSEA RMSEA_LL RMSEA_UL
         PROBCLFIT SRMR CFI BENTLERNNFI AIC CAIC SBC]
         noindextype;
run;
```

Instead of using the REFMODEL specifications, to show the parameters clearly in this invariance model, a strategy that explicates all non-default parameters is employed here. First, the *1-paths* for intercepts in both models have the same set of parameters Intercept01–Intercept22. Next, the *1-paths* for factor means of the 3 latent factors in Model 1 are set to fixed zeros for identifications. Note that these factor means would have been set to zeros by default even if you do not specify them. Because of the established identification of the mean structures in Model 1 and the invariance of intercepts across models (groups), the factor means in Model 2 are no longer under-identified and therefore they can be set as free parameters FactMeans01-FactMeans03. Specifications of loading and error variance parameters are the same for the two models. Because the two models are specified under distinct MODEL statements, all other un-named parameters are unconstrained across groups (models). Hence, the *intercept-invariance* model is specified.

The following table shows the fit summary of the "intercept-invariance" model and replicates similar results shown by running the *lavaan* code:

Fit Summary	
Chi-Square	1246.7688
Chi-Square DF	417
Pr > Chi-Square	<.0001
Standardized RMR (SRMR)	0.0531
RMSEA Estimate	0.0560
RMSEA Lower 90% Confidence Limit	0.0524
RMSEA Upper 90% Confidence Limit	0.0596
Probability of Close Fit	0.0031
Akaike Information Criterion	1512.7688
Bozdogan CAIC	2330.4988
Schwarz Bayesian Criterion	2197.4988
Bentler Comparative Fit Index	0.9282
Bentler-Bonett Non-normed Index	0.9205

To fit the *loading-invariance* mode, you can use the following code:

```
proc calis;
   group 1/ name='Elementary' data='c:\mbielm1';
   group 2/ name='Secondary' data='c:\mbisec1';
   model 1/ group = 1;
      path
         1  ==> I01-I22,
         EE ==> I01 I02 I03 I06 I08 I13 I14 I16 I20 = 1
                EEload02-EEload09,
         PA ==> I04 I07 I09 I12 I17 I18 I19 I21 = 1
                PAload02-PAload08,
         DP ==> I05 I10 I11 I15 I22 = 1 DPLoad02-DPLoad05,
         EE ==> I11 I12 = EELoad10-EELoad11;
      pcov
         I01 I02, I06 I16, I10 I11, I04 I07, I09 I19;
   model 2/ group = 2;
      path
         1  ==> I01-I22,
         EE ==> I01 I02 I03 I06 I08 I13 I14 I16 I20 = 1
                EEload02-EEload09,
         PA ==> I04 I07 I09 I12 I17 I18 I19 I21 = 1
                PAload02-PAload08,
         DP ==> I05 I10 I11 I15 I22 = 1 DPLoad02-DPLoad05,
         EE ==> I11 I12 = EELoad10-EELoad11;
      pcov
         I01 I02, I06 I16, I10 I11, I04 I07, I09 I19;
   fitindex on(only)=[chisq df probchi RMSEA RMSEA_LL RMSEA_UL
           PROBCLFIT SRMR CFI BENTLERNNFI AIC CAIC SBC]
           noindextype;
run;
```

The strategy of specifying the *loading-invariance* model is similar to that
of the *intercept-invariance* model. Models 1 and 2 are specified separately
and are configurally invariant. Intercept parameters are specified as 1-paths
for the two models but are not constrained. Factor means for the two models
are set to zeros by default for identification. The sets of loadings in the two
models are now constrained to be the same by using the same set of parameter names (or labels): EEload02–EEload11, PAload02–PAload08, and
DPLoad02–DPLoad05. Other parameters are unconstrained across groups
(models).

The following table shows the fit summary of the "loading-invariance"
model:

Fit Summary	
Chi-Square	1208.5994
Chi-Square DF	419
Pr > Chi-Square	<.0001
Standardized RMR (SRMR)	0.0539
RMSEA Estimate	0.0545
RMSEA Lower 90% Confidence Limit	0.0509
RMSEA Upper 90% Confidence Limit	0.0581
Probability of Close Fit	0.0200
Akaike Information Criterion	1470.5994
Bozdogan CAIC	2276.0326
Schwarz Bayesian Criterion	2145.0326
Bentler Comparative Fit Index	0.9317
Bentler-Bonett Non-normed Index	0.9247

Finally, to fit a multi-group model with invariance in both intercepts and loadings, you can *combine* the specifications of the previous two multi-group model. The strategy is to achieve the intercept and loading invariance by using the same names (labels) for the target parameters in the two models for the groups. This is left as an exercise for readers.

Once all model fits are obtained from all multi-group models that have various degrees of invariance, you can compute chi-square different tests by hand or by some simple coding. This is also left as an exercise for readers.

6.5   Discussions

In this chapter, we illustrated the test for multi-group invariance using the *MBI* data. As discussed in Rosseel (2012) and Gana and Broc (2019), this can be accomplished by using *lavaan* option *group.equal* to provide invariance in the SEM model components with:

- *loadings* for all factor loadings from observed variables to the latent variables

- *intercepts* for the intercepts of the observed variables

- *means* for the intercepts/means of the latent variables

- *residuals* for the residual variances of the observed variables

- *residual.covariances* for the residual covariances of the observed variables

- *lv.variances* for the (residual) variances of the latent variables

- *lv.covariances* for the (residual) covariances of the latent variables

- *regressions* for all regression coefficients in the model.

If the *group.equal* is omitted, all parameters are freely estimated in each group, which is the *configural* model. We used the *MBI* data and illustrated these selections with detailed steps.

One option left without illustration is to constrain one or a few of the parameters in each group with the model structure unchanged. For example, in the factor loading invariance model *fitMBI.Loadings*, suppose that we have substantive knowledge that the factor loadings from *I02* and *I03* to the latent variable *EE* are different between these two groups of teachers, we would need to relax the invariance constraints to freely estimate these two factor loadings for these two groups of teachers. In order to implement this, we can make use of the *lavaan* option *group.partial* to specify these two factor loadings as follows (not run to save pages):

```
fitMBI.Loadings2 = cfa(CFA4MBI, data=dMBI, group="School",
                group.equal="loadings",
                group.partial=c("EE =~ I02","EE =~ I03"))
```

More analysis of this data on multi-group invariance can be found in Byrne (2012). In Chapter 7 of this book, multi-group invariance was established using *MI* analysis to search for the specific paths.

6.6 Exercises

1. Using data *selfEsteem.txt* to test measurement invariance for latent factor *Self-Esteem* measured by ten items for two groups (i.e., *Sex*) using model:

   ```
   mod4SelfEsteem <-
   'F1 =~ se1 + se2 + se3 + se4 + se5 + se6 + se7 + se8
       + se9 + se10'
   ```

   Note: Data was downloaded from: https://raw.githubusercontent. com/rafavsbastos/data/main/measurement%20invariance.dat and used to illustrate the measurement invariance (https:// towardsdatascience.com/measurement-invariance-definition- and-example-in-r-15b4efcab351).

2. Read Chapter 7 (Byrne, 2012) and establish multi-group invariance using *group.equal* and *group.partial*.

3. Using the data in the main text, fit the *intercept and loading invariance* multi-group model for the elementary and secondary school teachers by using CALIS (Hint: Start with *intercept* invariance model and add more constraints to the model setup). Verify all the chi-square test results for comparing the *unconstrained, intercept-invariance, loading-invariance, intercept and loading invariance,* and *ideal* models.

7

Multi-Group Data Analysis: Categorical Data

Continuing from Chapter 6, this chapter discusses the multi-group analysis with categorical data. Due to the categorical characteristics of the data, some of the discussions in Chapter 5 will be linked in this chapter, such as the estimation of categorical *threshold* and the specification of *parameterization*. In this sense, this chapter is a continuation of both chapters.

We will make use of the data from Ren et al. (2021) for the illustration of this chapter. Using convenience sampling, the data was collected from 327 ($n = 192$ for boys and 135 for girls) children of Chinese rural-to-urban migrant workers in Hangzhou, China, to study their depression and anxiety using Kessler's 10-item Psychological Distress Scale (K10). The sample was selected in a local middle school where migrant children were congregated. Paper-and-pencil questionnaires on psychological distress and demographic variables were administered in a classroom setting.

Ren et al. (2021) have established the invariance with a series of analyses using *mplus*. The associated mplus program and data are included as *Ren.et.al.Mplus.zip* for interested readers to reproduce the analysis results with *mplus*. In this chapter, we will re-analyze this data using *R/lavaan* to establish the CFA measurement invariances between *boys* and *girls* and test the mean differences between *boys* and *girls* on their *Depression* and *Anxiety*, which are measured as latent factors.

7.1 Data and Preliminary Analysis

The data file is saved as *mig_chdn_k10.dta.dat* with 11 columns and 327 observations from these 327 school children. The format of this data set is that the first 10 columns are labeled as *b1-b10*, which are the 10 items on the K10 scale, and the last column is their *sex* where 0 represents for *girl* and 1 for *boy*:

Variable	Description	Short Description
b01	Feel tired out for no good reason	Tired
b02	Feel nervous	Nervous
b03	Feel so nervous that nothing could calm you down	Severely nervous
b04	Feel hopeless	Hopeless
b05	Feel restless or fidgety	Restless
b06	Feel so restless that could not sit still	Severely restless
b07	Feel depressed	Depressed
b08	Feel that everything was an effort	Effort
b09	Feel so sad that nothing could cheer up	Severely depressed
b10	Feel worthless	Worthless

We can read the data into *R* as follows:

```
# Read the data into R using 'read.table'
dmig = read.table("data/mig_chdn_k10.dta.dat",
                  na.strings="-9999", sep=",",header=F)
# Name the variables
colnames(dmig) = c("b01","b02","b03","b04","b05","b06",
                   "b07", "b08","b09","b10","sex")
# make "sex" as factor
dmig$sex = as.factor(dmig$sex)
# Check the data dimension
dim(dmig)
```

```
## [1] 327   11
```

```
# Print the data summary
summary(dmig)
```

```
##       b01             b02             b03
##  Min.   :1.00   Min.   :1.00   Min.   :1.00
##  1st Qu.:1.00   1st Qu.:1.00   1st Qu.:1.00
##  Median :1.00   Median :1.00   Median :1.00
##  Mean   :1.21   Mean   :1.16   Mean   :1.09
##  3rd Qu.:1.00   3rd Qu.:1.00   3rd Qu.:1.00
##  Max.   :3.00   Max.   :3.00   Max.   :3.00
##  NA's   :1                     NA's   :2
```

```
##         b04               b05               b06
##   Min.    :1.00   Min.    :1.00   Min.    :1.00
##   1st Qu.:1.00   1st Qu.:1.00   1st Qu.:1.00
##   Median :1.00   Median :1.00   Median :1.00
##   Mean    :1.14   Mean    :1.18   Mean    :1.09
##   3rd Qu.:1.00   3rd Qu.:1.00   3rd Qu.:1.00
##   Max.    :3.00   Max.    :3.00   Max.    :3.00
##   NA's    :1
##         b07               b08               b09
##   Min.    :1.00   Min.    :1.00   Min.    :1.00
##   1st Qu.:1.00   1st Qu.:1.00   1st Qu.:1.00
##   Median :1.00   Median :1.00   Median :1.00
##   Mean    :1.12   Mean    :1.14   Mean    :1.15
##   3rd Qu.:1.00   3rd Qu.:1.00   3rd Qu.:1.00
##   Max.    :3.00   Max.    :3.00   Max.    :3.00
##   NA's    :1     NA's    :1     NA's    :1
##         b10           sex
##   Min.    :1.00   0:135
##   1st Qu.:1.00   1:192
##   Median :1.00
##   Mean    :1.13
##   3rd Qu.:1.00
##   Max.    :3.00
##   NA's    :1
```

As noted, there are a few missing values coded as *-9999* in the data and they are loaded into *R* with option *na.strings="-9999"*.

We can see whether these ten items are significantly different by *sex* using t-test to reproduce the results in Table 1 from Ren et al. (2021) with following *R* code chunk:

```
t.test(b01~sex, dmig)
```

```
##
##   Welch Two Sample t-test
##
## data:  b01 by sex
## t = 1, df = 231, p-value = 0.1
## alternative hypothesis: true difference in means
## between group 0 and group 1 is not equal to 0
## 95 percent confidence interval:
##   -0.0275  0.2005
## sample estimates:
## mean in group 0 mean in group 1
##             1.26             1.17
```

```
t.test(b02~sex, dmig)
```

```
##
##   Welch Two Sample t-test
##
## data:  b02 by sex
## t = 2, df = 245, p-value = 0.04
## alternative hypothesis: true difference in means
## between group 0 and group 1 is not equal to 0
## 95 percent confidence interval:
##   0.00721 0.19765
## sample estimates:
## mean in group 0 mean in group 1
##             1.22             1.12
```

```
t.test(b03~sex, dmig)
```

```
##
##   Welch Two Sample t-test
##
## data:  b03 by sex
## t = 1, df = 258, p-value = 0.2
## alternative hypothesis: true difference in means
## between group 0 and group 1 is not equal to 0
## 95 percent confidence interval:
##   -0.0272  0.1150
## sample estimates:
## mean in group 0 mean in group 1
##             1.11             1.07
```

```
t.test(b04~sex, dmig)
```

```
##
##   Welch Two Sample t-test
##
## data:  b04 by sex
## t = 2, df = 239, p-value = 0.08
## alternative hypothesis: true difference in means
## between group 0 and group 1 is not equal to 0
## 95 percent confidence interval:
##   -0.0102  0.1816
## sample estimates:
## mean in group 0 mean in group 1
##             1.19             1.10
```

```
t.test(b05~sex, dmig)
```

```
##
##   Welch Two Sample t-test
##
## data:  b05 by sex
## t = 1, df = 253, p-value = 0.2
## alternative hypothesis: true difference in means
## between group 0 and group 1 is not equal to 0
## 95 percent confidence interval:
##   -0.0318  0.1741
## sample estimates:
## mean in group 0 mean in group 1
##             1.22            1.15
```

```
t.test(b06~sex, dmig)
```

```
##
##   Welch Two Sample t-test
##
## data:  b06 by sex
## t = 1, df = 259, p-value = 0.3
## alternative hypothesis: true difference in means
## between group 0 and group 1 is not equal to 0
## 95 percent confidence interval:
##   -0.0316  0.1184
## sample estimates:
## mean in group 0 mean in group 1
##             1.11            1.07
```

```
t.test(b07~sex, dmig)
```

```
##
##   Welch Two Sample t-test
##
## data:  b07 by sex
## t = 2, df = 211, p-value = 0.02
## alternative hypothesis: true difference in means
## between group 0 and group 1 is not equal to 0
## 95 percent confidence interval:
##   0.0176 0.1947
## sample estimates:
## mean in group 0 mean in group 1
##             1.18            1.07
```

```
t.test(b08~sex, dmig)
```

```
##
##   Welch Two Sample t-test
##
## data:  b08 by sex
## t = 0.9, df = 262, p-value = 0.4
## alternative hypothesis: true difference in means
## between group 0 and group 1 is not equal to 0
## 95 percent confidence interval:
##  -0.0562  0.1456
## sample estimates:
## mean in group 0 mean in group 1
##             1.17             1.13
```

```
t.test(b09~sex, dmig)
```

```
##
##   Welch Two Sample t-test
##
## data:  b09 by sex
## t = 2, df = 241, p-value = 0.07
## alternative hypothesis: true difference in means
## between group 0 and group 1 is not equal to 0
## 95 percent confidence interval:
##  -0.00698  0.19143
## sample estimates:
## mean in group 0 mean in group 1
##             1.21             1.12
```

```
t.test(b10~sex, dmig)
```

```
##
##   Welch Two Sample t-test
##
## data:  b10 by sex
## t = 2, df = 223, p-value = 0.1
## alternative hypothesis: true difference in means
## between group 0 and group 1 is not equal to 0
## 95 percent confidence interval:
##  -0.0177  0.1734
## sample estimates:
## mean in group 0 mean in group 1
##             1.17             1.09
```

As seen from these t-tests, only *b02* and *b07* are statistically significantly different between *boys* and *girls*. We also need to caution the readers that these t-tests could be unreliable since the data are recorded in 3 Likert scale of 1 and 3, and they are not normally distributed as required by the t-test as the fundamental assumption. The differences between *boys* and *girls* can be further graphically examined in Figure 7.1.

```
# Call tidyverse to reshape the data
library(tidyverse)
dmiglong <- dmig %>%
  gather("b01","b02","b03","b04","b05",
         "b06","b07","b08","b09","b10",
         key="Item", value="Value")
# Call ggplot2 to make the density plot
library(ggplot2)
ggplot(dmiglong, aes(Value, col=sex))+
  geom_density(size=1.5)+
  facet_wrap(~Item, nrow=4, scales="free")+
  labs(x=" ", y="Density")
```

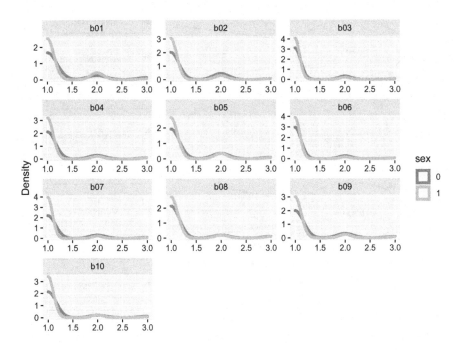

FIGURE 7.1
The Density Plot of all Ten Items by Boys and Girls.

7.2   CFA Models

Ren et al. (2021) tested three CFA models for K10 based on the psychological distress theory, including the second-order factor CFA model, two-factor CFA model and unidimentional CFA model. The final two-factor CFA model was accepted for K10, which is shown graphically in Figure 7.2.

This two-factor CFA model can be coded in *R/lavaan* as follows:

```
#  CFA model for K10
CFA4K10 <- '
# Latent variables: meaurement model
Depression =~ b04 + b07 + b08 + b09 + b10
Anxiety    =~ b01 + b02 + b03 + b05 + b06
'
```

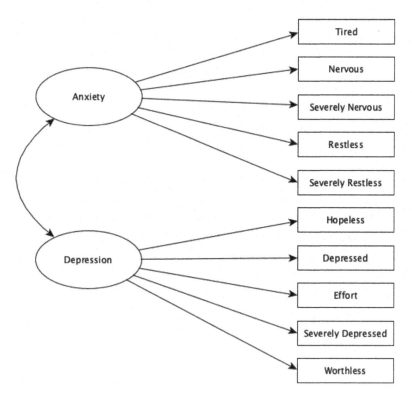

FIGURE 7.2
Final Two-Factor CFA Model for K10.

7.2.1 Step One: The Group-Specific Model

As the first step for group-invariance, we need to establish the group-specific model. Specifically for this application, we need to establish this two-factor CFA model for both *boys* and *girls*.

This can be done by calling the *cfa* to fit both *boys* and *girls* groups. Since this is a categorical data, we need to *order* the data categories using *ordered*. As discussed in Chapter 5, the default parameterization is *delta*, but for multi-group analysis, it is preferred to use *theta parameterization*. Therefore, the *R/lavaan* implementation is as follows:

```
#
# Fit the CFA model for Girls
#
fitK10.Girl = cfa(CFA4K10, data=dmig[dmig$sex==0,],
                parameterization='theta',
                ordered=c("b01","b02","b03","b04","b05",
                      "b06","b07","b08","b09","b10"))
# Print the model summary
summary(fitK10.Girl, fit.measures=TRUE)


## lavaan 0.6-12 ended normally after 93 iterations
##
##   Estimator                                    DWLS
##   Optimization method                        NLMINB
##   Number of model parameters                     31
##
##                                              Used       Total
##   Number of observations                      132         135
##
## Model Test User Model:
##                                          Standard      Robust
##   Test Statistic                           20.048      32.470
##   Degrees of freedom                           34          34
##   P-value (Chi-square)                      0.972       0.543
##   Scaling correction factor                             0.932
##   Shift parameter                                      10.952
##     simple second-order correction
##
## Model Test Baseline Model:
##
##   Test statistic                         1622.305     849.329
##   Degrees of freedom                           45          45
##   P-value                                   0.000       0.000
##   Scaling correction factor                             1.961
##
```

```
## User Model versus Baseline Model:
##
##   Comparative Fit Index (CFI)                           1.000      1.000
##   Tucker-Lewis Index (TLI)                              1.012      1.003
##
##   Robust Comparative Fit Index (CFI)                                  NA
##   Robust Tucker-Lewis Index (TLI)                                     NA
##
## Root Mean Square Error of Approximation:
##
##   RMSEA                                                 0.000      0.000
##   90 Percent confidence interval - lower                0.000      0.000
##   90 Percent confidence interval - upper                0.000      0.060
##   P-value RMSEA <= 0.05                                 0.998      0.893
##
##   Robust RMSEA                                                       NA
##   90 Percent confidence interval - lower                           0.000
##   90 Percent confidence interval - upper                             NA
##
## Standardized Root Mean Square Residual:
##
##   SRMR                                                  0.071      0.071
##
## Parameter Estimates:
##
##   Standard errors                              Robust.sem
##   Information                                    Expected
##   Information saturated (h1) model          Unstructured
##
## Latent Variables:
##                   Estimate  Std.Err  z-value  P(>|z|)
##   Depression =~
##     b04             1.000
##     b07             0.670    0.216    3.093    0.002
##     b08             0.600    0.250    2.397    0.017
##     b09             0.832    0.303    2.747    0.006
##     b10             0.660    0.230    2.872    0.004
##   Anxiety =~
##     b01             1.000
##     b02             0.654    0.188    3.476    0.001
##     b03             0.788    0.302    2.610    0.009
##     b05             1.111    0.432    2.570    0.010
##     b06             0.630    0.199    3.172    0.002
##
## Covariances:
##                   Estimate  Std.Err  z-value  P(>|z|)
```

```
## Depression ~~
##     Anxiety          3.075    1.281    2.399    0.016
##
## Intercepts:
##                     Estimate  Std.Err  z-value  P(>|z|)
##     .b04             0.000
##     .b07             0.000
##     .b08             0.000
##     .b09             0.000
##     .b10             0.000
##     .b01             0.000
##     .b02             0.000
##     .b03             0.000
##     .b05             0.000
##     .b06             0.000
##     Depression       0.000
##     Anxiety          0.000
##
## Thresholds:
##                     Estimate  Std.Err  z-value  P(>|z|)
##     b04|t1           2.275    0.588    3.870    0.000
##     b04|t2           3.922    0.769    5.101    0.000
##     b07|t1           1.704    0.319    5.335    0.000
##     b07|t2           3.105    0.388    8.012    0.000
##     b08|t1           1.752    0.311    5.630    0.000
##     b08|t2           2.616    0.396    6.609    0.000
##     b09|t1           1.857    0.398    4.669    0.000
##     b09|t2           3.408    0.496    6.872    0.000
##     b10|t1           1.856    0.372    4.995    0.000
##     b10|t2           2.771    0.442    6.272    0.000
##     b01|t1           1.783    0.445    4.003    0.000
##     b01|t2           3.018    0.575    5.248    0.000
##     b02|t1           1.300    0.238    5.465    0.000
##     b02|t2           3.052    0.373    8.179    0.000
##     b03|t1           2.207    0.471    4.686    0.000
##     b03|t2           4.155    0.713    5.825    0.000
##     b05|t1           2.061    0.515    4.000    0.000
##     b05|t2           3.904    0.698    5.594    0.000
##     b06|t1           1.927    0.331    5.822    0.000
##     b06|t2           3.236    0.525    6.158    0.000
##
## Variances:
##                     Estimate  Std.Err  z-value  P(>|z|)
##     .b04             1.000
##     .b07             1.000
##     .b08             1.000
```

```
##      .b09                    1.000
##      .b10                    1.000
##      .b01                    1.000
##      .b02                    1.000
##      .b03                    1.000
##      .b05                    1.000
##      .b06                    1.000
##       Depression            3.879     2.267     1.711     0.087
##       Anxiety               3.104     1.680     1.848     0.065
##
## Scales y*:
##                           Estimate  Std.Err   z-value   P(>|z|)
##      b04                    0.453
##      b07                    0.604
##      b08                    0.646
##      b09                    0.521
##      b10                    0.610
##      b01                    0.494
##      b02                    0.656
##      b03                    0.584
##      b05                    0.455
##      b06                    0.669
```

```
#
# Fit the CFA for Boys
#
fitK10.Boy  = cfa(CFA4K10,data=dmig[dmig$sex==1,],
                  parameterization='theta',
                  ordered=c("b01","b02","b03","b04","b05",
                            "b06","b07","b08","b09","b10"))
summary(fitK10.Boy, fit.measures=TRUE)
```

```
## lavaan 0.6-12 ended normally after 127 iterations
##
##    Estimator                              DWLS
##    Optimization method                   NLMINB
##    Number of model parameters              31
##
##                                          Used       Total
##    Number of observations                 188         192
##
## Model Test User Model:
##                                       Standard      Robust
##    Test Statistic                      23.695      35.808
##    Degrees of freedom                      34          34
##    P-value (Chi-square)                 0.907       0.384
```

```
## Scaling correction factor                                    0.992
## Shift parameter                                             11.934
##    simple second-order correction
##
## Model Test Baseline Model:
##
## Test statistic                              2778.510      1302.765
## Degrees of freedom                                45            45
## P-value                                        0.000         0.000
## Scaling correction factor                                    2.173
##
## User Model versus Baseline Model:
##
## Comparative Fit Index (CFI)                    1.000         0.999
## Tucker-Lewis Index (TLI)                       1.005         0.998
##
## Robust Comparative Fit Index (CFI)                              NA
## Robust Tucker-Lewis Index (TLI)                                 NA
##
## Root Mean Square Error of Approximation:
##
## RMSEA                                          0.000         0.017
## 90 Percent confidence interval - lower         0.000         0.000
## 90 Percent confidence interval - upper         0.022         0.057
## P-value RMSEA <= 0.05                          0.997         0.900
##
## Robust RMSEA                                                    NA
## 90 Percent confidence interval - lower                       0.000
## 90 Percent confidence interval - upper                         NA
##
## Standardized Root Mean Square Residual:
##
## SRMR                                           0.071         0.071
##
## Parameter Estimates:
##
## Standard errors                         Robust.sem
## Information                               Expected
## Information saturated (h1) model      Unstructured
##
##
## Latent Variables:
##                   Estimate  Std.Err  z-value  P(>|z|)
## Depression =~
##    b04              1.000
##    b07              1.132    0.449    2.519    0.012
```

```
##      b08            1.067    0.337    3.170    0.002
##      b09            1.401    0.424    3.307    0.001
##      b10            0.654    0.165    3.958    0.000
##   Anxiety =~
##      b01            1.000
##      b02            1.201    0.394    3.045    0.002
##      b03            1.382    0.538    2.566    0.010
##      b05            1.105    0.376    2.940    0.003
##      b06            1.203    0.407    2.955    0.003
##
## Covariances:
##                   Estimate  Std.Err  z-value  P(>|z|)
##   Depression ~~
##      Anxiety        2.607    0.990    2.634    0.008
##
## Intercepts:
##                   Estimate  Std.Err  z-value  P(>|z|)
##      .b04           0.000
##      .b07           0.000
##      .b08           0.000
##      .b09           0.000
##      .b10           0.000
##      .b01           0.000
##      .b02           0.000
##      .b03           0.000
##      .b05           0.000
##      .b06           0.000
##      Depression     0.000
##      Anxiety        0.000
##
## Thresholds:
##                   Estimate  Std.Err  z-value  P(>|z|)
##      b04|t1         3.154    0.623    5.060    0.000
##      b04|t2         4.432    0.687    6.454    0.000
##      b07|t1         3.680    0.899    4.092    0.000
##      b07|t2         5.563    1.488    3.740    0.000
##      b08|t1         3.007    0.651    4.616    0.000
##      b08|t2         4.267    0.787    5.425    0.000
##      b09|t1         3.789    1.104    3.433    0.001
##      b09|t2         5.882    1.495    3.934    0.000
##      b10|t1         2.275    0.375    6.075    0.000
##      b10|t2         3.471    0.508    6.829    0.000
##      b01|t1         1.805    0.332    5.443    0.000
##      b01|t2         3.887    0.608    6.392    0.000
##      b02|t1         2.642    0.511    5.172    0.000
##      b02|t2         4.445    0.821    5.412    0.000
```

```
##       b03|t1              3.737    1.027    3.639    0.000
##       b03|t2              5.331    1.393    3.826    0.000
##       b05|t1              2.213    0.452    4.900    0.000
##       b05|t2              3.946    0.761    5.186    0.000
##       b06|t1              3.351    0.653    5.134    0.000
##       b06|t2              4.452    0.836    5.323    0.000
##
## Variances:
##                        Estimate  Std.Err  z-value  P(>|z|)
##       .b04                1.000
##       .b07                1.000
##       .b08                1.000
##       .b09                1.000
##       .b10                1.000
##       .b01                1.000
##       .b02                1.000
##       .b03                1.000
##       .b05                1.000
##       .b06                1.000
##       Depression          3.775    1.735    2.176    0.030
##       Anxiety             2.283    1.003    2.275    0.023
##
## Scales y*:
##                        Estimate  Std.Err  z-value  P(>|z|)
##       b04                 0.458
##       b07                 0.414
##       b08                 0.434
##       b09                 0.345
##       b10                 0.618
##       b01                 0.552
##       b02                 0.483
##       b03                 0.432
##       b05                 0.514
##       b06                 0.482
```

As seen from the model fitting measures, these K10 CFA models can fit the data for both *boys* and *girls* extremely well. In addition, all the factor loadings to the two latent factors *Depression* and *Anxiety* are statistically significant.

This process established the group-specific K10 CFA model for both *boys* and *girls* groups.

7.2.2   Step Two: The Configural Model

With the established group-specific K10 CFA model, step 2 is to establish the *configural model* to test *configural invariance* by fitting the K10 CFA model

for the data from both groups simultaneously. This can be done in *R/lavaan*
using *group* option as follows:

```
# Fit the configural model
fitK10.Configural = cfa(CFA4K10, data=dmig, group="sex",
                parameterization='theta',
                ordered=c("b01","b02","b03","b04","b05",
                    "b06","b07","b08","b09","b10"))
# Print the model summary with the fitting measures
summary(fitK10.Configural, fit.measures=TRUE)
```

```
## lavaan 0.6-12 ended normally after 247 iterations
##
##   Estimator                                      DWLS
##   Optimization method                          NLMINB
##   Number of model parameters                       62
##
##   Number of observations per group:             Used        Total
##     1                                             188          192
##     0                                             132          135
##
## Model Test User Model:
##                                            Standard       Robust
##   Test Statistic                             43.743       68.384
##   Degrees of freedom                             68           68
##   P-value (Chi-square)                        0.990        0.464
##   Scaling correction factor                                0.963
##   Shift parameter for each group:
##     1                                                     13.477
##     0                                                      9.462
##     simple second-order correction
##   Test statistic for each group:
##     1                                          23.695       38.093
##     0                                          20.048       30.291
##
## Model Test Baseline Model:
##
##   Test statistic                           4400.815     2172.634
##   Degrees of freedom                             90           90
##   P-value                                     0.000        0.000
##   Scaling correction factor                                2.070
##
## User Model versus Baseline Model:
##
##   Comparative Fit Index (CFI)                 1.000        1.000
##   Tucker-Lewis Index (TLI)                    1.007        1.000
```

```
##
##    Robust Comparative Fit Index (CFI)                            NA
##    Robust Tucker-Lewis Index (TLI)                               NA
##
## Root Mean Square Error of Approximation:
##
##    RMSEA                                             0.000      0.006
##    90 Percent confidence interval - lower            0.000      0.000
##    90 Percent confidence interval - upper            0.000      0.047
##    P-value RMSEA <= 0.05                             1.000      0.967
##
##    Robust RMSEA                                                 NA
##    90 Percent confidence interval - lower                      0.000
##    90 Percent confidence interval - upper                      NA
##
## Standardized Root Mean Square Residual:
##
##    SRMR                                              0.071      0.071
##
## Parameter Estimates:
##
##    Standard errors                               Robust.sem
##    Information                                     Expected
##    Information saturated (h1) model            Unstructured
##
##
## Group 1 [1]:
##
## Latent Variables:
##                     Estimate  Std.Err  z-value  P(>|z|)
##    Depression =~
##      b04              1.000
##      b07              1.132    0.449    2.518    0.012
##      b08              1.067    0.337    3.168    0.002
##      b09              1.401    0.424    3.306    0.001
##      b10              0.654    0.165    3.956    0.000
##    Anxiety =~
##      b01              1.000
##      b02              1.201    0.395    3.044    0.002
##      b03              1.382    0.539    2.565    0.010
##      b05              1.105    0.376    2.939    0.003
##      b06              1.203    0.407    2.954    0.003
##
## Covariances:
##                     Estimate  Std.Err  z-value  P(>|z|)
##    Depression ~~
```

```
##      Anxiety           2.607    0.990   2.632   0.008
##
## Intercepts:
##                      Estimate  Std.Err  z-value  P(>|z|)
##      .b04              0.000
##      .b07              0.000
##      .b08              0.000
##      .b09              0.000
##      .b10              0.000
##      .b01              0.000
##      .b02              0.000
##      .b03              0.000
##      .b05              0.000
##      .b06              0.000
##      Depression        0.000
##      Anxiety           0.000
##
## Thresholds:
##                      Estimate  Std.Err  z-value  P(>|z|)
##      b04|t1            3.154    0.624    5.058    0.000
##      b04|t2            4.432    0.687    6.451    0.000
##      b07|t1            3.680    0.900    4.090    0.000
##      b07|t2            5.563    1.488    3.738    0.000
##      b08|t1            3.007    0.652    4.614    0.000
##      b08|t2            4.267    0.787    5.423    0.000
##      b09|t1            3.788    1.104    3.431    0.001
##      b09|t2            5.882    1.496    3.932    0.000
##      b10|t1            2.275    0.375    6.072    0.000
##      b10|t2            3.471    0.508    6.826    0.000
##      b01|t1            1.805    0.332    5.441    0.000
##      b01|t2            3.887    0.608    6.389    0.000
##      b02|t1            2.642    0.511    5.170    0.000
##      b02|t2            4.445    0.822    5.410    0.000
##      b03|t1            3.737    1.027    3.638    0.000
##      b03|t2            5.331    1.394    3.824    0.000
##      b05|t1            2.213    0.452    4.898    0.000
##      b05|t2            3.946    0.761    5.183    0.000
##      b06|t1            3.351    0.653    5.132    0.000
##      b06|t2            4.452    0.837    5.321    0.000
##
## Variances:
##                      Estimate  Std.Err  z-value  P(>|z|)
##      .b04              1.000
##      .b07              1.000
##      .b08              1.000
##      .b09              1.000
```

```
##    .b10              1.000
##    .b01              1.000
##    .b02              1.000
##    .b03              1.000
##    .b05              1.000
##    .b06              1.000
##    Depression        3.775    1.736    2.175    0.030
##    Anxiety           2.283    1.004    2.274    0.023
##
## Scales y*:
##                    Estimate  Std.Err  z-value  P(>|z|)
##    b04               0.458
##    b07               0.414
##    b08               0.434
##    b09               0.345
##    b10               0.618
##    b01               0.552
##    b02               0.483
##    b03               0.432
##    b05               0.514
##    b06               0.482
##
##
## Group 2 [0]:
##
## Latent Variables:
##                    Estimate  Std.Err  z-value  P(>|z|)
##    Depression =~
##    b04               1.000
##    b07               0.670    0.216    3.095    0.002
##    b08               0.600    0.250    2.398    0.016
##    b09               0.832    0.303    2.749    0.006
##    b10               0.660    0.229    2.874    0.004
##    Anxiety =~
##    b01               1.000
##    b02               0.654    0.188    3.478    0.001
##    b03               0.788    0.302    2.612    0.009
##    b05               1.111    0.432    2.572    0.010
##    b06               0.630    0.198    3.174    0.002
##
## Covariances:
##                    Estimate  Std.Err  z-value  P(>|z|)
##    Depression ~~
##    Anxiety           3.075    1.280    2.401    0.016
##
## Intercepts:
```

```
##                          Estimate  Std.Err  z-value  P(>|z|)
##      .b04                   0.000
##      .b07                   0.000
##      .b08                   0.000
##      .b09                   0.000
##      .b10                   0.000
##      .b01                   0.000
##      .b02                   0.000
##      .b03                   0.000
##      .b05                   0.000
##      .b06                   0.000
##      Depression             0.000
##      Anxiety                0.000
##
## Thresholds:
##                          Estimate  Std.Err  z-value  P(>|z|)
##      b04|t1                 2.275    0.587    3.873    0.000
##      b04|t2                 3.922    0.768    5.105    0.000
##      b07|t1                 1.704    0.319    5.339    0.000
##      b07|t2                 3.105    0.387    8.018    0.000
##      b08|t1                 1.752    0.311    5.634    0.000
##      b08|t2                 2.616    0.396    6.613    0.000
##      b09|t1                 1.857    0.397    4.672    0.000
##      b09|t2                 3.408    0.496    6.877    0.000
##      b10|t1                 1.856    0.371    4.999    0.000
##      b10|t2                 2.771    0.442    6.276    0.000
##      b01|t1                 1.783    0.445    4.006    0.000
##      b01|t2                 3.018    0.575    5.251    0.000
##      b02|t1                 1.300    0.238    5.469    0.000
##      b02|t2                 3.052    0.373    8.185    0.000
##      b03|t1                 2.207    0.471    4.689    0.000
##      b03|t2                 4.155    0.713    5.829    0.000
##      b05|t1                 2.061    0.515    4.002    0.000
##      b05|t2                 3.904    0.697    5.598    0.000
##      b06|t1                 1.927    0.331    5.826    0.000
##      b06|t2                 3.236    0.525    6.162    0.000
##
## Variances:
##                          Estimate  Std.Err  z-value  P(>|z|)
##      .b04                   1.000
##      .b07                   1.000
##      .b08                   1.000
##      .b09                   1.000
##      .b10                   1.000
##      .b01                   1.000
##      .b02                   1.000
```

```
##    .b03              1.000
##    .b05              1.000
##    .b06              1.000
##    Depression        3.879    2.265    1.712    0.087
##    Anxiety           3.104    1.678    1.850    0.064
##
## Scales y*:
##                   Estimate  Std.Err  z-value  P(>|z|)
##    b04              0.453
##    b07              0.604
##    b08              0.646
##    b09              0.521
##    b10              0.610
##    b01              0.494
##    b02              0.656
##    b03              0.584
##    b05              0.455
##    b06              0.669
```

Again as seen from the model fitting measures of CFI/TLI/RMSEA/SRMR, this model fitting is very satisfactory and this established the *configural invariance*. Notice that in this *configural invariance* model, all the parameters (such as factor loadings, variances, and thresholds) are group-specific and they are different between these two groups.

7.2.3   Step Three: The *Ideal* Invariance

With the established *configural invariance* model, we can test the most strict invariance model to constrain the measurement invariance, which includes factor loadings, thresholds, residual variances, latent variable variances and convariance. This can be implemented in detail as follows using *R/lavaan*:

```
# CFA model for K10
CFA4K10 <- '
# Constrain latent variables: meaurement model
Depression =~ 1*b04+c(L7,L7)*b07+c(L8,L8)*b08+c(L9,L9)*b09
              +c(L10,L10)*b10
Anxiety     =~ 1*b01+c(L2,L2)*b02+c(L3,L3)*b03+c(L5,L5)*b05
              +c(L6,L6)*b06

# Constrain thresholds
b01|c(t11,t11)*t1+c(t12.t12)*t2; b02|c(t21,t21)*t1+c(t22.t22)*t2
b03|c(t31,t31)*t1+c(t32.t32)*t2; b04|c(t41,t41)*t1+c(t42.t42)*t2
b05|c(t51,t51)*t1+c(t52.t52)*t2; b06|c(t61,t61)*t1+c(t62.t62)*t2
b07|c(t71,t71)*t1+c(t72.t72)*t2; b08|c(t81,t81)*t1+c(t82.t82)*t2
```

```
b09|c(t91,t91)*t1+c(t92.t92)*t2; b10|c(t101,t101)*t1+c(t102.t102)*t2

# Constrain latent variances and covariance
Depression   ~~ c(lvd, lvd)*Depression
Anxiety      ~~ c(lva, lva)*Anxiety
Depression   ~~ c(lvda,lvda)*Anxiety


# Unique Variances
b01 ~~ 1*b01; b02 ~~ 1*b02;
b03 ~~ 1*b03; b04 ~~ 1*b04;
b05 ~~ 1*b05; b06 ~~ 1*b06;
b07 ~~ 1*b07; b08 ~~ 1*b08;
b09 ~~ 1*b09; b10 ~~ 1*b10;
'
```

We can then call *R/lavaan* function *cfa* to fit the *ideal* invariance model for factor loadings, residual variance, residual.covariances, lv.variances (i.e., latent variance), and lv.covariances (i.e., latent covariance). This is implemented as follows:

```
# Fit the ideal invariance model
fitK10.ideal = cfa(CFA4K10, data=dmig, group="sex",
            parameterization='theta',
            ordered=c("b01","b02","b03","b04","b05",
                    "b06","b07","b08","b09","b10"))


# Print the model summary with the fitting measures
summary(fitK10.ideal, fit.measures=TRUE)

## lavaan 0.6-12 ended normally after 112 iterations
##
##    Estimator                                    DWLS
##    Optimization method                        NLMINB
##    Number of model parameters                     62
##    Number of equality constraints                 31
##
##    Number of observations per group:      Used       Total
##    1                                        188         192
##    0                                        132         135
##
## Model Test User Model:
##                                        Standard      Robust
##    Test Statistic                      117.308     119.152
##    Degrees of freedom                       99          99
##    P-value (Chi-square)                  0.101       0.082
```

```
## Scaling correction factor                                        1.529
## Shift parameter for each group:
##      1                                                          24.923
##      0                                                          17.499
##    simple second-order correction
## Test statistic for each group:
##      1                                          58.310          63.063
##      0                                          58.998          56.089
##
## Model Test Baseline Model:
##
## Test statistic                               4400.815        2172.634
## Degrees of freedom                                 90              90
## P-value                                         0.000           0.000
## Scaling correction factor                                       2.070
##
## User Model versus Baseline Model:
##
## Comparative Fit Index (CFI)                     0.996           0.990
## Tucker-Lewis Index (TLI)                        0.996           0.991
##
## Robust Comparative Fit Index (CFI)                                 NA
## Robust Tucker-Lewis Index (TLI)                                    NA
##
## Root Mean Square Error of Approximation:
##
## RMSEA                                           0.034           0.036
## 90 Percent confidence interval - lower          0.000           0.000
## 90 Percent confidence interval - upper          0.056           0.057
## P-value RMSEA <= 0.05                           0.871           0.847
##
## Robust RMSEA                                                       NA
## 90 Percent confidence interval - lower                          0.000
## 90 Percent confidence interval - upper                             NA
##
## Standardized Root Mean Square Residual:
##
## SRMR                                            0.088           0.088
##
## Parameter Estimates:
##
## Standard errors                           Robust.sem
## Information                                 Expected
## Information saturated (h1) model        Unstructured
##
##
```

```
## Group 1 [1]:
##
## Latent Variables:
##                        Estimate  Std.Err  z-value  P(>|z|)
##    Depression =~
##      b04                  1.000
##      b07        (L7)      0.927    0.223    4.153    0.000
##      b08        (L8)      0.897    0.236    3.803    0.000
##      b09        (L9)      1.203    0.292    4.121    0.000
##      b10        (L10)     0.671    0.140    4.788    0.000
##    Anxiety =~
##      b01                  1.000
##      b02        (L2)      0.897    0.191    4.698    0.000
##      b03        (L3)      1.009    0.257    3.931    0.000
##      b05        (L5)      1.113    0.292    3.813    0.000
##      b06        (L6)      0.905    0.203    4.453    0.000
##
## Covariances:
##                        Estimate  Std.Err  z-value  P(>|z|)
##    Depression ~~
##      Anxiety  (lvda)     2.698    0.748    3.609    0.000
##
## Intercepts:
##                        Estimate  Std.Err  z-value  P(>|z|)
##     .b04                 0.000
##     .b07                 0.000
##     .b08                 0.000
##     .b09                 0.000
##     .b10                 0.000
##     .b01                 0.000
##     .b02                 0.000
##     .b03                 0.000
##     .b05                 0.000
##     .b06                 0.000
##      Depression          0.000
##      Anxiety             0.000
##
## Thresholds:
##                        Estimate  Std.Err  z-value  P(>|z|)
##      b01|t1    (t11)     1.794    0.267    6.708    0.000
##      b01|t2    (t12.)    3.257    0.396    8.218    0.000
##      b02|t1    (t21)     1.872    0.231    8.119    0.000
##      b02|t2    (t22.)    3.650    0.376    9.709    0.000
##      b03|t1    (t31)     2.768    0.429    6.456    0.000
##      b03|t2    (t32.)    4.479    0.609    7.352    0.000
##      b04|t1    (t41)     2.636    0.403    6.535    0.000
```

```
##       b04|t2  (t42.)    4.062    0.473    8.591    0.000
##       b05|t1  (t51)     2.150    0.344    6.244    0.000
##       b05|t2  (t52.)    3.899    0.541    7.213    0.000
##       b06|t1  (t61)     2.566    0.318    8.076    0.000
##       b06|t2  (t62.)    3.809    0.440    8.665    0.000
##       b07|t1  (t71)     2.542    0.369    6.894    0.000
##       b07|t2  (t72.)    4.132    0.560    7.379    0.000
##       b08|t1  (t81)     2.416    0.358    6.747    0.000
##       b08|t2  (t82.)    3.497    0.447    7.816    0.000
##       b09|t1  (t91)     2.838    0.534    5.317    0.000
##       b09|t2  (t92.)    4.720    0.719    6.568    0.000
##       b10|t1  (t101)    2.061    0.260    7.932    0.000
##       b10|t2  (t102)    3.032    0.322    9.405    0.000
##
## Variances:
##                       Estimate  Std.Err  z-value  P(>|z|)
##     Depressn (lvd)     3.575    1.258    2.843    0.004
##     Anxiety  (lva)     2.596    0.887    2.926    0.003
##    .b01                1.000
##    .b02                1.000
##    .b03                1.000
##    .b04                1.000
##    .b05                1.000
##    .b06                1.000
##    .b07                1.000
##    .b08                1.000
##    .b09                1.000
##    .b10                1.000
##
## Scales y*:
##                       Estimate  Std.Err  z-value  P(>|z|)
##     b04                0.468
##     b07                0.496
##     b08                0.508
##     b09                0.402
##     b10                0.619
##     b01                0.527
##     b02                0.569
##     b03                0.524
##     b05                0.487
##     b06                0.565
##
##
## Group 2 [0]:
##
## Latent Variables:
```

```
##                      Estimate  Std.Err  z-value  P(>|z|)
##    Depression =~
##      b04                1.000
##      b07      (L7)      0.927    0.223    4.153    0.000
##      b08      (L8)      0.897    0.236    3.803    0.000
##      b09      (L9)      1.203    0.292    4.121    0.000
##      b10      (L10)     0.671    0.140    4.788    0.000
##    Anxiety =~
##      b01                1.000
##      b02      (L2)      0.897    0.191    4.698    0.000
##      b03      (L3)      1.009    0.257    3.931    0.000
##      b05      (L5)      1.113    0.292    3.813    0.000
##      b06      (L6)      0.905    0.203    4.453    0.000
##
## Covariances:
##                      Estimate  Std.Err  z-value  P(>|z|)
##    Depression ~~
##      Anxiety (lvda)   2.698    0.748    3.609    0.000
##
## Intercepts:
##                      Estimate  Std.Err  z-value  P(>|z|)
##     .b04               0.000
##     .b07               0.000
##     .b08               0.000
##     .b09               0.000
##     .b10               0.000
##     .b01               0.000
##     .b02               0.000
##     .b03               0.000
##     .b05               0.000
##     .b06               0.000
##      Depression        0.000
##      Anxiety           0.000
##
## Thresholds:
##                      Estimate  Std.Err  z-value  P(>|z|)
##      b01|t1   (t11)    1.794    0.267    6.708    0.000
##      b01|t2   (t12.)   3.257    0.396    8.218    0.000
##      b02|t1   (t21)    1.872    0.231    8.119    0.000
##      b02|t2   (t22.)   3.650    0.376    9.709    0.000
##      b03|t1   (t31)    2.768    0.429    6.456    0.000
##      b03|t2   (t32.)   4.479    0.609    7.352    0.000
##      b04|t1   (t41)    2.636    0.403    6.535    0.000
##      b04|t2   (t42.)   4.062    0.473    8.591    0.000
##      b05|t1   (t51)    2.150    0.344    6.244
```

```
## b05|t2  (t52.)  3.899  0.541  7.213  0.000
## b06|t1  (t61)   2.566  0.318  8.076  0.000
## b06|t2  (t62.)  3.809  0.440  8.665  0.000
## b07|t1  (t71)   2.542  0.369  6.894  0.000
## b07|t2  (t72.)  4.132  0.560  7.379  0.000
## b08|t1  (t81)   2.416  0.358  6.747  0.000
## b08|t2  (t82.)  3.497  0.447  7.816  0.000
## b09|t1  (t91)   2.838  0.534  5.317  0.000
## b09|t2  (t92.)  4.720  0.719  6.568  0.000
## b10|t1  (t101)  2.061  0.260  7.932  0.000
## b10|t2  (t102)  3.032  0.322  9.405  0.000
##
## Variances:
##                      Estimate  Std.Err  z-value  P(>|z|)
##    Depressn (lvd)    3.575     1.258    2.843    0.004
##    Anxiety  (lva)    2.596     0.887    2.926    0.003
##    .b01              1.000
##    .b02              1.000
##    .b03              1.000
##    .b04              1.000
##    .b05              1.000
##    .b06              1.000
##    .b07              1.000
##    .b08              1.000
##    .b09              1.000
##    .b10              1.000
##
## Scales y*:
##                      Estimate  Std.Err  z-value  P(>|z|)
##    b04               0.468
##    b07               0.496
##    b08               0.508
##    b09               0.402
##    b10               0.619
##    b01               0.527
##    b02               0.569
##    b03               0.524
##    b05               0.487
##    b06               0.565
```

From this *ideal* invariance model, we can see from the model fitting measures of CFI/TLI/RMSEA/SRMR that this model fits the data very satisfactorily. This process established the *ideal* invariance model where all the factor loadings, residual variances, latent variances and covariance are all equal.

7.2.4   Testing Latent Mean Difference between Sex

Since we have now established the *ideal* invariance model, we can then turn
to test the latent mean differences by group. In this case, the mean difference
would be solely contributed by the *sex* difference with all other effects being
equivalent. These latent mean differences would be contributed by the *sex* on
Depression and *Anxiety* between *boys* and *girls*.

We can just add the constraints of Depression c(0, DDif)*1 and
Anxiety   c(0, ADif)*1 where *c(0, DDif)* is used to constrain the latent factor
mean for group 1 (i.e., boys) to be zero and freely estimate the latent mean
(i.e., *DDif*) for the second group 2 (i.e., girls). This *DDif* is then the latent
mean difference. The same *R* coding in *c(0, ADif)* for *Anxiety* with *ADif* as
the latent mean difference between *boys* and *girls*:

```
# CFA model for K10
CFA4K10meandif <- '
# Constrain latent variables: meaurement model
Depression =~ 1*b04+c(L7,L7)*b07+c(L8,L8)*b08+c(L9,L9)*b09
              +c(L10,L10)*b10
Anxiety    =~ 1*b01+c(L2,L2)*b02+c(L3,L3)*b03+c(L5,L5)*b05
              +c(L6,L6)*b06

# Constrain thresholds
b01|c(t11,t11)*t1+c(t12.t12)*t2; b02|c(t21,t21)*t1+c(t22.t22)*t2
b03|c(t31,t31)*t1+c(t32.t32)*t2; b04|c(t41,t41)*t1+c(t42.t42)*t2
b05|c(t51,t51)*t1+c(t52.t52)*t2; b06|c(t61,t61)*t1+c(t62.t62)*t2
b07|c(t71,t71)*t1+c(t72.t72)*t2; b08|c(t81,t81)*t1+c(t82.t82)*t2
b09|c(t91,t91)*t1+c(t92.t92)*t2; b10|c(t101,t101)*t1+c(t102.t102)*t2

# Constrain latent variances and covariance
Depression   ~~ c(1vd, 1vd)*Depression
Anxiety      ~~ c(1va, 1va)*Anxiety
Depression   ~~ c(1vda,1vda)*Anxiety

# Unique Variances
b01 ~~ 1*b01; b02 ~~ 1*b02;
b03 ~~ 1*b03; b04 ~~ 1*b04;
b05 ~~ 1*b05; b06 ~~ 1*b06;
b07 ~~ 1*b07; b08 ~~ 1*b08;
b09 ~~ 1*b09; b10 ~~ 1*b10;

# Factor Mean
Depression ~ c(0, DDif)*1
Anxiety    ~ c(0, ADif)*1
'
```

We can then call *R/lavaan* function *cfa* to fit the *ideal* invariance model for factor loadings, residual variance, residual.covariances, lv.variances (i.e., latent variance), and lv.covariances (i.e., latent covariance). This is implemented as follows:

```
# Fit the ideal invariance model
fitK10.meandif = cfa(CFA4K10meandif, data=dmig, group="sex",
            parameterization='theta',
            ordered=c("b01","b02","b03","b04","b05",
                "b06","b07","b08","b09","b10"))


# Print the model summary with the fitting measures
summary(fitK10.meandif)


## lavaan 0.6-12 ended normally after 97 iterations
##
##    Estimator                                      DWLS
##    Optimization method                          NLMINB
##    Number of model parameters                       66
##    Number of equality constraints                   33
##
##    Number of observations per group:           Used          Total
##    1                                            188            192
##    0                                            132            135
##
## Model Test User Model:
##                                           Standard         Robust
##    Test Statistic                          117.308        116.948
##    Degrees of freedom                           97             97
##    P-value (Chi-square)                      0.079          0.082
##    Scaling correction factor                                1.545
##    Shift parameter for each group:
##    1                                                       24.086
##    0                                                       16.911
##      simple second-order correction
##    Test statistic for each group:
##    1                                        58.310         61.838
##    0                                        58.998         55.110
##
## Parameter Estimates:
##
##    Standard errors                        Robust.sem
##    Information                              Expected
##    Information saturated (h1) model     Unstructured
##
##
```

```
## Group 1 [1]:
##
## Latent Variables:
##                        Estimate  Std.Err  z-value  P(>|z|)
##   Depression =~
##     b04                   1.000
##     b07        (L7)       0.927    0.223    4.153    0.000
##     b08        (L8)       0.897    0.236    3.803    0.000
##     b09        (L9)       1.203    0.292    4.121    0.000
##     b10       (L10)       0.671    0.140    4.788    0.000
##   Anxiety =~
##     b01                   1.000
##     b02        (L2)       0.897    0.191    4.698    0.000
##     b03        (L3)       1.009    0.257    3.931    0.000
##     b05        (L5)       1.113    0.292    3.813    0.000
##     b06        (L6)       0.905    0.203    4.453    0.000
##
## Covariances:
##                        Estimate  Std.Err  z-value  P(>|z|)
##   Depression ~~
##     Anxiety (lvda)        2.698    0.748    3.609    0.000
##
## Intercepts:
##                        Estimate  Std.Err  z-value  P(>|z|)
##     Deprssn (DDif)      -0.942    0.268   -3.520    0.000
##     Anxiety (ADif)      -0.787    0.198   -3.970    0.000
##    .b04                  0.000
##    .b07                  0.000
##    .b08                  0.000
##    .b09                  0.000
##    .b10                  0.000
##    .b01                  0.000
##    .b02                  0.000
##    .b03                  0.000
##    .b05                  0.000
##    .b06                  0.000
##
## Thresholds:
##                        Estimate  Std.Err  z-value  P(>|z|)
##     b01|t1     (t11)     1.007    0.169    5.970    0.000
##     b01|t2    (t12.)     2.469    0.279    8.837    0.000
##     b02|t1     (t21)     1.166    0.154    7.583    0.000
##     b02|t2    (t22.)     2.944    0.283   10.386    0.000
##     b03|t1     (t31)     1.974    0.248    7.956    0.000
##     b03|t2    (t32.)     3.684    0.414    8.898    0.000
```

```
## b04|t1  (t41)   1.694   0.225   7.532   0.000
## b04|t2  (t42.)  3.120   0.282  11.059   0.000
## b05|t1  (t51)   1.274   0.183   6.975   0.000
## b05|t2  (t52.)  3.022   0.361   8.380   0.000
## b06|t1  (t61)   1.853   0.196   9.469   0.000
## b06|t2  (t62.)  3.096   0.308  10.063   0.000
## b07|t1  (t71)   1.669   0.198   8.422   0.000
## b07|t2  (t72.)  3.259   0.411   7.925   0.000
## b08|t1  (t81)   1.571   0.203   7.754   0.000
## b08|t2  (t82.)  2.653   0.284   9.329   0.000
## b09|t1  (t91)   1.704   0.224   7.610   0.000
## b09|t2  (t92.)  3.587   0.400   8.966   0.000
## b10|t1  (t101)  1.428   0.161   8.873   0.000
## b10|t2  (t102)  2.400   0.218  10.997   0.000
##
## Variances:
##                     Estimate  Std.Err  z-value  P(>|z|)
## Depressn (lvd)      3.575     1.258    2.843    0.004
## Anxiety  (lva)      2.596     0.887    2.926    0.003
## .b01                1.000
## .b02                1.000
## .b03                1.000
## .b04                1.000
## .b05                1.000
## .b06                1.000
## .b07                1.000
## .b08                1.000
## .b09                1.000
## .b10                1.000
##
## Scales y*:
##                     Estimate  Std.Err  z-value  P(>|z|)
## b04                 0.468
## b07                 0.496
## b08                 0.508
## b09                 0.402
## b10                 0.619
## b01                 0.527
## b02                 0.569
## b03                 0.524
## b05                 0.487
## b06                 0.565
##
##
```

```
## Group 2 [0]:
##
## Latent Variables:
##                      Estimate  Std.Err  z-value  P(>|z|)
##   Depression =~
##     b04                 1.000
##     b07        (L7)      0.927    0.223    4.153    0.000
##     b08        (L8)      0.897    0.236    3.803    0.000
##     b09        (L9)      1.203    0.292    4.121    0.000
##     b10        (L10)     0.671    0.140    4.788    0.000
##   Anxiety =~
##     b01                 1.000
##     b02        (L2)      0.897    0.191    4.698    0.000
##     b03        (L3)      1.009    0.257    3.931    0.000
##     b05        (L5)      1.113    0.292    3.813    0.000
##     b06        (L6)      0.905    0.203    4.453    0.000
##
## Covariances:
##                      Estimate  Std.Err  z-value  P(>|z|)
##   Depression ~~
##     Anxiety (lvda)      2.698    0.748    3.609    0.000
##
## Intercepts:
##                      Estimate  Std.Err  z-value  P(>|z|)
##     Deprssn (DDif)     -0.942    0.268   -3.520    0.000
##     Anxiety (ADif)     -0.787    0.198   -3.970    0.000
##    .b04                 0.000
##    .b07                 0.000
##    .b08                 0.000
##    .b09                 0.000
##    .b10                 0.000
##    .b01                 0.000
##    .b02                 0.000
##    .b03                 0.000
##    .b05                 0.000
##    .b06                 0.000
##
## Thresholds:
##                      Estimate  Std.Err  z-value  P(>|z|)
##     b01|t1    (t11)     1.007    0.169    5.970    0.000
##     b01|t2    (t12.)    2.469    0.279    8.837    0.000
##     b02|t1    (t21)     1.166    0.154    7.583    0.000
##     b02|t2    (t22.)    2.944    0.283   10.386    0.000
```

```
## b03|t1  (t31)   1.974   0.248   7.956   0.000
## b03|t2  (t32.)  3.684   0.414   8.898   0.000
## b04|t1  (t41)   1.694   0.225   7.532   0.000
## b04|t2  (t42.)  3.120   0.282  11.059   0.000
## b05|t1  (t51)   1.274   0.183   6.975   0.000
## b05|t2  (t52.)  3.022   0.361   8.380   0.000
## b06|t1  (t61)   1.853   0.196   9.469   0.000
## b06|t2  (t62.)  3.096   0.308  10.063   0.000
## b07|t1  (t71)   1.669   0.198   8.422   0.000
## b07|t2  (t72.)  3.259   0.411   7.925   0.000
## b08|t1  (t81)   1.571   0.203   7.754   0.000
## b08|t2  (t82.)  2.653   0.284   9.329   0.000
## b09|t1  (t91)   1.704   0.224   7.610   0.000
## b09|t2  (t92.)  3.587   0.400   8.966   0.000
## b10|t1  (t101)  1.428   0.161   8.873   0.000
## b10|t2  (t102)  2.400   0.218  10.997   0.000
##
## Variances:
##                    Estimate Std.Err z-value  P(>|z|)
##    Depressn (lvd)    3.575   1.258   2.843    0.004
##    Anxiety  (lva)    2.596   0.887   2.926    0.003
##    .b01              1.000
##    .b02              1.000
##    .b03              1.000
##    .b04              1.000
##    .b05              1.000
##    .b06              1.000
##    .b07              1.000
##    .b08              1.000
##    .b09              1.000
##    .b10              1.000
##
## Scales y*:
##                    Estimate Std.Err z-value  P(>|z|)
##    b04               0.468
##    b07               0.496
##    b08               0.508
##    b09               0.402
##    b10               0.619
##    b01               0.527
##    b02               0.569
##    b03               0.524
##    b05               0.487
##    b06               0.565
```

From the output, we can see that

```
Intercepts:
                    Estimate  Std.Err  z-value  P(>|z|)
    Deprssn (DDif)    -0.942    0.268   -3.520    0.000
    Anxiety (ADif)    -0.787    0.198   -3.970    0.000
```

and this means that there is a significantly higher *Depression* and *Anxiety* for *girls* than *boys*.

7.3   Data Analysis Using SAS

Although *PROC CALIS* supports the multiple-group analysis, at present there is no estimation method implemented specifically for analyzing categorical responses.

7.4   Discussions

In this chapter, we illustrated the test for multi-group invariance for categorical data using data that we collected from Chinese migrants children living in Hangzhou, China, to study their nonspecific psychological distress.

We used *R/lavaan* to fit the two-factor K10 CFA model and concluded that the K10 was cross-gender invariance and equivalent, which was used further used to test the difference for latent factor defined as *Anxiety* and *Depression* between *boys* and *girls*. These results indicated that, on average, girls experienced higher levels of depression and anxiety than boys. This analysis controlled all the potential confounding factors from factor loadings, residual variances, categorical data thresholds, latent variances, and covariances.

8

Pain-Related Disability for People with Temporomandibular Disorder: Full Structural Equation Modeling

In this chapter, we illustrate a full structural equation model (SEM) to analyze the relationship between features associated with pain-related disability (PRD) in people and painful temporomandibular disorder (TMD) (Miller et al., 2020).

As described by Miller et al. (2020), painful temporomandibular disorders (TMDs) are characterized by localized pain in the jaw joints and/or masticatory muscles in conjunction with limited jaw function. According to the National Health Interview Survey, TMD affects 4.6% of the US adult population, with prevalence higher among women than men. The experience of ongoing jaw pain can evoke psychological distress as pain imposes restrictions on work and social activity. Pain-related disability (PRD) is a multifaceted construct that refers to the impact of pain on an individual's capacity to fulfill their self-defined social roles. The research by Miller et al. (2020) examined the relationship between clinical, psychological, and pain sensitivity factors and pain-related disability among adults with chronic temporomandibular disorder (TMD). Data from a cross-sectional community-based sample of 1088 men and women with chronic TMD enrolled in study *Orofacial Pain: Prospective Evaluation and Risk Assessment (OPPERA)* were used for this secondary data analysis. Details of the *OPPERA* study, protocol and procedures can be found in Slade et al. (2011), Greenspan et al. (2011), Fillingim et al. (2011), and Ohrbach et al. (2011).

The main purpose of this chapter is to demonstrate the two-stage process of structural equation modeling. First, we establish the measurement model. Then, we fit a full structural equation model that builds upon the measurement model. Several structural equation models are possible, but we will illustrate only one structural equation model by using *R* and *SAS*. The associated *Mplus* code is included as an appendix. Another purpose of this chapter is to make the data from Miller et al. (2020) publicly available for interested researchers and readers to conduct further analysis.

Note to install the *R* packages *lavaan* for *sem* analysis, *semPlot* for SEM path diagrams, *tidyverse* for data processing, and *ggplot2* for plotting.

```
# Load the lavaan package into R session
library(lavaan)
# Load the library for SEM plot
library(semPlot)
# Load tidyverse for data processing
library(tidyverse)
# Load ggplot2 for plotting
library(ggplot2)
```

8.1   Data Description

Data set is available in *Pain.csv*, which is edited to de-identify the information of the study participants. It includes data from 1088 participants with 17 variables as follows:

- *Site* is the identifier for the four study sites (1 = University at Buffalo, NY, 2 = University of Florida in Gainesville, FL, 3 = University of Maryland in Baltimore, MD, and 4 = University of North Carolina at Chapel Hill, NC),

- *Age* is the age for study participants,

- *Gender* is the gender information with 1 = "Male" and 2 = "Female",

- *CPI1* is the rating of current facial pain (0–10 scale),

- *CPI2* is the rating of average facial pain in the last 6 months (0–10 scale),

- *CPI3* is the rating of worst facial pain in the last 6 months (0–10 scale),

- *DD1* is the number of days in the past 6 months which have been kept from usual activities because of facial pain,

- *DD2* is number of days efficiency dropped below 50% of what consider normal because of facial pain,

- *I1* is the interference in daily activities due to facial pain (0–10 scale),

- *I2* is the interference in social activities due to facial pain (0–10 scale),

- *I3* is the interference in work due to facial pain (0–10 scale),

- *JF1* is the *Jaw Functional Limitation Scale (JFLS)* measuring the chewing limitation (0–10 scale),

- *JF2* is the *Jaw Functional Limitation Scale (JFLS)* measuring the Opening Limitation (0–10 scale),

- *JF3* is the *Jaw Functional Limitation Scale (JFLS)* measuring Verbal and Emotional Expression Limitation (0–10 scale),

- *P8* is the overall negative affect score (30–120 scale),

- *CSQ1* is the *Coping Strategies Questionnaire (CSQ)* catastrophizing subscale (0–6 scale),

- *SCL905* is the *Symptoms Checklist 90—Revised (SCL-90R)* somatization full scale.

The data can be loaded into *R* as follows:

```
# Read the data into R
dPain = read.csv("data/pain.csv", na.strings="-1234",header=TRUE)
# Check the data dimension
dim(dPain)
```

```
## [1] 1088    17
```

```
# Print the data summary
summary(dPain)
```

```
##       Site            Age            Gender
##   Min.   :1.00   Min.   :18.0   Min.   :1.00
##   1st Qu.:1.00   1st Qu.:23.0   1st Qu.:2.00
##   Median :2.00   Median :28.0   Median :2.00
##   Mean   :2.35   Mean   :29.2   Mean   :1.77
##   3rd Qu.:3.00   3rd Qu.:36.0   3rd Qu.:2.00
##   Max.   :4.00   Max.   :44.0   Max.   :2.00
##
##        CPI1            CPI2            CPI3
##   Min.   : 0.00   Min.   : 0.00   Min.   : 0.00
##   1st Qu.: 2.00   1st Qu.: 6.00   1st Qu.: 4.00
##   Median : 4.00   Median : 8.00   Median : 5.00
##   Mean   : 3.76   Mean   : 7.51   Mean   : 5.42
##   3rd Qu.: 6.00   3rd Qu.: 9.00   3rd Qu.: 7.00
##   Max.   :10.00   Max.   :10.00   Max.   :10.00
##   NA's   :26      NA's    :26     NA's    :26
##        DD1             DD2             I1
##   Min.   :  0.0   Min.   :  0     Min.   : 0.00
##   1st Qu.:  0.0   1st Qu.:  0     1st Qu.: 0.00
##   Median :  0.0   Median :  5     Median : 2.00
##   Mean   : 17.3   Mean   : 29     Mean   : 2.54
##   3rd Qu.: 12.0   3rd Qu.: 35     3rd Qu.: 4.00
##   Max.   :180.0   Max.   :180     Max.   :10.00
##   NA's   :34      NA's    :39     NA's    :26
```

```
##            I2                I3                JF1
## Min.   : 0.00    Min.   : 0.00    Min.   :0.00
## 1st Qu.: 0.00    1st Qu.: 0.00    1st Qu.:1.00
## Median : 2.00    Median : 1.00    Median :2.33
## Mean   : 2.35    Mean   : 2.23    Mean   :2.49
## 3rd Qu.: 4.00    3rd Qu.: 4.00    3rd Qu.:3.67
## Max.   :10.00    Max.   :10.00    Max.   :9.00
## NA's   :26       NA's   :25       NA's   :19
##           JF2               JF3               P8
## Min.   :0.00     Min.   :0.00     Min.   : 30.0
## 1st Qu.:1.00     1st Qu.:0.00     1st Qu.: 43.0
## Median :2.50     Median :0.44     Median : 55.0
## Mean   :2.74     Mean   :1.24     Mean   : 58.6
## 3rd Qu.:4.25     3rd Qu.:1.78     3rd Qu.: 72.0
## Max.   :9.75     Max.   :9.50     Max.   :111.0
## NA's   :19       NA's   :19       NA's   :6
##          CSQ1             SCL905
## Min.   :0.00     Min.   :0.00
## 1st Qu.:0.50     1st Qu.:0.25
## Median :1.17     Median :0.58
## Mean   :1.49     Mean   :0.74
## 3rd Qu.:2.17     3rd Qu.:1.00
## Max.   :6.00     Max.   :3.41
## NA's   :9        NA's   :15
```

Note that there are missing values which is coded by *-1234*. These missing values are read into *R* with option *na.strings= "-1234"* in *R* function *read.csv*.

8.2   Theoretical Model and Estimation

8.2.1   The Hypothesized Model for Pain-Related Disability

The hypothesized model of pain-related disability (PRD) and its contributing constructs were presented in Figure 1 of Miller et al. (2020). With extensive analyses and model selection, the final model and standardized parameter estimates of *pain-related disability, psychological unease, and jaw limitation* were presented in Figure 2 of Miller et al. (2020). This figure is reproduced in Figure 8.1 with some slight simplifications that remove some *Control variables* and the residual correlation between *Interference* and *Restriction*.

As seen in Figure 8.1, the final structural equation model to be fitted for *PRD* include several model components as follows:

1. A first-order CFA measurement model for latent variable *Psychological unease* (denoted by UNEASE), which is measured by three observed variables of P8, SCL905, and CSQ1.

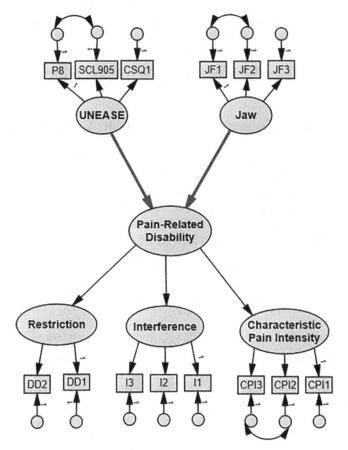

FIGURE 8.1
Pain-Related Disability and its Associated Constructs.

2. A first-order CFA measurement model for latent variable *Jaw Limitation* (denoted by *Jaw*), which is measured by three observed variables of *JF1*, *JF2*, and *JF3*.

3. A second-order CFA measurement model with a level-2 latent variable, *Pain-Related Disability*, which is measured by three level-1 latent variables—*Restriction, Interference,* and *Characteristic Pain Intensity*. At level 1, the latent variable *Restriction* is measured by two observed variables (i.e., *DD1* and *DD2*), *Interference* is measured by three observed variables (i.e., *I1, I2,* and *I3*), and *Characteristic Pain Intensity* is measured by three observed variables (i.e., *CPI1, CPI2,* and *CPI3*).

4. A structural equation model with regression paths from latent variables *UNEASE* and *Jaw* to *Pain-Related Disability*.

Note that in SEM convention, the latent variables are shown in *circles*, whereas observed variables are shown in *rectangles*. Arrows from latent variables onto observed variables indicate that the variables were measured to reflect the latent constructs. Arrows between latent variables (in bold) represent hypothesized regression relationships to be estimated and tested.

8.2.2   The Mathematical Model for Pain-Related Disability

Corresponding to the hypothesized model in Figure 8.1, the entire mathematical model can be described by the following four components:

1. A first-order CFA measurement model for latent variable *Psychological unease* (denoted by UNEASE), which is measured by three observed variables of *P8*, *SCL905* and *CSQ1*. The mathematical formulation can be written as follows:

$$P8 = a_{P8} + b_{P8} \times \text{UNEASE} + e_{P8}$$
$$SCL905 = a_{SCL905} + b_{SCL905} \times \text{UNEASE} + e_{SCL905}$$
$$CSQ1 = a_{CSQ1} + b_{CSQ1} \times \text{UNEASE} + e_{CSQ1}.$$

Following Miller et al. (2020), e_{P8} and e_{SCL905} are correlated.

2. A first-order CFA measurement model for latent variable *Jaw Limitation* (denoted by *Jaw*), which is measured by three observed variables of *JF1*, *JF2* and *JF3*. The mathematical formulation can be written as follows:

$$JF1 = a_{JF1} + b_{JF1} \times \text{Jaw} + e_{JF1}$$
$$JF2 = a_{JF2} + b_{JF2} \times \text{Jaw} + e_{JF2}$$
$$JF3 = a_{JF3} + b_{JF3} \times \text{Jaw} + e_{JF3}.$$

Following Miller et al. (2020), e_{JF1} and e_{JF2} are correlated.

3. A second-order CFA measurement model with level-2 latent variable *Pain-Related Disability* that is measured by three level-1 latent variables of *Restriction, Interference,* and *Characteristic Pain Intensity*. At level-1, the latent variable *Restriction* is measured by two observed variables (i.e., *DD1* and *DD2*), *Interference* is measured by three observed variables (i.e., *I1, I2,* and *I3*), and *Characteristic Pain Intensity* is measured by three observed variables (i.e., *CPI1, CPI2,* and *CPI3*). The corresponding mathematical formulation can be written as follows:

- At Level-1

$$DD1 = a_{DD1} + b_{DD1} \times \text{Restriction} + e_{DD1}$$
$$DD2 = a_{DD2} + b_{DD2} \times \text{Restriction} + e_{DD2}$$

$$I1 = a_{I1} + b_{I1} \times \text{Interference} + e_{I1}$$
$$I2 = a_{I2} + b_{I2} \times \text{Interference} + e_{I2}$$
$$I3 = a_{I3} + b_{I3} \times \text{Interference} + e_{I3}$$

$$CPI1 = a_{CPI1} + b_{CPI1} \times \text{Characteristic pain intensity} + e_{CPI1}$$
$$CPI2 = a_{CPI2} + b_{CPI2} \times \text{Characteristic pain intensity} + e_{CPI2}$$
$$CPI3 = a_{CPI3} + b_{CPI3} \times \text{Characteristic pain intensity} + e_{CPI3}.$$

Following Miller et al. (2020), e_{CPI2} and e_{JCPI3} are correlated.

- At Level-2

$$CPI = a_{CPI} + b_{CPI} \times \text{Pain-related disability} + e_{CPI}$$
$$\text{Interference} = a_{\text{Interference}} + b_{\text{Interference}} \times \text{Pain-related disability} + e_{\text{Interference}}$$
$$\text{Restriction} = a_{\text{Restriction}} + b_{\text{Restriction}} \times \text{Pain-related disability} + e_{\text{Restriction}}.$$

4. A structural equation model with regression paths from latent variables *UNEASE* and *Jaw* to *Pain-Related Disability*. The corresponding mathematical formulation can be written as follows:

$$\text{Pain-related disability} = \beta_0 + \beta_1 \times \text{UNEASE} + \beta_2 \times \text{Jaw} + e. \quad (8.1)$$

In each of the first three measurement components, the as are the intercepts, bs are the factor-loadings, es are the error terms whose variances are being estimated. In the fourth component for the structural equation, β_0 is the intercept, and β_1 and β_2 are the regression path parameters.

Note that for the identification of model parameters in the measurement models, one factor loading in each of the six measurement models for latent factors must be fixed to 1. Although *lavaan* will automatically fix the first loading to 1 for each of the latent factors, you must specify these fixed values at the desired locations when you use *PROC CALIS* of *SAS* to analyze the model. Additional identification constraints on the intercepts and means are also needed. These constraints depend on the structural model in the fourth component and will be described in the next section.

8.2.3   Measurement and Structural Models

In the previous section, the first three components are essentially confirmatory factor models for measuring latent factors. The fourth component is a structural model, as it specifies the functional relationships among the latent factors involved. Some variations of the fourth component are of interest here.

Model 1: No functional relationships among the latent factors involved. In this case, the latent factors *UNEASE, Jaw,* and *Pain-Related Disability* are merely correlated with each other and there are no directed arrows among them. Figure 8.2 represents this model and Section 8.3.1 fits this model by *lavaan.* As a result, the entire model is a measurement model of the latent factors, with one second-order factor. The factor means of *UNEASE, Jaw,* and *Pain-Related Disability* and the intercepts for *Restriction, Interference,* and *Characteristic Pain Intensity* are all fixed to zeros for the identification of mean structures. Once this measurement model is built and validated, we might explore other structural models and compare them with this "baseline" measurement model.

Model 2: The functional relationships among the latent factors are represented in diagram in Figure 8.1. In this case, *UNEASE* and *JAW* would predict *Pain-Related Disability* and the predictors themselves are uncorrelated.

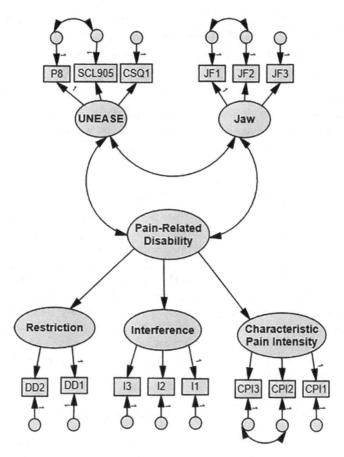

FIGURE 8.2
The CFA Models for Pain-Related Disability.

The factor means of *UNEASE* and *Jaw* and the intercepts for *Pain-Related Disability, Restriction, Interference,* and *Characteristic Pain Intensity* are all fixed to zeros for the identification of mean structures.

Model 3: Adding the correlation or covariance between *UNEASE* and *JAW* to Model 2. This model would be represented by Figure 8.1 with an additional double-headed arrow between *UNEASE* and *JAW*. All other parameterizations of this model are similar to that of Model 2.

Note that for the current example, Models 1 and 3 turn out to be equivalent models, which means that they will have the same statistical fit for the same data set. While Model 1 saturates the relationships among the three exogenous latent factors by six variance and covariance parameters, Model 3 simply transforms those six parameters into another set of six parameters—that is, two variances and one covariance for latent factors *UNEASE* and *JAW*, two regression coefficients, and one error variance for *Pain-Related Disability*. Similarly, the three mean parameters for the exogenous latent factors in Model 1 transform into two-factor means of *UNEASE* and *Jaw* and one intercept for *Pain-Related Disability* in Model 3. Consequently, Models 1 and 3 explain the same data equally well in the current context. In general, however, such an equivalence might not be expected when you have more than one endogenous latent factor and their residuals are not all correlated in a structural model similar to the current Model 3.

Model 4: Adding observed variables as predictor or control variables of some latent factors. For example, after establishing the measurement models for the latent factors, Miller et al. (2020) included *Age, Gender*, and other observed variables as predictor or control variables of *Pain-Related Disability*, in addition to the hypothesized relationships among latent factors, such as those described in Model 2 or 3.

8.3 Data Analysis Using *R*

8.3.1 Validation of Measurement Models

To many practitioners of SEM, the first logical step in building a model is to validate the measurement model that is associated with their full hypothesized structural equation model. This procedure is demonstrated here.

Corresponding to the full SEM model described in Figure 8.1, the three measurement models can be constructed as seen in Figure 8.2, where the structural paths from the two latent variables *UNEASE* and *Jaw* to the latent variable *Pain-Related Disability* are not considered. Instead of the two structural paths for describing the relationships of the three latent factors involved, the three correlations among these three exogenous latent variables are included by default in *lavaan*. This is Model 1 described in Section 8.2.3.

As an exercise, interested readers can validate each of these three-component CFA models separately. We will, however, validate simultaneously all three measurement models that are represented in Figure 8.2 for illustration. These three measurement models can be programmed in *lavaan* as follows:

```
# The three CFA measurement models
CFA4Pain <- '
#

# The two first-order CFA models
#
# Latent variables for 1st-order CFA "Jaw Limitation",
  denoted by "JAW"
Jaw     =~ JF1 + JF2 + JF3
# Latent variable for 1st-order CFA "Psychological unease",
  denoted by "UNEASE"
UNEASE =~ P8 + SCL905 + CSQ1

#
# The second-order CFA model
#
# Latent variable for level-1 CFA "Characteristic Pain Intensity",
  denoted by "CPI"
CPI =~ CPI1 + CPI2 + CPI3
# Latent variable for level-1 CFA "Interference",
  denoted by "INTER"
INTER =~ I1 + I2 + I3
# Latent variable for level-1 CFA disability "Restriction",
  denoted by "DIS"
DIS =~ DD1 + DD2
# Latent variable for level-2 CFA "PRD"
PRD =~  CPI + INTER + DIS

#
# residual covariances
#
JF1      ~~ JF2
SCL905 ~~ P8
CPI2     ~~ CPI3
'
```

We can then call *lavaan* function *sem* to fit the model as follows:

```
# fit the CFA model
fitCFA.Pain = sem(CFA4Pain, data=dPain, estimator ="MLR",
                  missing="ml")
```

```
## Warning in lav_data_full(data = data, group = group,
## cluster = cluster, : lavaan WARNING: some observed
## variances are (at least) a factor 1000 times larger
## than others; use varTable(fit) to investigate

## Warning in lav_data_full(data = data, group = group,
##   cluster = cluster, :
## lavaan WARNING: some cases are empty and will be ignored:
##   426 925 940
```

```
# Print the model summary with the fitting measures
summary(fitCFA.Pain, fit.measures=TRUE)
```

```
## lavaan 0.6-12 ended normally after 178 iterations
##
## Estimator                                     ML
## Optimization method                       NLMINB
## Number of model parameters                    51
##
##                                             Used       Total
## Number of observations                      1085        1088
## Number of missing patterns                    18
##
## Model Test User Model:
##                                         Standard      Robust
## Test Statistic                           322.289     277.081
## Degrees of freedom                            68          68
## P-value (Chi-square)                       0.000       0.000
## Scaling correction factor                              1.163
##    Yuan-Bentler correction (Mplus variant)
##
## Model Test Baseline Model:
##
## Test statistic                          8151.480    6120.541
## Degrees of freedom                            91          91
## P-value                                    0.000       0.000
## Scaling correction factor                              1.332
##
## User Model versus Baseline Model:
##
## Comparative Fit Index (CFI)                0.968       0.965
## Tucker-Lewis Index (TLI)                   0.958       0.954
##
## Robust Comparative Fit Index (CFI)                     0.970
## Robust Tucker-Lewis Index (TLI)                        0.959
##
```

```
## Loglikelihood and Information Criteria:
##
##    Loglikelihood user model (H0)           -35598.026  -35598.026
##    Scaling correction factor                             1.513
##        for the MLR correction
##    Loglikelihood unrestricted model (H1)   -35436.882  -35436.882
##    Scaling correction factor                             1.313
##        for the MLR correction
##
##    Akaike (AIC)                             71298.052   71298.052
##    Bayesian (BIC)                          71552.509   71552.509
##    Sample-size adjusted Bayesian (BIC)     71390.522   71390.522
##
## Root Mean Square Error of Approximation:
##
##    RMSEA                                       0.059       0.053
##    90 Percent confidence interval - lower      0.052       0.047
##    90 Percent confidence interval - upper      0.065       0.059
##    P-value RMSEA <= 0.05                        0.013       0.183
##
##    Robust RMSEA                                            0.057
##    90 Percent confidence interval - lower                  0.050
##    90 Percent confidence interval - upper                  0.065
##
## Standardized Root Mean Square Residual:
##
##    SRMR                                        0.042       0.042
##
## Parameter Estimates:
##
##    Standard errors                          Sandwich
##    Information bread                        Observed
##    Observed information based on             Hessian
##
## Latent Variables:
##                     Estimate  Std.Err  z-value  P(>|z|)
##    Jaw =~
##      JF1              1.000
##      JF2              1.397    0.067   20.856    0.000
##      JF3              1.390    0.111   12.483    0.000
##    UNEASE =~
##      P8               1.000
##      SCL905           0.047    0.004   10.708    0.000
##      CSQ1             0.122    0.010   11.748    0.000
##    CPI =~
##      CPI1             1.000
```

```
## CPI2                 0.751   0.043   17.445   0.000
## CPI3                 0.954   0.052   18.383   0.000
## INTER =~
## I1                   1.000
## I2                   1.082   0.024   44.385   0.000
## I3                   1.061   0.027   39.423   0.000
## DIS =~
## DD1                  1.000
## DD2                  1.465   0.086   17.019   0.000
## PRD =~
## CPI                  1.000
## INTER                1.330   0.101   13.215   0.000
## DIS                 16.805   1.590   10.570   0.000
##
## Covariances:
##                   Estimate Std.Err z-value  P(>|z|)
## .JF1 ~~
## .JF2                 0.925   0.123    7.499   0.000
## .P8 ~~
## .SCL905              2.962   0.376    7.883   0.000
## .CPI2 ~~
## .CPI3                0.511   0.130    3.934   0.000
## Jaw ~~
## UNEASE               4.567   0.555    8.224   0.000
## PRD                  0.992   0.126    7.863   0.000
## UNEASE ~~
## PRD                  8.368   0.995    8.410   0.000
##
## Intercepts:
##                   Estimate Std.Err z-value  P(>|z|)
## .JF1                 2.493   0.054   46.000   0.000
## .JF2                 2.747   0.066   41.709   0.000
## .JF3                 1.247   0.053   23.352   0.000
## .P8                 58.667   0.580  101.105   0.000
## .SCL905              0.741   0.019   39.193   0.000
## .CSQ1                1.490   0.039   38.408   0.000
## .CPI1                3.757   0.075   50.070   0.000
## .CPI2                7.510   0.060  124.861   0.000
## .CPI3                5.414   0.067   80.279   0.000
## .I1                  2.544   0.076   33.639   0.000
## .I2                  2.358   0.080   29.523   0.000
## .I3                  2.231   0.079   28.351   0.000
## .DD1                17.508   1.192   14.684   0.000
## .DD2                29.589   1.486   19.910   0.000
## Jaw                  0.000
## UNEASE               0.000
```

```
##    .CPI          0.000
##    .INTER        0.000
##    .DIS          0.000
##     PRD          0.000
##
## Variances:
##                Estimate  Std.Err  z-value  P(>|z|)
##    .JF1           1.920    0.125   15.383    0.000
##    .JF2           2.250    0.164   13.726    0.000
##    .JF3           0.679    0.142    4.763    0.000
##    .P8          299.020   14.131   21.160    0.000
##    .SCL905        0.240    0.017   14.098    0.000
##    .CSQ1          0.657    0.084    7.832    0.000
##    .CPI1          2.540    0.214   11.891    0.000
##    .CPI2          1.895    0.149   12.737    0.000
##    .CPI3          1.692    0.193    8.771    0.000
##    .I1            1.313    0.143    9.180    0.000
##    .I2            1.190    0.130    9.152    0.000
##    .I3            1.212    0.136    8.900    0.000
##    .DD1         643.512   77.705    8.282    0.000
##    .DD2         508.351  104.762    4.852    0.000
##     Jaw           1.220    0.150    8.149    0.000
##     UNEASE       65.412   11.844    5.523    0.000
##    .CPI           1.523    0.167    9.105    0.000
##    .INTER         1.349    0.197    6.834    0.000
##    .DIS         292.076   56.062    5.210    0.000
##     PRD           1.951    0.231    8.436    0.000
```

Two important *options* that we used in the *lavaan* function *sem* are explained as follows:

1. The *missing="ml"* option: Due to the missing values (coded in -1234 in this data), we fit the data with missing values by using the *full information maximum likelihood (fiml)* estimation, which assumes that the missing mechanism is MCAR (missing completely at random) or MAR (missing at random). The *fiml* estimation computes the individual likelihoods using all available data from complete or incomplete observations alike and it is an optimal choice when certain statistical conditions are satisfied. Alternative strategies dealing with missing data exist. According to Rosseel (2012), the default missing data treatment in *lavaan* is *listwise* (deletion), which would remove all cases with missing values prior to the model fitting and estimation. The *listwise* deletion is valid if the data are missing completely at random (MCAR), but it might produce biased estimates in the MAR situation. In the MAR situation, unlike the *full information maximum likelihood* method, the *listwise* deletion is also not optimal (though it can be useful for a first run).

Hence, if the estimator belongs to the ML family and the missing mechanism is MAR, the *ml* (alias: *fiml* or *direct*) option, where *ml* corresponds to the full information maximum likelihood (*fiml*) method, is a better choice.

2. The *estimator = "MLR"* option: This fits the data by the maximum likelihood estimation with robust (Huber-White) standard error and a scaled test statistic that is asymptotically equal to the Yuan-Bentler test statistic. This estimator adjusts for the incomplete (i.e., missing) data as well as a possible non-normality in the data. The non-normality issue has been discussed in Chapter 4.

Due to the *estimator = "MLR"* option, the output of model fitting has an additional column labeled as *Robust* with the scaling factor estimated. As seen from the output, the Chi-square test statistic from the *Robust* column is smaller, although it is still significant. The other overall model fitting measures are quite satisfactory as indicated by the Comparative Fit Index (CFI) = 0.970, the Tucker-Lewis Index (TLI) = 0.959, the Root Mean Square Error of Approximation (RMSEA) = 0.053 and the p-value for RMSEA ≤ 0.05 is 0.183, and the Standardized Root Mean Square Residual (SRMR) = 0.042. All the factor loadings, intercepts, and variances/covariances are highly statistically significant. This validates the measurement components of the full model (Model 1).

8.3.2   Fit the Full Structural Equation Model for Pain-Related Disability

The next step is to fit and validate the full SEM in Figure 8.1, which is Model 2 described in Section 8.2.3. However, we found that the fit of this model was not entirely satisfactory. For example, SRMR is 0.1, RMSEA is 0.068, and the p-value for testing close fit is 0, indicating that you should reject the null hypothesis that the population RMSEA is < 0.05. For this reason, we will allow the correlation/covariance between *UNEASE* and *Jaw* to be freely estimated, which becomes Model 3 described in Section 8.2.3. The *lavaan* code for this model is as follows:

```
# The full SEM for PRD
SEM4Pain <- '
#
# The two first-order CFA models
#
# Latent variables for 1st-order CFA "Jaw Limitation",
    denoted by "JAW"
Jaw    =~ JF1 + JF2 + JF3
# Latent variable for 1st-order CFA "Psychological unease",
    denoted by "UNEASE"
UNEASE =~ P8 + SCL905 + CSQ1
```

```
#
# The second-order CFA model
#
# Latent variable for level-1 CFA "Characteristic Pain Intensity",
      denoted by "CPI"
CPI =~ CPI1 + CPI2 + CPI3
# Latent variable for level-1 CFA "Interference",
      denoted by "INTER"
INTER =~ I1 + I2 + I3
# Latent variable for level-1 CFA disability "Restriction",
      denoted by "DIS"
DIS =~ DD1 + DD2
# Latent variable for level-2 CFA "PRD"
PRD =~ CPI + INTER + DIS

#
# residual covariances
#
JF1     ~~ JF2
SCL905 ~~ P8
CPI2    ~~ CPI3

#
# Uncorrelated exogenous constructs
#
# UNEASE ~~ 0*Jaw

#
# Add the structural regression paths
#
PRD ~ UNEASE
PRD ~ Jaw
'
```

Notice that we have commented out the specification "UNEASE ~~ 0*Jaw" in the code. Were this not commented out, we would have fitted Model 2 instead of the intended Model 3. We now call *lavaan* function *sem* to fit the full SEM (Model 3) as follows:

```
# fit the SEM model
fitSEM.Pain = sem(SEM4Pain, data=dPain, estimator ="MLR",
                     missing="ml")

## Warning in lav_data_full(data = data, group = group,
## cluster = cluster, : lavaan WARNING: some observed
```

```
## variances are (at least) a factor 1000 times larger
## than others; use varTable(fit) to investigate

## Warning in lav_data_full(data = data, group = group,
   cluster = cluster, :
## lavaan WARNING: some cases are empty and will be ignored:
##    426 925 940
```

# Print the model summary with the fitting measures
summary(fitSEM.Pain, fit.measures=TRUE, estimates=FALSE)

```
## lavaan 0.6-12 ended normally after 172 iterations
##
##   Estimator                                         ML
##   Optimization method                           NLMINB
##   Number of model parameters                        51
##
##                                                   Used      Total
##   Number of observations                          1085       1088
##   Number of missing patterns                        18
##
## Model Test User Model:
##                                               Standard     Robust
##   Test Statistic                               322.289    277.081
##   Degrees of freedom                                68         68
##   P-value (Chi-square)                           0.000      0.000
##   Scaling correction factor                                 1.163
##     Yuan-Bentler correction (Mplus variant)
##
## Model Test Baseline Model:
##
##   Test statistic                              8151.480   6120.541
##   Degrees of freedom                                91         91
##   P-value                                        0.000      0.000
##   Scaling correction factor                                 1.332
##
## User Model versus Baseline Model:
##
##   Comparative Fit Index (CFI)                    0.968      0.965
##   Tucker-Lewis Index (TLI)                       0.958      0.954
##
##   Robust Comparative Fit Index (CFI)                        0.970
##   Robust Tucker-Lewis Index (TLI)                           0.959
##
## Loglikelihood and Information Criteria:
##
```

```
## 	Loglikelihood user model (H0)         -35598.026  -35598.026
## 	Scaling correction factor                          1.513
## 	    for the MLR correction
## 	Loglikelihood unrestricted model (H1) -35436.882  -35436.882
## 	Scaling correction factor                          1.313
## 	    for the MLR correction
##
## 	Akaike (AIC)                            71298.052   71298.052
## 	Bayesian (BIC)                          71552.509   71552.509
## 	Sample-size adjusted Bayesian (BIC)     71390.522   71390.522
##
## Root Mean Square Error of Approximation:
##
## 	RMSEA                                        0.059       0.053
## 	90 Percent confidence interval - lower       0.052       0.047
## 	90 Percent confidence interval - upper       0.065       0.059
## 	P-value RMSEA <= 0.05                         0.013       0.183
##
## 	Robust RMSEA                                             0.057
## 	90 Percent confidence interval - lower                   0.050
## 	90 Percent confidence interval - upper                   0.065
##
## Standardized Root Mean Square Residual:
##
## 	SRMR                                         0.042       0.042
```

As we have discussed in Section 8.2.3, we expect that Model 3 should fit as well as Model 1 (measurement models for latent factors) because they are equivalent models. As Model 1 has a good fit, the current model should have the same good fit. The model equivalence is verified here: Robust Comparative Fit Index (CFI) = 0.970, Robust Tucker-Lewis Index (TLI) = 0.959, Root Mean Square Error of Approximation (RMSEA) = 0.053 and the *p*-value for testing RMSEA ≤ 0.05 is 0.183, and the Standardized Root Mean Square Residual (SRMR) = 0.042. All these values are the same as that of Model 1. This validates the full structural equation model (Model 3, in which *UNEASE* and *Jaw* are correlated) for the pain-related disability.

Note that we used the option *estimates=FALSE* to disable the output on parameter estimates because we would like to compare the estimates and their standardized estimates with those obtained from Miller et al. (2020). The standardized estimates can be accessed using the option *standardized=TRUE* from *lavaan* function *parameterEstimates* as follows:

```
# Get the standardized parameter estimates
estStandard = parameterEstimates(fitSEM.Pain, standardized=TRUE)
# Print only the estimates, both unstandardized and standardized
estStandard[,c("lhs","op","rhs","est","std.all")]
```

```
##          lhs op    rhs      est std.all
## 1        Jaw =~    JF1    1.000   0.623
## 2        Jaw =~    JF2    1.397   0.717
## 3        Jaw =~    JF3    1.390   0.881
## 4     UNEASE =~     P8    1.000   0.424
## 5     UNEASE =~ SCL905    0.047   0.613
## 6     UNEASE =~   CSQ1    0.122   0.771
## 7        CPI =~   CPI1    1.000   0.760
## 8        CPI =~   CPI2    0.751   0.713
## 9        CPI =~   CPI3    0.954   0.807
## 10     INTER =~     I1    1.000   0.886
## 11     INTER =~     I2    1.082   0.908
## 12     INTER =~     I3    1.061   0.904
## 13       DIS =~    DD1    1.000   0.753
## 14       DIS =~    DD2    1.465   0.884
## 15       PRD =~    CPI    1.000   0.749
## 16       PRD =~  INTER    1.330   0.848
## 17       PRD =~    DIS   16.805   0.808
## 18       JF1 ~~    JF2    0.925   0.445
## 19        P8 ~~ SCL905    2.962   0.350
## 20      CPI2 ~~   CPI3    0.511   0.285
## 21       PRD  ~ UNEASE    0.096   0.558
## 22       PRD  ~    Jaw    0.453   0.358
## 23       JF1 ~~    JF1    1.920   0.611
## 24       JF2 ~~    JF2    2.250   0.486
## 25       JF3 ~~    JF3    0.679   0.223
## 26        P8 ~~     P8  299.020   0.821
## 27    SCL905 ~~ SCL905    0.240   0.624
## 28      CSQ1 ~~   CSQ1    0.657   0.405
## 29      CPI1 ~~   CPI1    2.540   0.422
## 30      CPI2 ~~   CPI2    1.895   0.492
## 31      CPI3 ~~   CPI3    1.692   0.349
## 32        I1 ~~     I1    1.313   0.215
## 33        I2 ~~     I2    1.190   0.175
## 34        I3 ~~     I3    1.212   0.183
## 35       DD1 ~~    DD1  643.512   0.433
## 36       DD2 ~~    DD2  508.349   0.219
## 37       Jaw ~~    Jaw    1.220   1.000
## 38    UNEASE ~~ UNEASE   65.412   1.000
## 39       CPI ~~    CPI    1.523   0.438
## 40     INTER ~~  INTER    1.349   0.281
## 41       DIS ~~    DIS  292.076   0.346
## 42       PRD ~~    PRD    0.696   0.357
## 43       Jaw ~~ UNEASE    4.567   0.511
## 44       JF1 ~1           2.493   1.407
```

```
## 45     JF2 ~1      2.747   1.276
## 46     JF3 ~1      1.247   0.716
## 47      P8 ~1     58.667   3.073
## 48 SCL905 ~1      0.741   1.194
## 49    CSQ1 ~1      1.490   1.169
## 50    CPI1 ~1      3.757   1.532
## 51    CPI2 ~1      7.510   3.826
## 52    CPI3 ~1      5.414   2.457
## 53      I1 ~1      2.544   1.029
## 54      I2 ~1      2.358   0.903
## 55      I3 ~1      2.231   0.868
## 56     DD1 ~1     17.508   0.454
## 57     DD2 ~1     29.589   0.615
## 58     Jaw ~1      0.000   0.000
## 59  UNEASE ~1      0.000   0.000
## 60     CPI ~1      0.000   0.000
## 61   INTER ~1      0.000   0.000
## 62     DIS ~1      0.000   0.000
## 63     PRD ~1      0.000   0.000
```

The standardized estimates here are very close to the values shown in Figure 2 from Miller et al. (2020). They are not exactly equal because we did not adjust for *control variables* here in the current illustration. Fitting a structural equation model with adjustment of control variables will be left to interested readers as an exercise.

8.3.3   Plotting of the Fitted Model

Even its graphical capability is not perfect at this point, an *R* package called *semPlot* can be used to plot the fitted models. For example, to make a similar figure like in Figure 8.1, we can call the function *semPlots* with the fitted model object *fitSEM.Pain*. The function will produce a diagram like the one in Figure 8.3.

```
# Load the library
library(semPlot)
# Call "semPlots" to plot the estimated model
semPaths(fitSEM.Pain, whatLabels = "est")
```

Except for the additional double-headed path between *UNEASE* and *Jaw* and the display of estimates in Figure 8.3, you can see that this figure has the same set of graphical elements as that of Figure 8.1, which has a better graphical illustration. Therefore, more plotting options in *semPlots* should be explored and sought out to make Figure 8.3 more presentable. We will let the interested readers to discover these potential graphical improvements.

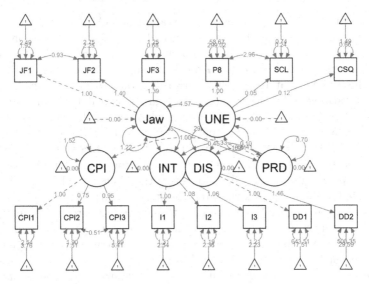

FIGURE 8.3
Estimated Structural Model for Pain-Related Disability.

8.4 Data Analysis Using SAS

A *SAS* data set *Dpain* has been prepared for all analyses in this section. The missing values in the original CSV file were recoded to '.', which is a standard coding of missing value in the *SAS* system. In the following code, the *CALIS* procedure fits a composite measurement model for the data:

```
/******* First step to establish the measurement model **********/
proc calis data='c:\Dpain' method=FIML;
   path
      Unease       ==> P8 SCL905 CSQ1 = 1.,
      Jaw          ==> JF1 JF2 JF3    = 1.,
      DIS          ==> DD1 DD2        = 1.,
      Inter        ==> I1 I2 I3       = 1.,
      CPI          ==> CPI1 CPI2 CPI3 = 1.,
      PRD          ==> CPI Inter DIS  = 1.;
   pcov
      P8 SCL905, JF1 JF2, CPI2 CPI3;
   fitindex on(only)=[chisq df probchi RMSEA RMSEA_LL RMSEA_UL
           PROBCLFIT SRMR CFI BENTLERNNFI] noindextype;
run;
```

In the first five entries of the PATH statement, relationships between the sets of latent variables and observed variables are specified. The last path entry specifies a second-order factor model. Fixed ones are applied to the first factor loading in each path entry. All other loadings are free parameters.

In the PCOV statement, three error covariances are specified for the observed variables. All other error terms are assumed to be uncorrelated. Because *UNEASE, Jaw*, and *PRD* are exogenous variables in the model, the covariances among them are free parameters by default.

The METHOD = FIML option in the PROC CALIS statement requests the full information maximum likelihood method for estimation. As discussed, this method uses all available data for estimation. When the FIML estimation is requested, default mean structures are applied to the variables and hence there is no need to specify additionally the mean or intercept parameters in the code. Explicit specifications of mean and intercept parameters might be needed when these parameters are subject to constraints (such as equality constraints or fixed numbers).

The following table shows the fit summary. Not surprisingly, these fit statistics are the same as those obtained from the *lavaan* output and they indicate that the measurement model has a decent fit.

Fit Summary	
Chi-Square	322.2892
Chi-Square DF	68
Pr > Chi-Square	<.0001
Standardized RMR (SRMR)	0.0452
RMSEA Estimate	0.0587
RMSEA Lower 90% Confidence Limit	0.0523
RMSEA Upper 90% Confidence Limit	0.0652
Probability of Close Fit	0.0127
Bentler Comparative Fit Index	0.9685
Bentler-Bonett Non-normed Index	0.9578

NOTE: Saturated mean structure parameters are excluded from the computations of fit indices.

Now, to fit the full structural model, a path entry is added to PATH statement to represent the paths from *UNEASE* and *Jaw*, respectively, to *PRD*, as shown in the following:

```
/******* Second step to fit the structural model ***********/
proc calis data='c:\Dpain' method=FIMLSB;
    path
        Unease        ==> P8 SCL905 CSQ1  = 1.,
        Jaw           ==> JF1 JF2 JF3     = 1.,
        DIS           ==> DD1 DD2         = 1.,
        Inter         ==> I1 I2 I3        = 1.,
        CPI           ==> CPI1 CPI2 CPI3  = 1.,
        PRD           ==> CPI Inter DIS   = 1.,
        Unease Jaw    ==> PRD;
```

```
pcov
   P8 SCL905, JF1 JF2, CPI2 CPI3;
fitindex on(only)=[chisq df probchi RMSEA RMSEA_LL RMSEA_UL
       PROBCLFIT SRMR CFI BENTLERNNFI] noindextype;
run;
```

Fit Summary	
Chi-Square	322.2893
Chi-Square DF	68
Pr > Chi-Square	<.0001
Standardized RMR (SRMR)	0.0453
RMSEA Estimate	0.0587
RMSEA Lower 90% Confidence Limit	0.0523
RMSEA Upper 90% Confidence Limit	0.0652
Probability of Close Fit	0.0127
Bentler Comparative Fit Index	0.9685
Bentler-Bonett Non-normed Index	0.9578
NOTE: Saturated mean structure parameters are excluded from the computations of fit indices.	

As discussed previously, this structural model is equivalent to the measurement model and therefore the sets of model fit statistics for the two models are almost identical.

PROC CALIS computes the standardized estimates by default. The following output shows standardized estimates for loadings, paths, variances, and covariances in the model. These standardized estimates match those from the *lavaan* output.

Path			Parameter	Estimate	Standard Error	t Value	Pr > |t|
Unease	===>	P8		0.42384	0.03199	13.2488	<.0001
Unease	===>	SCL905	_Parm01	0.61325	0.02731	22.4552	<.0001
Unease	===>	CSQ1	_Parm02	0.77155	0.02736	28.1970	<.0001
Jaw	===>	JF1		0.62339	0.02656	23.4682	<.0001
Jaw	===>	JF2	_Parm03	0.71710	0.02246	31.9305	<.0001
Jaw	===>	JF3	_Parm04	0.88126	0.02238	39.3770	<.0001
DIS	===>	DD1		0.75309	0.01849	40.7207	<.0001
DIS	===>	DD2	_Parm05	0.88353	0.01650	53.5403	<.0001
Inter	===>	I1		0.88616	0.00837	105.8	<.0001
Inter	===>	I2	_Parm06	0.90848	0.00735	123.5	<.0001
Inter	===>	I3	_Parm07	0.90372	0.00754	119.8	<.0001
CPI	===>	CPI1		0.75999	0.02013	37.7453	<.0001
CPI	===>	CPI2	_Parm08	0.71278	0.02350	30.3272	<.0001
CPI	===>	CPI3	_Parm09	0.80716	0.02027	39.8209	<.0001
PRD	===>	CPI		0.74934	0.02383	31.4477	<.0001
PRD	===>	Inter	_Parm10	0.84799	0.01676	50.5844	<.0001
PRD	===>	DIS	_Parm11	0.80850	0.02060	39.2498	<.0001
Unease	===>	PRD	_Parm12	0.55759	0.04196	13.2874	<.0001
Jaw	===>	PRD	_Parm13	0.35801	0.04130	8.6683	<.0001

Standardized Results for PATH List

Although the structural model is equivalent to the measurement model, it offers another set of estimates to interpret the relationships among the variables. For example, in the previous table, the standardized coefficients for paths from *UNEASE* and *Jaw*, respectively, to *PRD* are both significant, indicating strong functional relationships among these latent factors.

The next table shows the error variance estimates when all the variances of latent factors and observed variables are standardized to one.

Standardized Results for Variance Parameters						
Variance Type	Variable	Parameter	Estimate	Standard Error	t Value	Pr > |t|
Exogenous	Unease	_Add01	1.00000			
	Jaw	_Add02	1.00000			
Error	CPI1	_Add03	0.42241	0.03060	13.8025	<.0001
	CPI2	_Add04	0.49195	0.03350	14.6830	<.0001
	CPI3	_Add05	0.34849	0.03272	10.6499	<.0001
	DD1	_Add06	0.43285	0.02786	15.5392	<.0001
	DD2	_Add07	0.21938	0.02916	7.5231	<.0001
	I1	_Add08	0.21473	0.01484	14.4712	<.0001
	I2	_Add09	0.17467	0.01336	13.0716	<.0001
	I3	_Add10	0.18329	0.01363	13.4460	<.0001
	JF1	_Add11	0.61138	0.03312	18.4604	<.0001
	JF2	_Add12	0.48577	0.03221	15.0815	<.0001
	JF3	_Add13	0.22339	0.03944	5.6633	<.0001
	P8	_Add14	0.82036	0.02712	30.2509	<.0001
	CSQ1	_Add15	0.40471	0.04222	9.5850	<.0001
	SCL905	_Add16	0.62393	0.03350	18.6275	<.0001
	CPI	_Add17	0.43848	0.03571	12.2786	<.0001
	Inter	_Add18	0.28091	0.02843	9.8802	<.0001
	DIS	_Add19	0.34633	0.03331	10.3980	<.0001
	PRD	_Add20	0.35687	0.03606	9.8965	<.0001

The final tables in the following show the standardized results for covariances among latent factors and errors. The correlation between *UNEASE* and *Jaw* is 0.51 ($p < 0.0001$), indicating a strong association between the two latent factors.

Standardized Results for Covariances Among Exogenous Variables						
Var1	Var2	Parameter	Estimate	Standard Error	t Value	Pr > |t|
Jaw	Unease	_Add21	0.51112	0.03548	14.4054	<.0001

Standardized Results for Covariances Among Errors						
Error of	Error of	Parameter	Estimate	Standard Error	t Value	Pr > |t|
P8	SCL905	_Parm14	0.25008	0.02679	9.3340	<.0001
JF1	JF2	_Parm15	0.24262	0.02836	8.5561	<.0001
CPI2	CPI3	_Parm16	0.11813	0.02722	4.3403	<.0001

8.5 Discussions

In this chapter, we used a real data set from 1088 adults with painful chronic TMD enrolled in the *Orofacial Pain: Prospective Evaluation and Risk Assessment (OPPERA)* study to illustrate the process of building a structural equation model that tests the relationship between features associated with pain-related disability in people with painful TMD.

We used the *R* package *lavaan* and *SAS PROC CALIS* to illustrate the analysis with detailed step-by-step implementations. We first validated a measurement model that was composed of component CFA models for jaw limitation as measured from the *Jaw Functional Limitation Scale (JFLS)*, psychological unease (negative affect, somatic symptoms, and catastrophizing), and pain-related disability (which is a second-order CFA model). Then, a full structural equation model for the pain-related disability was built based on its hypothesized functional relationships with jaw limitation and psychological unease. The final structural model included the covariance (correlation) between the factors for jaw limitation and psychological unease fitted well and they were found to be strongly related to pain-related disability.

8.6 Exercises

1. Fit the full structural equation model for the pain-related disability in Figure 8.1 with zero covariance between *Jaw* and *UNEASE* (Model 2 in Section 8.2.3). Does this model fit well? Is this model equivalent to Model 1 described in Section 8.2.3?

2. Fit the full structural equation model for the pain-related disability in Figure 8.1 with control variable *Age* to see whether *Age* is statistically significant. Let the covariance between *Jaw* and *UNEASE* be a free parameter in the model.

3. Fit the full structural equation model for the pain-related disability in Figure 8.1 with control variables *Gender* to test whether there is a *gender* difference. Let the covariance between *Jaw* and *UNEASE* be a free parameter in the model.

4. Fit the full structural equation model for the pain-related disability in Figure 8.1 with control variables *Site* to test whether there is a *Site* difference. Let the covariance between *Jaw* and *UNEASE* be a free parameter in the model.

Appendix: Mplus Programs

We will include the *mplus* programs as an appendix. Note that you will need to have *mplus* installed to run the program

1.   Mplus for Measurement Model

```
TITLE:
  The three CFA models for the Pain SEM model

DATA:
  FILE IS "Pain.dat";

VARIABLE:
NAMES =
  Site   Age Gender CPI1    CPI2    CPI3    DD1 DD2
  I1     I2   I3   JF1 JF2 JF3 P8   CSQ1     SCL905;

MISSING = ALL (-1234);

ANALYSIS:
  ESTIMATOR IS MLR;

MODEL:
 ! 2nd-order CFA for PRD
   CPI BY CPI1 CPI2 CPI3;
   INTER BY I1 I2 I3;
   DIS BY DD1 DD2;
   PRD BY CPI Inter DIS;

 ! 1st-order
   JAW BY JF1 JF2 JF3 ;
   UNEASE BY SCL905 CSQ1 P8;

 ! Covariances
   JF1 WITH JF2;
   SCL905 WITH P8;
   CPI2 WITH CPI3;

OUTPUT:
  STDYX TECH4 MODINDICES;
```

2. Mplus for Structural Equation Model

```
TITLE:
  The full SEM model for PRD

DATA:
  FILE IS "Pain.dat";

VARIABLE:
NAMES =
  Site  Age Gender  CPI1    CPI2    CPI3    DD1 DD2
  I1    I2  I3  JF1 JF2 JF3 P8  CSQ1    SCL905;

MISSING = ALL (-1234);

ANALYSIS:
  ESTIMATOR IS MLR;

MODEL:
 ! 2nd-order CFA for PRD
   CPI BY CPI1 CPI2 CPI3;
   INTER BY I1 I2 I3;
   DIS BY DD1 DD2;
   PRD BY CPI Inter DIS;

 ! 1st-order
   JAW BY JF1 JF2 JF3 ;
   UNEASE BY SCL905 CSQ1 P8;

 ! Covariances
   JF1 WITH JF2;
   SCL905 WITH P8;
   CPI2 WITH CPI3;

 ! regression
   PRD ON JAW UNEASE;

OUTPUT:
  STDYX TECH4 MODINDICES;
```

9

Breast-Cancer Post-Surgery Assessment—Latent Growth-Curve Modeling

In this chapter, we illustrate latent growth-curve (LGC) modeling with a longitudinal data collected from 405 Hong Kong women on breast-cancer post-surgery assessment presented in Byrne (2012). In this data set, two outcome variables of their mood (i.e., MOOD) and their social adjustment (i.e., SOCADJ) were measured three times longitudinally at 1, 4, and 8 months post-surgery.

Latent growth-curve modeling is one of the most powerful models in structural equation modeling to analyze longitudinal data. It models longitudinal growth trajectories over time and tests differences between groups (such as, in different intervention groups, age groups, etc.). As a general model for longitudinal growth, this model can incorporate both the within-individual and between-individual variations along with the longitudinal trajectories. Corresponding to the breast-cancer data, for example, we can model the within-woman (i.e., within-individual) longitudinal change on both their mood (MOOD) and social adjustment (SOCADJ) over 8 months as well as the between-woman (i.e., between-individual) variations of their longitudinal trajectories.

Note to install the *R* packages *lavaan* for *sem* analysis (https://www.lavaan.ugent.be/), *semPlot* for SEM path diagrams, *tidyverse* for data processing, and *ggplot2* for plotting.

```
# Load the lavaan package into R session
library(lavaan)
# Load the library for SEM plot
library(semPlot)
# Load tidyverse for data processing
library(tidyverse)
# Load ggplot2 for plotting
library(ggplot2)
```

DOI: 10.1201/9781003365860-9

9.1 Data and Preliminary Analysis

9.1.1 Data Description

The data are available in the *hkcancer.csv* file and there are 405 women with 10 variables as follows:

- *ID* is the identification of each woman.

- $MOOD_1$, $MOOD_4$ and $MOOD_8$ are the self-rated *mood* measured at 1, 4, and 8 months after surgery. Note that these variables measure *bad mood*— the higher the value of MOOD, the worse their mood after surgery.

- $SOCADJ_1$, $SOCADJ_4$ and $SOCADJ_8$ are the measured *social adjustment* at 1, 4 and 8 months after surgery—the higher the value of SOCADJ, the better their social adjustment after surgery.

- *Age* is the age of each woman at surgery.

- *AgeGrp* is the dichotomized age group from their *Age*, where *Younger* is for woman's age < 50 and *Older* is for woman's age > 50.

- *SurgTX* is to denote their surgery treatment of *lumpectomy* and *mastectomy*.

Let us load the data into *R* as follows:

```
# Read the data into R
dCancer <- read.csv("data/hkcancer.csv", na.strings="*",header=T)
# Check the data dimension
dim(dCancer)
```

```
## [1] 405   10
```

```
# Print the first 6 observations
head(dCancer)
```

```
##   ID MOOD1 MOOD4 MOOD8 SOCADJ1 SOCADJ4 SOCADJ8 Age
## 1  1    15    NA    NA    95.9      NA      NA  70
## 2  2    16    25    22   114.9   105.1    90.4  47
## 3  3    37    26    25    80.7    95.3    95.3  47
## 4  4    19    16    15   112.8   108.6    99.0  52
## 5  5    13    16    14   115.0   105.0   101.0  43
## 6  6    21    28    19   106.5   115.0   107.5  34
##        SurgTx  AgeGrp
## 1 Mastectomy   Older
## 2 Mastectomy Younger
```

```
## 3 Mastectomy Younger
## 4 Mastectomy   Older
## 5 Mastectomy Younger
## 6 Lumpectomy Younger
```

Note that there are some missing values coded as ∗, which was specified as na.strings = "∗" in *hkcancer.csv*.

9.1.2   Exploratory Data Analysis

The exploratory data analysis is to investigate whether the longitudinal trajectories change linearly or nonlinear over time. If the trajectories are linear, we can use a linear growth-curve model where only the intercept and slope parameters are needed to be focused on in the latent growth-curve modeling. However, if the growth trajectories are not linear, a nonlinear growth-curve model would have to be specified. There are many nonlinear models to be selected, depending on the substantive knowledge about the data collected. Mostly practically, an exploratory data analysis with graphical plotting is usually the first step for this investigation.

9.1.2.1   Changes in Means and Variances

We can investigate the changes in means and variances for the two outcomes along with the three time points, which can be calculated by calling the *R* function *apply* as follows:

```
# Get the Means for numerical data
apply(dCancer[,2:8],2, function(x) mean(x, na.rm=T))
```

```
##    MOOD1    MOOD4   MOOD8 SOCADJ1 SOCADJ4 SOCADJ8
##    21.4     21.0    20.0   100.9   100.4   100.6
##     Age
##    51.6
```

```
# Get the variances
apply(dCancer[,2:8],2, function(x) var(x, na.rm=T))
```

```
##    MOOD1    MOOD4   MOOD8 SOCADJ1 SOCADJ4 SOCADJ8
##    40.3     41.5    34.7    95.0   100.1    76.0
##     Age
##    114.1
```

As seen from the output, the means for the MOOD changed from 21.370 at 1-month to 20.985 at 4-month and 20.000 at 8-month after surgery, which indicated a slight positive mood for these women after surgery (recall that MOOD measures bad mood) even though the decline was very small. The means for the SOCADJ changed from 100.911 at 1-month to 100.424 at 4-month and 100.599 at 8-month after surgery, where there are no clear patterns and the changes are very minimal. The changes in variances are more apparent at 8 months, where both $MOOD_8$ and $SOCADJ_8$ show smaller variances than their previous measurements. Finally, the mean is 51 with variance of 114 for the *Age* variable.

To see these patterns graphically, we can plot the data with *boxplot* to see their distributions at each time point. We illustrate how to do this in *R* for MOOD and leave the SOCADJ to interested readers to explore.

We select the outcome MOOD and *reshape* the data from the *wide* format (i.e., person-format, where the data are presented for each woman with one row and three columns of $MOOD_1$, $MOOD_2$ and $MOOD_3$) to *long* format (i.e., person-time format, where the data are presented for each woman at each time point with just one column) using *R* library *tidyverse*. With the *long* format, we can then use *ggplot2* to make the *geom_boxplot* for the MOOD, which is shown in Figure 9.1.

```
# Call tidyverse to reshape the data
library(tidyverse)
```

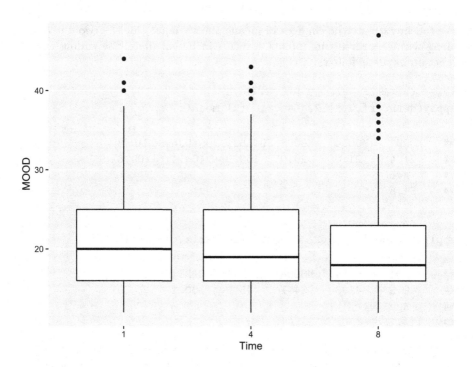

FIGURE 9.1
Illustration on Linear Growth-Curve from Boxplots.

```
# Get the "Mood" data
dMOOD <- dCancer[,c("ID","MOOD1","MOOD4","MOOD8", "AgeGrp",
                    "SurgTx")] %>%
# Reshape the data to long format
  gather("MOOD1","MOOD4","MOOD8",key="Var", value="Value",
         na.rm=TRUE) %>%
# Create the "Time" variable by "separate" the "Var"
  separate(Var, into=c("tmp","Time"), sep=4, convert=TRUE)
# Call ggplot2 to make the plot
library(ggplot2)
ggplot(dMOOD, aes(as.factor(Time),Value)) +
  # Add boxplot
  geom_boxplot() +
  labs(x="Time", y="MOOD")
```

As seen from Figure 9.1, the means for the MOOD changed from 21.370 at 1-month to 20.985 at 4-month and 20.000 at 8-month after surgery with their standard deviations from 40.255 at 1-month to 41.543 at 4-month and 34.659 at 8-month.

9.1.3   Changes in Longitudinal Trajectories

We can further investigate the individual changes for the 405 women graphically. This can be done similarly using *ggplot*, but changing *geom_boxplot* to *geom_point* for plotting the observed data points, and *geom_line* for linking the observed data points with *line*, and *geom_smooth* for adding a *linear regression* line to the plot. In addition, we use *facet_wrap* to show this longitudinal trajectories for each surgical and age group using $\sim SurgTX * AgeGrp$. This is shown in Figure 9.2.

```
# Call ggplot for Spaghetti plot
ggplot(dMOOD, aes(Time,Value)) +
  # Add points
  geom_point() +
  # Add lines
  geom_line(aes(group=ID)) +
  geom_smooth(method="lm", se=FALSE, size=2, col="red") +
  # By SurgTX and AgeGrp
  facet_wrap(~SurgTx*AgeGrp,nrow=2, scales="free")+
  # Add the labs for both x-axis and y-axis
  labs(x="Time", y="MOOD")
```

```
## `geom_smooth()` using formula 'y ~ x'
```

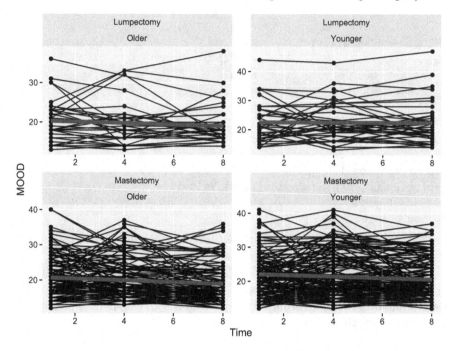

FIGURE 9.2
Illustration on Linear Growth-Curve Trajectories.

In each of the four plots in Figure 9.2, a linear growth (or the lack of growth) trend is fitted for the corresponding data. These linear trends describe the overall changes of mood over the 8-month period. However, it can also be seen in Figure 9.2 that individual growth-curves might deviate from these overall trends. That is, individual growths follow the overall trends in different degrees. Thus, the variability of these individual growth trends are important elements to be included in modeling. As we will see, the latent growth-curve modeling, which we apply to analyze the current data, can capture the individual variability in modeling.

9.2 Latent Growth-Curve Modeling

Latent growth-curve (LGC) models can be vastly extensive. We start with a simple LGC model, which is for a single outcome, and then extend to multiple outcomes. For example, there are two outcomes in the cancer data that we will use to illustrate the growth-curve analysis.

9.2.1 Latent Growth-Curve Model with a Single Outcome

For the ease of interpretation, we describe the LGC model along with the *hkcancer* data. We use the MOOD measures as the outcome variables that were measured at three time points: 1-, 4-, and 8-month after surgery. A linear growth-curve model for the repeated measure is fitted:

$$y_{it} = \beta_{i0} + \beta_{i1}\text{Time}_{it} + \epsilon_{it} \tag{9.1}$$

where y_{it} denotes the outcome of MOOD at time t for woman i, β_{i0} and β_{i1} represent the *intercept* and *slope* of the longitudinal linear model for the ith woman, and ϵ_{it} is the error term that is assumed to be normally distributed with mean of 0 and within-woman residual variance σ_t^2. This residual standard deviation σ_t can be time-specific or time-independent.

In standard linear regression models, *intercepts* and *slopes* are treated as fixed parameters in the model. These fixed parameters are estimated in modeling. However, in the latent growth-curve model described here, the *intercepts* and *slopes* are not fixed quantities across subjects (women). Instead, the subject index i for intercepts β_{i0} and slopes β_{i1} signifies that these quantities can vary across women. Mathematically, these intercepts and slopes are treated as unobserved (latent) random quantities that are modeled as:

$$\begin{aligned} \beta_{i0} &= \beta_0 + \nu_{i0} \\ \beta_{i1} &= \beta_1 + \nu_{i1} \end{aligned} \tag{9.2}$$

where β_0 and β_1 are the overall *intercept* and *slope* (or *effect*) parameters, respectively, and νs are random variables that generate random intercepts (β_{i0}s) and slopes (β_{i1}s) for women. Random variables νs are assumed to be distributed as a bivariate normal distribution such that:

$$\begin{pmatrix} \nu_{i0} \\ \nu_{i1} \end{pmatrix} \sim N\left(\begin{pmatrix} 0 \\ 0 \end{pmatrix}, \begin{pmatrix} \tau_0^2 & \tau_0\tau_1\rho_{01} \\ \tau_0\tau_1\rho_{01} & \tau_1^2 \end{pmatrix} \right) \tag{9.3}$$

where τ_0^2 and τ_1^2 are the between-women (between-subject) variances for the intercepts and slopes, respectively, and ρ_{01} is the correlation between random intercepts and slopes.

To summarize, as compared with the linear regression models, the formulation in Equations (9.1)–(9.3) have the following novel features:

(1) Intercepts β_{i0} and slopes β_{i1} for women (individuals or subjects) are **random** quantities that are formulated in the model to account for between-subject differences, and hence for different trajectories in subjects. They are **not** parameters to be estimated in the model. Instead, they are assumed to have a joint bivariate normal distribution, which is described next.

(2) The overall intercept β_0 and slope β_1 and the covariance matrix elements τ_0^2, τ_1^2, and ρ_{01} in Equation (9.3) are model parameters that

are used to characterize the distribution of the random intercepts and slopes—that is,

$$\begin{pmatrix} \beta_{i0} \\ \beta_{i1} \end{pmatrix} \sim N\left(\begin{pmatrix} \beta_0 \\ \beta_1 \end{pmatrix}, \begin{pmatrix} \tau_0^2 & \tau_0\tau_1\rho_{01} \\ \tau_0\tau_1\rho_{01} & \tau_1^2 \end{pmatrix} \right) \qquad (9.4)$$

In SEM software, however, the covariance matrix in Equation (9.4) is usually replaced with the following equivalent covariance matrix parameterization:

$$\begin{pmatrix} \tau_0^2 & \tau_{01} \\ \tau_{01} & \tau_1^2 \end{pmatrix} \qquad (9.5)$$

where τ_{01} represents the covariance between random intercepts and slopes and it replaces ρ_{01} (correlation) as a parameter to be estimated in the model.

9.2.2   Latent Growth-Curve Model with Multiple Outcomes

The LGC model with a single outcome can be extended to the case with P $(P > 1)$ multiple outcomes. For example, $P = 2$ for the Hong Kong cancer data because MOOD and SOCADJ are the target outcome variables whose trajectories are of interest.

In the multiple-outcome LGC model, the model for the p-th $(p = 1, \ldots, P)$ outcome can be written as follows:

$$y_{pit} = \beta_{pi0} + \beta_{pi1}\text{Time}_{pit} + \epsilon_{pit} \qquad (9.6)$$

where y_{pit} denotes the p-th outcome at time t for woman i, β_{pi0} and β_{pi1} represent the random *intercept* and *slope* for the p-th outcome of the i-th woman, and ϵ_{pit} is the error term that is assumed to be normally distributed with a mean of 0 and within-woman variance σ_{pt}^2.

The coefficients of β_{pi0} and β_{pi1} in Equation (9.6) are random quantities for modeling *between-individual* variation in growth trajectory and themselves are modeled as follows:

$$\begin{aligned} \beta_{pi0} &= \beta_{p0} + \nu_{pi0} \\ \beta_{pi1} &= \beta_{p1} + \nu_{pi1} \end{aligned} \qquad (9.7)$$

where β_{p0} and β_{p1} are the overall *intercept* and *slope* parameters for the p-th outcome and νs are random variables that generate random intercepts (β_{pi0}s) and slopes (β_{pi1}s) for the p-th outcome of the women.

In the multiple-outcome LGC model, the random variables νs can be correlated within or between outcomes. As a result, random intercepts and slopes can be correlated within or between outcomes. When the between-outcome correlations are not assumed for the νs, a straightforward extension of Equation (9.4) for the joint distribution of within-outcome random intercepts and slopes for the p-th outcome variable $(p = 1, \ldots, P)$ is:

$$\begin{pmatrix} \beta_{pi0} \\ \beta_{pi1} \end{pmatrix} \sim N\left(\begin{pmatrix} \beta_{p0} \\ \beta_{p1} \end{pmatrix}, \begin{pmatrix} \tau_{p0}^2 & \tau_{p0}\tau_{p1}\rho_{p01} \\ \tau_{p0}\tau_{p1}\rho_{p01} & \tau_{p1}^2 \end{pmatrix} \right) \qquad (9.8)$$

In addition, zero correlations of νs between outcomes imply that all between-outcome correlations or covariances among the random intercepts and slopes are also zeros. In this case of uncorrelated intercepts and slopes between outcomes, the set of parameters that characterize the random intercepts and slopes include P sets (one set for each outcome variable) of overall intercept (i.e., β_{p0}), slope (i.e., β_{p1}s), and P sets of covariance elements that include τ_{p0}^2, τ_{p1}^2, and ρ_{p01} (or covariance denoted as τ_{p01} instead as correlation ρ_{p01}).

An obvious limitation of the assumption of zero correlations of νs between outcomes is that all outcome variable and their trajectories would be uncorrelated under the hypothesized model. This is certainly restrictive and often impractical. In general, to allow for a more flexible modeling of the covariation among outcome variables, all νs, within or between P outcomes, are assumed to be correlated in the model. In addition, a multinormal distribution, with a null mean vector and a $2P \times 2P$ covariance matrix Σ_ν, is assumed for the $2P$ νs. Consequently, the random intercepts and slopes are multivariate-normally distributed—that is, the distribution is specified as:

$$
\begin{pmatrix} \beta_{1i0} \\ \beta_{1i1} \\ \vdots \\ \beta_{pi0} \\ \beta_{pi1} \\ \vdots \\ \beta_{Pi0} \\ \beta_{Pi1} \end{pmatrix} \sim N \left(\begin{pmatrix} \beta_{10} \\ \beta_{11} \\ \vdots \\ \beta_{p0} \\ \beta_{p1} \\ \vdots \\ \beta_{P0} \\ \beta_{P1} \end{pmatrix}, \Sigma_\nu \right)
\tag{9.9}
$$

In this case of correlated intercepts and slopes between outcomes, the set of parameters that characterize the random intercepts and slopes include P sets (one set for each outcome variable) of overall intercept (i.e., β_{p0}) and slope (i.e., β_{p1}s), and the $2P \times (2P + 1)/2$ covariance elements in Σ_ν.

9.2.3   Latent Growth-Curve Model with Covariates

The LGC modeling can further incorporate covariates at different levels. For example, if there are covariates at the *outcome model* level, say a vector of covariates X_{pit}, we can extend Equation (9.6) as follows:

$$
y_{pit} = \beta_{pi0} + \beta_{pi1}\text{Time}_{pit} + \beta'_{pi2}X_{pit} + \epsilon_{pit}
\tag{9.10}
$$

where an extra parameter vector β_{pi2} is included.

If there are covariates at the random intercept or slope level (or the latent-factor level, to be explained in the next section), say a vector of covariates Z_{pi}, we can extend Equation (9.7) as follows:

$$
\begin{aligned}
\beta_{pi0} &= \beta_{p0} + \beta'_{p01}Z_{pi} + \nu_{pi0} \\
\beta_{pi1} &= \beta_{p1} + \beta'_{p11}Z_{pi} + \nu_{pi1}
\end{aligned}
\tag{9.11}
$$

where β_{p01} and β_{p11} are the extra parameter vectors to be estimated.

Which level the covariates should be specified depends on applications. For example, the SurgTX (surgical type) and *AgeGrp* (age group) in the Hong Kong cancer data would be incorporated as covariates for modeling the random intercepts and slopes (i.e., Equation (9.11)) rather than as covariates for modeling the outcome directly (i.e., Equation (9.10)). The reason is that surgical type and age are treated as determining factors of the growth trajectories of MOOD and *SOCIADJ*. They occur prior to the observations of growth trends and are supposed to explain individuals' different initial response levels (i.e., through the random intercept) and their differential effects of the trajectories (i.e., through the random slopes).

For covariates that are considered to be haphazard factors that affect the measurement of an outcome variable, modeling by Equation (9.10) might be more appropriate. For example, if X indicates the good/bad weather of the day when the outcomes were measured, it would be reasonable to include it as a covariate for the modeling of MOOD.

9.2.4   Modeling of Latent Growth Curves by Structural Equation Modeling

To estimate the parameters of a latent growth-curve model by using structural equation modeling techniques, we start with the simplest model with a single outcome. The data model in Equation (9.1) is recapitulated here:

$$y_{it} = \beta_{i0} + \beta_{i1}\text{Time}_{it} + \epsilon_{it} \tag{9.12}$$

Suppose now we want to model the trajectory of the MOOD variables of the *hkcancer* data. Because the mood was measured at three different time points, Equation (9.12) expands to three equations for the MOOD outcomes at three time points:

$$\text{MOOD}_{i1} = \beta_{i0} + \beta_{i1}\text{Time}_{i1} + \epsilon_{i1}$$
$$\text{MOOD}_{i4} = \beta_{i0} + \beta_{i1}\text{Time}_{i4} + \epsilon_{i4}$$
$$\text{MOOD}_{i8} = \beta_{i0} + \beta_{i1}\text{Time}_{i8} + \epsilon_{i8} \tag{9.13}$$

In this set of equations, the t index has been replaced with the specific months for measuring the outcomes.

Next, by treating the measurement in the first month as the baseline time, we can code $\text{Time}_{i1} = 0$. For the next time point of the fourth month, we can code $\text{Time}_{i4} = 1$ so that a unit of the time code reflects a lapse of three months. Finally, for the last time point of the eighth month, $\text{Time}_{i8} = (8 - 1)/3 = 2.33$. With these fixed time coding, the set of equations in Equation (9.13) is rewritten as:

$$\text{MOOD}_{i1} = 1 \times \beta_{i0} + 0 \times \beta_{i1} + \epsilon_{i1}$$
$$\text{MOOD}_{i4} = 1 \times \beta_{i0} + 1 \times \beta_{i1} + \epsilon_{i4}$$
$$\text{MOOD}_{i8} = 1 \times \beta_{i0} + 2.33 \times \beta_{i1} + \epsilon_{i8} \tag{9.14}$$

The set of equations in Equation (9.14) still represents a data model because i is used as an index of observations. We can instead write (9.14) as a variable model as follows:

$$MOOD_1 = 1 \times Intercept_M + 0 \times Slope_M + \epsilon_{M1}$$
$$MOOD_4 = 1 \times Intercept_M + 1 \times Slope_M + \epsilon_{M4}$$
$$MOOD_8 = 1 \times Intercept_M + 2.33 \times Slope_M + \epsilon_{M8} \qquad (9.15)$$

Now, $MOOD_1$, $MOOD_4$, $MOOD_8$, ϵ_{M1}, ϵ_{M4}, and ϵ_{M8} are all random variables in Equation (9.15). The observation index i for these variables can be eliminated here because $Intercept_M$ and $Slope_M$ are now treated as random variables, which have realized (though unobserved) values β_{i0}s and β_{i1}s for the individuals in the data model (Equation 9.14). More importantly, the set of equations in Equation (9.15) now has the same form as that of the confirmatory factor-analysis (CFA) model (see Chapter 2), with the following three distinctive features of parameterization:

(1) In the context of CFA, random intercepts $Intercept_M$ and random slopes $Slope_M$ are now treated as latent factors for the mood measurements. Deriving from the joint distribution of the random intercept and slope in Equation (9.8), the distribution of the latent factors $Intercept_M$ and $Slope_M$ in the CFA can be described as:

$$\begin{pmatrix} Intercept_M \\ Slope_M \end{pmatrix} \sim N \left(\begin{pmatrix} \beta_{M0} \\ \beta_{M1} \end{pmatrix}, \begin{pmatrix} \tau_{M0}^2 & \tau_{M0,M1} \\ \tau_{M0,M1} & \tau_{M1}^2 \end{pmatrix} \right) \qquad (9.16)$$

(2) The means of these two latent factors in Equation (9.16), representing the overall intercept and slope, are generally nonzero in this CFA formulation. Therefore, the mean structures of this CFA model must be fitted simultaneously with the covariance structures in the modeling.

(3) All loadings in such a CFA model are fixed values (either fixed 1's or other values for reflecting time lapses)—that is, no free loadings are estimated in this CFA model (although *free* loadings can be specified in more advanced modeling).

Consequently, by recasting the latent growth-curve model as a confirmatory factor model with some specific features of parameterization, one can fit a LGC model by the SEM technique for CFA, which has been discussed in Chapter 2 and needs not be repeated here.

9.2.4.1 Modeling of Multiple Outcomes

In the *hkcancer* data, the social adjustment variables were also measured as outcomes at different time points. To model both the mood and social adjustment outcomes together, the following set of equations must now be included in the CFA model, in addition to the equations in Equation (9.15):

$$\text{SOCADJ}_1 = 1 \times \text{Intercept}_S + 0 \times \text{Slope}_S + \epsilon_{S1}$$
$$\text{SOCADJ}_4 = 1 \times \text{Intercept}_S + 1 \times \text{Slope}_S + \epsilon_{S4}$$
$$\text{SOCADJ}_8 = 1 \times \text{Intercept}_S + 2.33 \times \text{Slope}_S + \epsilon_{S8} \tag{9.17}$$

The variances and covariances among all (exogenous) latent factors Intercept_M, Slope_M, Intercept_S and Slope_S are now free parameters in the CFA model. Adapting Equation (9.9) to the current model, the distributional assumption for the latent intercepts and slopes is stated as:

$$\begin{pmatrix} \text{Intercept}_M \\ \text{Slope}_M \\ \text{Intercept}_S \\ \text{Slope}_S \end{pmatrix} \sim N \left(\begin{pmatrix} \beta_{M0} \\ \beta_{M1} \\ \beta_{S0} \\ \beta_{S1} \end{pmatrix}, \begin{pmatrix} \tau_{M0}^2 & \tau_{M0,M1} & \tau_{M0,S0} & \tau_{M0,S1} \\ \tau_{M0,M1} & \tau_{M1}^2 & \tau_{M1,S0} & \tau_{M1,S1} \\ \tau_{M0,S0} & \tau_{M1,S0} & \tau_{S0}^2 & \tau_{S0,S1} \\ \tau_{M0,S1} & \tau_{M1,S1} & \tau_{S0,S1} & \tau_{S1}^2 \end{pmatrix} \right) \tag{9.18}$$

9.2.4.2 Modeling with Covariates

For the *hkcancer* data, the random intercepts and slopes of the mood and social adjustment variables might be dependent on the surgery type (SurgTX) and the age group (AgeGrp) of the women. If these dependencies are modeled, the following four model equations are specified (instead of the distributional assumption (Equation 9.18)):

$$\text{Intercept}_M = \beta_{M0} + \beta_{M01} \times \text{SurgTX} + \beta_{M02} \times \text{AgeGrp} + \nu_{M0}$$
$$\text{Slope}_M = \beta_{M1} + \beta_{M11} \times \text{SurgTX} + \beta_{M12} \times \text{AgeGrp} + \nu_{M1}$$
$$\text{Intercept}_S = \beta_{S0} + \beta_{S01} \times \text{SurgTX} + \beta_{S02} \times \text{AgeGrp} + \nu_{S0}$$
$$\text{Slope}_S = \beta_{S1} + \beta_{S11} \times \text{SurgTX} + \beta_{S12} \times \text{AgeGrp} + \nu_{S1} \tag{9.19}$$

where νs are random error terms, β_{M0}, β_{M1}, β_{S0}, and β_{S1} are (nonrandom) intercepts, and all other βs are (nonrandom) effects or slopes in these equations. With covariates SurgTX and AgeGrp in Equations (9.19), the latent factors Intercept_M, Slope_M, Intercept_S and Slope_S are endogenous and so their means and covariances would be implied from the model. As a result, parameter specifications of the covariance matrix for these latent factors like those in Equation (9.18) would be unnecessary. Instead, the means and covariances of SurgTX and AgeGrp would be model parameters. The means, variances, and covariances of exogenous observed variables such as SurgTX and AgeGrp are usually set by default as free parameters in SEM software.

9.3 Data Analysis Using *R*

It is easy to implement the LGC model in *R*. Conventionally, to address model identification we assign a fixed coefficient of *1* attached to the random *intercept* factor in each of its related outcome variable equations. This reflects the fact that the intercept value remains the same across time for each individual, even though individuals might have different intercept values.

In addition, we fix the coefficients attached to the random *slope* factors according to the observed time code values for the outcome variables. For example, in modeling the Hong Kong breast-cancer data, we assign the 0 value to the coefficient attached to the random *slope* factor in the model equation for outcome $MOOD_1$, which was measured at one month after surgery. Next, we assign *1* to the coefficient attached to the random *slope* factor in the model equation for outcome $MOOD_4$, which was measured in the fourth month after surgery. Hence, these assignments assume that a unit of time code represents a lapse of 3 months. Because the next measurement was four months (at the eighth month after surgery) after the time code 1 for $MOOD_4$, the time code for $MOOD_8$ is therefore $2.33 = 1 + 4/3$. So, finally we assign 2.33 to the coefficient attached to the random *slope* factor in the model equation for outcome $MOOD_8$.

This LGC model specification is depicted in Figure 9.3 as a CFA for the MOOD measurements. A similar figure can be produced for *Social Adjustment*. In Figure 9.3, there are two *latent* variables of *Intercept* and *Slope* to describe the linear growth-curve trajectories from $MOOD_1$ to $MOOD_4$ and

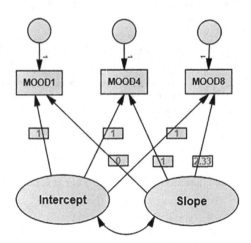

FIGURE 9.3
The LGC Model for MOOD.

then to MOOD$_8$. As explained previously, in this CFA specification, the loadings for the *Intercept* factor are all fixed at 1 and the loadings for the *Slope* factor are fixed at 0, 1, 2.33, respectively, for different time points of mood measurements.

9.3.1   Latent Growth-Curve Modeling for MOOD

9.3.1.1   The Basic Model for the Trajectory of Mood

With the above explanation, the LGC model for MOOD as depicted in Figure 9.3 can be easily implemented as follows:

```
# Model for Mood
mod4MOOD1 <- '
  # Intercept and Slope with fixed-coefficients
  iMOOD =~ 1*MOOD1 + 1*MOOD4 + 1*MOOD8
  sMOOD =~ 0*MOOD1 + 1*MOOD4 + 2.33*MOOD8
'
```

We can then call *lavaan* function *growth* to fit the LGC model. Since there are missing values in the data, we use the option *missing = "ml"* for full-information maximum likelihood estimation. Also due to the skewness of the data illustrated in Figure 9.2, we make use of *estimator = "MLR"* to get the robust standard errors.

```
library(lavaan)
fitMOOD1 <- growth(mod4MOOD1, data = dCancer,estimator ="MLR",
                   missing="ml")
```

```
## Warning in lav_data_full(data = data, group = group,
   cluster = cluster, :
   lavaan WARNING: some cases are empty and will be ignored:
##    88 90 111 113 121 135 137 158 171 192 199 210 219 245 253
      301 320 332 357 364 394
```

```
summary(fitMOOD1)
```

```
## lavaan 0.6-12 ended normally after 72 iterations
##
##    Estimator                                         ML
##    Optimization method                           NLMINB
##    Number of model parameters                         8
##
##                                               Used      Total
##    Number of observations                      384        405
##    Number of missing patterns                    7
```

```
## 
## Model Test User Model:
##                                            Standard      Robust
##    Test Statistic                             1.717       1.778
##    Degrees of freedom                             1           1
##    P-value (Chi-square)                       0.190       0.182
##    Scaling correction factor                              0.965
##       Yuan-Bentler correction (Mplus variant)
## 
## Parameter Estimates:
## 
##    Standard errors                          Sandwich
##    Information bread                         Observed
##    Observed information based on             Hessian
## 
## Latent Variables:
##                   Estimate  Std.Err  z-value  P(>|z|)
##    iMOOD =~
##       MOOD1          1.000
##       MOOD4          1.000
##       MOOD8          1.000
##    sMOOD =~
##       MOOD1          0.000
##       MOOD4          1.000
##       MOOD8          2.330
## 
## Covariances:
##                   Estimate  Std.Err  z-value  P(>|z|)
##    iMOOD ~~
##       sMOOD         -2.109    1.654   -1.275    0.202
## 
## Intercepts:
##                   Estimate  Std.Err  z-value  P(>|z|)
##    .MOOD1           0.000
##    .MOOD4           0.000
##    .MOOD8           0.000
##     iMOOD          21.474    0.319   67.330    0.000
##     sMOOD          -0.590    0.133   -4.437    0.000
## 
## Variances:
##                   Estimate  Std.Err  z-value  P(>|z|)
##    .MOOD1          14.305    3.449    4.148    0.000
##    .MOOD4          18.612    2.497    7.454    0.000
##    .MOOD8           6.822    4.029    1.693    0.090
##     iMOOD          25.833    3.609    7.158    0.000
##     sMOOD           2.269    1.306    1.737    0.082
```

As seen from the measures of model goodness-of-fit, we obtained a very satisfactory model fit for this data. The estimated variance for the slope sMOOD and the error variance for $MOOD_8$ are only marginally significant (i.e., p-values are 0.082 and 0.90, respectively). Also, the estimated covariance between *intercept* and *slope* (i.e., iMOOD $\sim$ $\sim$ sMOOD) is not statistically significant (i,e., p-value = 0.202).

Insignificant variance estimate for sMOOD means that women's slopes in their trajectories are not much different from each other. Insignificant error variance for $MOOD_8$ means that the mood in the eighth month could be determined almost completely by the systematic sources (i.e., the random intercepts and slopes). Insignificant covariance between the *intercept* and *slope* factors means that there is no association between the initial mood and random slopes in women's trajectories.

To improve the model parsimony, We first fixed the covariance between the *intercept* (iMOOD) and *slope* (sMOOD) to zero and then refit the model. Because the refitted model still showed a highly insignificant variance estimate for the slope sMOOD, this variance was also fixed to zero for the final model fitting.

9.3.1.2   Final Model for the Trajectory of Mood

With the two parameters fixed to zero, the final model is specified as follows:

```
# Final Model for Mood
mod4MOOD <- '
  # Intercept and Slope with fixed-coefficients
  iMOOD =~ 1*MOOD1 + 1*MOOD4 + 1*MOOD8
  sMOOD =~ 0*MOOD1 + 1*MOOD4 + 2.33*MOOD8
  # Constrains to fix non-significant parameter estimates
  iMOOD ~~ 0*sMOOD
  sMOOD ~~ 0*sMOOD
'
```

```
library(lavaan)
fitMOOD <- growth(mod4MOOD, data = dCancer, estimator ="MLR",
                  missing="ml")
```

```
summary(fitMOOD, fit.measures=TRUE)
```

```
## lavaan 0.6-12 ended normally after 48 iterations
##
##   Estimator                                         ML
##   Optimization method                           NLMINB
##   Number of model parameters                         6
##
```

```
##                                                Used          Total
##    Number of observations                       384            405
##    Number of missing patterns                     7
##
## Model Test User Model:
##                                            Standard         Robust
##    Test Statistic                             4.771          4.204
##    Degrees of freedom                             3              3
##    P-value (Chi-square)                       0.189          0.240
##    Scaling correction factor                                 1.135
##       Yuan-Bentler correction (Mplus variant)
##
## Model Test Baseline Model:
##
##    Test statistic                           310.701        196.993
##    Degrees of freedom                             3              3
##    P-value                                    0.000          0.000
##    Scaling correction factor                                 1.577
##
## User Model versus Baseline Model:
##
##    Comparative Fit Index (CFI)                0.994          0.994
##    Tucker-Lewis Index (TLI)                   0.994          0.994
##
##    Robust Comparative Fit Index (CFI)                        0.996
##    Robust Tucker-Lewis Index (TLI)                           0.996
##
## Loglikelihood and Information Criteria:
##
##    Loglikelihood user model (H0)          -3232.262      -3232.262
##    Scaling correction factor                                 1.371
##       for the MLR correction
##    Loglikelihood unrestricted model (H1)  -3229.877      -3229.877
##    Scaling correction factor                                 1.292
##       for the MLR correction
##
##    Akaike (AIC)                            6476.524       6476.524
##    Bayesian (BIC)                          6500.228       6500.228
##    Sample-size adjusted Bayesian (BIC)     6481.191       6481.191
##
## Root Mean Square Error of Approximation:
##
##    RMSEA                                      0.039          0.032
##    90 Percent confidence interval - lower     0.000          0.000
##    90 Percent confidence interval - upper     0.102          0.093
##    P-value RMSEA <= 0.05                      0.522          0.600
```

```
##
##    Robust RMSEA                                               0.034
##    90 Percent confidence interval - lower                     0.000
##    90 Percent confidence interval - upper                     0.104
##
## Standardized Root Mean Square Residual:
##
##    SRMR                                            0.036        0.036
##
## Parameter Estimates:
##
##    Standard errors                                 Sandwich
##    Information bread                               Observed
##    Observed information based on                    Hessian
##
## Latent Variables:
##                      Estimate  Std.Err  z-value  P(>|z|)
##    iMOOD =~
##      MOOD1            1.000
##      MOOD4            1.000
##      MOOD8            1.000
##    sMOOD =~
##      MOOD1            0.000
##      MOOD4            1.000
##      MOOD8            2.330
##
## Covariances:
##                      Estimate  Std.Err  z-value  P(>|z|)
##    iMOOD ~~
##      sMOOD            0.000
##
## Intercepts:
##                      Estimate  Std.Err  z-value  P(>|z|)
##    .MOOD1            0.000
##    .MOOD4            0.000
##    .MOOD8            0.000
##     iMOOD           21.492     0.317    67.731    0.000
##     sMOOD           -0.585     0.134    -4.361    0.000
##
## Variances:
##                      Estimate  Std.Err  z-value  P(>|z|)
##     sMOOD            0.000
##    .MOOD1           18.790     2.304     8.154    0.000
##    .MOOD4           17.234     2.469     6.979    0.000
##    .MOOD8           13.123     1.797     7.303    0.000
##     iMOOD           22.726     2.666     8.523    0.000
```

As seen from the model fitting, the Chi-square test is not significant with p-value = 0.240, the other model fitting measures are very satisfactory as indicated by the Comparative Fit Index (CFI) = 0.996, the Tucker-Lewis Index (TLI) = 0.994, the Root Mean Square Error of Approximation (RMSEA) = 0.034, and the Standardized Root Mean Square Residual (SRMR) = 0.036.

With this latent growth-curve modeling, we can conclude that the estimated overall MOOD at 1-month is 21.492, which is highly statistically significant with p-value < 0.001. The estimated overall slope is −0.585, which is also highly statistically significant, indicating that the overall mood for women improved (again, MOOD measured negative mood) from 21.492 at the first month at the rate of 0.585 per 3 months (or per 1 unit of the time code). This finding implies that as time following surgery increases, women tend gradually to report a more positive mood.

9.3.2 Latent Growth-Curve Modeling for SOCADJ

9.3.2.1 The Basic Model for the Trajectory of Social Adjustment

We can similarly build an LGC model for SOCADJ and fit the model to the data as follows:

```
# Model for SOCADJ
mod4SOCADJ1 <- '
  # Intercept and Slope with fixed-coefficients
  iSOCADJ =~ 1*SOCADJ1 + 1*SOCADJ4 + 1*SOCADJ8
  sSOCADJ =~ 0*SOCADJ1 + 1*SOCADJ4 + 2.33*SOCADJ8
'
# Call "growth* to fit the model
fitSOCADJ <- growth(mod4SOCADJ1, data = dCancer,
                    estimator ="MLR", missing="ml" )
```

```
## Warning in lav_object_post_check(object): lavaan
## WARNING: some estimated ov variances are negative
```

```
summary(fitSOCADJ)
```

```
## lavaan 0.6-12 ended normally after 132 iterations
##
## Estimator                               ML
## Optimization method                 NLMINB
## Number of model parameters               8
##
##                                       Used       Total
## Number of observations                 386         405
## Number of missing patterns               7
```

```
##
## Model Test User Model:
##                                              Standard      Robust
##    Test Statistic                               4.145       3.835
##    Degrees of freedom                               1           1
##    P-value (Chi-square)                         0.042       0.050
##    Scaling correction factor                                1.081
##       Yuan-Bentler correction (Mplus variant)
##
## Parameter Estimates:
##
##    Standard errors                              Sandwich
##    Information bread                            Observed
##    Observed information based on                 Hessian
##
## Latent Variables:
##                    Estimate  Std.Err  z-value  P(>|z|)
##    iSOCADJ =~
##       SOCADJ1         1.000
##       SOCADJ4         1.000
##       SOCADJ8         1.000
##    sSOCADJ =~
##       SOCADJ1         0.000
##       SOCADJ4         1.000
##       SOCADJ8         2.330
##
## Covariances:
##                    Estimate  Std.Err  z-value  P(>|z|)
##    iSOCADJ ~~
##       sSOCADJ       -16.346    4.997   -3.271    0.001
##
## Intercepts:
##                    Estimate  Std.Err  z-value  P(>|z|)
##    .SOCADJ1          0.000
##    .SOCADJ4          0.000
##    .SOCADJ8          0.000
##     iSOCADJ        100.933    0.502  201.029    0.000
##     sSOCADJ         -0.265    0.205   -1.289    0.197
##
## Variances:
##                    Estimate  Std.Err  z-value  P(>|z|)
##    .SOCADJ1          9.153    8.217    1.114    0.265
##    .SOCADJ4         42.258    7.881    5.362    0.000
##    .SOCADJ8         -0.280   13.004   -0.022    0.983
##     iSOCADJ         85.765   14.271    6.010    0.000
##     sSOCADJ         13.062    4.021    3.248    0.001
```

Notice that there is a *lavaan WARNING* that *some estimated ov variances are negative.* Examining the output carefully, we can see that the estimated residual variance for $SOCADJ_8$ is -0.280. However, because the corresponding p-value is 0.983, this estimate is not statistically significantly different from zero. Therefore, we can tentatively interpret this insignificant estimate as a nearly complete determination of the $SOCADJ_8$ by the systematic sources in the model (but it happens to have a negative estimate due to sampling fluctuations).

Were the negative variance estimate statistically significant, it would be a clear example of the so-called *Heywood Cases* in structural equation modeling. It is a mathematical peculiarity because variance is not supposed to be below 0. In general, a Heywood case indicates model misspecifications (e.g., path misspecifications, too many factors, a wrong measurement model, and so on). Another kind of mathematical peculiarity that could be encountered in practical structural equation modeling is when the estimated correlation is outside the range of 0 to 1, which is also an indication of model misspecifications.

Fixing the negative variance estimate to zero merely hides the model misspecification problems and might not be the best way to deal with Heywood cases. Researchers must instead think about whether alternative models are more plausible for the data at hand. In the current example, we notice that the mean slope estimate (mean of sSOCADJ) is negative but not significant (p=0.197). This means that on average there is no change in social adjustment during the 8-month period. Hence, it would be useful to first refit the model with the mean slope fixed to zero and then to see if the negative variance estimate persists.

9.3.2.2   The Final Model for the Trajectory of Social Adjustment

After fixing the mean slope parameter to zero, the refitted model still showed a negative (but insignificant) estimate of the error variance for $SOCADJ_8$. We then fit a final model with both of these parameters fixed at zero, as shown in the following:

```
# Model for SOCADJ
mod4SOCADJ <- '
  # Intercept and Slope with fixed-coefficients
  iSOCADJ =~ 1*SOCADJ1 + 1*SOCADJ4 + 1*SOCADJ8
  sSOCADJ =~ 0*SOCADJ1 + 1*SOCADJ4 + 2.33*SOCADJ8
  # Due to Insignificance, fix them to zero
  sSOCADJ ~ 0
  SOCADJ8 ~~ 0*SOCADJ8
'

# Call "growth* to fit the model
fitSOCADJ <- growth(mod4SOCADJ, data = dCancer,
                    estimator ="MLR", missing="ml" )
```

```
summary(fitSOCADJ,fit.measures=TRUE)
```

```
## lavaan 0.6-12 ended normally after 91 iterations
##
## Estimator                                     ML
## Optimization method                       NLMINB
## Number of model parameters                     6
##
##                                            Used         Total
## Number of observations                      386           405
## Number of missing patterns                    7
##
## Model Test User Model:
##                                        Standard        Robust
## Test Statistic                            5.802         3.625
## Degrees of freedom                            3             3
## P-value (Chi-square)                      0.122         0.305
## Scaling correction factor                               1.601
##     Yuan-Bentler correction (Mplus variant)
##
## Model Test Baseline Model:
##
## Test statistic                          363.782       213.112
## Degrees of freedom                            3             3
## P-value                                   0.000         0.000
## Scaling correction factor                               1.707
##
## User Model versus Baseline Model:
##
## Comparative Fit Index (CFI)               0.992         0.997
## Tucker-Lewis Index (TLI)                  0.992         0.997
##
## Robust Comparative Fit Index (CFI)                      0.997
## Robust Tucker-Lewis Index (TLI)                         0.997
##
## Loglikelihood and Information Criteria:
##
## Loglikelihood user model (H0)         -3681.673     -3681.673
## Scaling correction factor                               2.047
##     for the MLR correction
## Loglikelihood unrestricted model (H1) -3678.771     -3678.771
## Scaling correction factor                               1.898
##     for the MLR correction
##
## Akaike (AIC)                           7375.345      7375.345
```

```
## Bayesian (BIC)                                    7399.080    7399.080
## Sample-size adjusted Bayesian (BIC)               7380.043    7380.043
##
## Root Mean Square Error of Approximation:
##
## RMSEA                                                0.049       0.023
## 90 Percent confidence interval - lower               0.000       0.000
## 90 Percent confidence interval - upper               0.109       0.078
## P-value RMSEA <= 0.05                                0.422       0.726
##
## Robust RMSEA                                                     0.029
## 90 Percent confidence interval - lower                           0.000
## 90 Percent confidence interval - upper                           0.116
##
## Standardized Root Mean Square Residual:
##
## SRMR                                                 0.027       0.027
##
## Parameter Estimates:
##
## Standard errors                                  Sandwich
## Information bread                                Observed
## Observed information based on                     Hessian
##
## Latent Variables:
##                   Estimate  Std.Err  z-value  P(>|z|)
## iSOCADJ =~
##    SOCADJ1           1.000
##    SOCADJ4           1.000
##    SOCADJ8           1.000
## sSOCADJ =~
##    SOCADJ1           0.000
##    SOCADJ4           1.000
##    SOCADJ8           2.330
##
## Covariances:
##                   Estimate  Std.Err  z-value  P(>|z|)
## iSOCADJ ~~
##    sSOCADJ         -16.223    4.715   -3.440    0.001
##
## Intercepts:
##                   Estimate  Std.Err  z-value  P(>|z|)
##    sSOCADJ           0.000
##    .SOCADJ1          0.000
##    .SOCADJ4          0.000
##    .SOCADJ8          0.000
```

```
##      iSOCADJ        100.586    0.433  232.200    0.000
##
## Variances:
##                     Estimate  Std.Err  z-value  P(>|z|)
##      .SOCADJ8          0.000
##      .SOCADJ1         10.038    6.221    1.613    0.107
##      .SOCADJ4         41.955    5.362    7.824    0.000
##      iSOCADJ          85.018   14.069    6.043    0.000
##      sSOCADJ          12.995    2.033    6.393    0.000
```

The final model for social adjustment now has a very good fit, as indicated by various measures: RMSEA = 0.023, CFI = 0.997, TLI = 0.997. There is no negative variance estimate anymore. But there is still one insignificant error variance estimate for $SOCADJ_1$. Again, this means that social adjustment at time 1 might be completely accounted for by the systematic sources in the model.

9.3.3   Multiple-Outcome Latent Growth-Curve Modeling

9.3.3.1   Joint LGC Models for MOOD and SOCADJ

Tooling with powerful structural equation modeling techniques, latent growth-curve analysis can easily incorporate multiple outcomes. For example, in the Hong Kong women cancer data, two outcomes, MOOD and SOCADJ, were measured at the 1-month, 4-month and 8-month time points after surgery. We now illustrate the multiple-outcome LGC modeling by applying the LGC model depicted in Figure 9.4 to the cancer data.

The multiple-outcome LGC modeling is a natural extension of the single outcome LGC modeling. With multiple outcomes, there are some novel modeling features to consider, as listed in the following:

1. The LGC or trajectories for all outcomes are modeled simultaneously. As seen from Figure 9.4, we can now model both outcomes MOOD and SOCADJ together with ease and flexibility.

2. The covariances (correlations) among the *latent* intercepts and slopes of all outcomes can be modeled. As seen from Figure 9.4, we can include the covariances among the random intercepts and slopes between the two outcomes MOOD and SOCADJ, as depicted by the four thin bidirectional paths connecting the random intercepts and slopes between the two outcomes.

3. The residual covariances (correlations) among the observed outcomes can potentially be modeled. For example, we can include bidirectional paths among the error terms of $MOOD_1$, $MOOD_4$, $MOOD_8$ and $SOCADJ_1$, $SOCADJ_4$, $SOCADJ_8$ to represent error covariances (correlations) if desired (even though they are not currently being modeled in Figure 9.4).

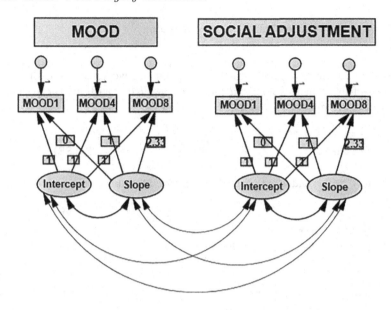

FIGURE 9.4
The LGC Model for Mood and Social Adjustment.

9.3.3.2 Initial Model Fitting of the Joint Model for MOOD and SOCADJ

The joint LGC model in Figure 9.4 can be implemented in *R lavaan* as follows:

```
# Joint LGC Model for Both MOOD and SOCADJ
mod4HK1 <- '
    # Intercept and Slope with MOOD
    iMOOD =~ 1*MOOD1 + 1*MOOD4 + 1*MOOD8
    sMOOD =~ 0*MOOD1 + 1*MOOD4 + 2.33*MOOD8
    # Intercept and Slope with SOCADJ
    iSOCADJ =~ 1*SOCADJ1 + 1*SOCADJ4 + 1*SOCADJ8
    sSOCADJ =~ 0*SOCADJ1 + 1*SOCADJ4 + 2.33*SOCADJ8
'
```

Note that we only need to code the *Intercept*s and *Slope*s associated with their observed measures. The correlations among them are by default included. With this joint LGC model, we can call *growth* to fit the model to the data with both outcomes as follows:

```
# Load the library lavaan
library(lavaan)
# Call growth to fit the model to the data
```

```
fitHK1 <- growth(mod4HK1, data = dCancer, estimator ="MLR",
                 missing="ml" )
```

For the purpose of model fitting diagnostics, we only print the model goodness-of-fit measures using *fit.measures = TRUE* without the parameter estimates using *estimates = FALSE* as follows:

```
# Print the model fitting
summary(fitHK1, fit.measures = TRUE, estimates = FALSE)
```

```
## lavaan 0.6-12 ended normally after 228 iterations
##
##   Estimator                                         ML
##   Optimization method                           NLMINB
##   Number of model parameters                        20
##
##                                                 Used       Total
##   Number of observations                         386         405
##   Number of missing patterns                      15
##
## Model Test User Model:
##                                             Standard      Robust
##   Test Statistic                              40.278      36.036
##   Degrees of freedom                               7           7
##   P-value (Chi-square)                         0.000       0.000
##   Scaling correction factor                                1.118
##     Yuan-Bentler correction (Mplus variant)
##
## Model Test Baseline Model:
##
##   Test statistic                             801.198     529.602
##   Degrees of freedom                              15          15
##   P-value                                      0.000       0.000
##   Scaling correction factor                                1.513
##
## User Model versus Baseline Model:
##
##   Comparative Fit Index (CFI)                  0.958       0.944
##   Tucker-Lewis Index (TLI)                     0.909       0.879
##
##   Robust Comparative Fit Index (CFI)                       0.958
##   Robust Tucker-Lewis Index (TLI)                          0.911
##
## Loglikelihood and Information Criteria:
##
##   Loglikelihood user model (H0)            -6865.430   -6865.430
```

```
## Scaling correction factor                                      1.687
##       for the MLR correction
## Loglikelihood unrestricted model (H1)    -6845.291    -6845.291
## Scaling correction factor                                      1.539
##       for the MLR correction
##
## Akaike (AIC)                            13770.860    13770.860
## Bayesian (BIC)                          13849.976    13849.976
## Sample-size adjusted Bayesian (BIC)     13786.519    13786.519
##
## Root Mean Square Error of Approximation:
##
## RMSEA                                       0.111        0.104
## 90 Percent confidence interval - lower      0.079        0.073
## 90 Percent confidence interval - upper      0.145        0.136
## P-value RMSEA <= 0.05                       0.001        0.003
##
## Robust RMSEA                                             0.110
## 90 Percent confidence interval - lower                   0.076
## 90 Percent confidence interval - upper                   0.146
##
## Standardized Root Mean Square Residual:
##
## SRMR                                        0.041        0.041
```

As seen from the model fitting, the joint LGC modeling does not yield better model fit than that of the single outcome LGC models. The Chi-square test is highly significant and furthermore the Root Mean Square Error of Approximation (RMSEA) is 0.104 and the p-value for testing RMSEA ≤ 0.05 (test of close fit) is 0.003, indicating an unaccepable model fit.

9.3.3.3 Model Fitting with Modification Indices

We can attempt to use the model modification indices to improve the model fit. We do not intent to print all the modification indices, but just for the larger *mi*s. This can be implemented as follows:

```
# Get the modification indices
MI.JM = modindices(fitHK1)
# Print the mi's with mi > 10
subset(MI.JM, mi > 10)
```

```
##       lhs op    rhs   mi   epc sepc.lv sepc.all
## 58 MOOD4 ~~ SOCADJ4 29.6 -9.64   -9.64   -0.361
##    sepc.nox
## 58   -0.361
```

We can see that a suggestion to add the correlation $MOOD_4 \sim \sim SOCADJ_4$ would decrease the model fit chi-square statistic by 29.565 approximately. Substantively, this means that residuals of MOOD and the SOCADJ in the fourth month could be correlated.

Accordingly, we can *modify* the above joint LGC model as follows:

```
# Joint LGC Model for Both MOOD and SOCADJ
mod4HK2 <- '
  # Intercept and Slope with MOOD
  iMOOD =~ 1*MOOD1 + 1*MOOD4 + 1*MOOD8
  sMOOD =~ 0*MOOD1 + 1*MOOD4 + 2.33*MOOD8
  # Intercept and Slope with SOCADJ
  iSOCADJ =~ 1*SOCADJ1 + 1*SOCADJ4 + 1*SOCADJ8
  sSOCADJ =~ 0*SOCADJ1 + 1*SOCADJ4 + 2.33*SOCADJ8
  # Add the correlation based on modification indices
  MOOD4 ~~ SOCADJ4
'
```

We can then call *growth* to fit the model to the data again as follows:

```
# Call growth to fit the model to the data
fitHK2 <- growth(mod4HK2, data = dCancer, estimator ="MLR",
                 missing="ml" )
```

```
# Print the model fitting
summary(fitHK2, fit.measures = TRUE)
```

```
## lavaan 0.6-12 ended normally after 229 iterations
##
##   Estimator                                    ML
##   Optimization method                      NLMINB
##   Number of model parameters                   21
##
##                                      Used       Total
##   Number of observations              386         405
##   Number of missing patterns           15
##
## Model Test User Model:
##                                   Standard      Robust
##   Test Statistic                     7.148       6.590
##   Degrees of freedom                     6           6
##   P-value (Chi-square)               0.307       0.360
##   Scaling correction factor                      1.085
##     Yuan-Bentler correction (Mplus variant)
##
## Model Test Baseline Model:
```

```
##
##    Test statistic                             801.198    529.602
##    Degrees of freedom                              15         15
##    P-value                                      0.000      0.000
##    Scaling correction factor                               1.513
##
## User Model versus Baseline Model:
##
##    Comparative Fit Index (CFI)                  0.999      0.999
##    Tucker-Lewis Index (TLI)                     0.996      0.997
##
##    Robust Comparative Fit Index (CFI)                      0.999
##    Robust Tucker-Lewis Index (TLI)                         0.998
##
## Loglikelihood and Information Criteria:
##
##    Loglikelihood user model (H0)            -6848.865  -6848.865
##    Scaling correction factor                               1.669
##        for the MLR correction
##    Loglikelihood unrestricted model (H1)    -6845.291  -6845.291
##    Scaling correction factor                               1.539
##        for the MLR correction
##
##    Akaike (AIC)                            13739.730  13739.730
##    Bayesian (BIC)                          13822.802  13822.802
##    Sample-size adjusted Bayesian (BIC)     13756.172  13756.172
##
## Root Mean Square Error of Approximation:
##
##    RMSEA                                        0.022      0.016
##    90 Percent confidence interval - lower       0.000      0.000
##    90 Percent confidence interval - upper       0.072      0.068
##    P-value RMSEA <= 0.05                        0.769      0.820
##
##    Robust RMSEA                                            0.017
##    90 Percent confidence interval - lower                  0.000
##    90 Percent confidence interval - upper                  0.072
##
## Standardized Root Mean Square Residual:
##
##    SRMR                                         0.023      0.023
##
## Parameter Estimates:
##
##    Standard errors                           Sandwich
##    Information bread                         Observed
```

```
##    Observed information based on              Hessian
##
## Latent Variables:
##                    Estimate  Std.Err  z-value  P(>|z|)
##    iMOOD =~
##      MOOD1          1.000
##      MOOD4          1.000
##      MOOD8          1.000
##    sMOOD =~
##      MOOD1          0.000
##      MOOD4          1.000
##      MOOD8          2.330
##    iSOCADJ =~
##      SOCADJ1        1.000
##      SOCADJ4        1.000
##      SOCADJ8        1.000
##    sSOCADJ =~
##      SOCADJ1        0.000
##      SOCADJ4        1.000
##      SOCADJ8        2.330
##
## Covariances:
##                    Estimate  Std.Err  z-value  P(>|z|)
##   .MOOD4 ~~
##     .SOCADJ4       -10.371    2.495   -4.156    0.000
##    iMOOD ~~
##      sMOOD          -1.958    1.642   -1.193    0.233
##      iSOCADJ       -22.352    4.002   -5.585    0.000
##      sSOCADJ         1.999    1.584    1.262    0.207
##    sMOOD ~~
##      iSOCADJ         1.917    1.737    1.103    0.270
##      sSOCADJ        -1.426    0.750   -1.902    0.057
##    iSOCADJ ~~
##      sSOCADJ       -15.482    4.530   -3.418    0.001
##
## Intercepts:
##                    Estimate  Std.Err  z-value  P(>|z|)
##    .MOOD1           0.000
##    .MOOD4           0.000
##    .MOOD8           0.000
##    .SOCADJ1         0.000
##    .SOCADJ4         0.000
##    .SOCADJ8         0.000
##     iMOOD          21.399     0.314   68.164    0.000
##     sMOOD          -0.560     0.133   -4.223    0.000
##     iSOCADJ       100.922     0.503  200.774    0.000
```

```
##      sSOCADJ          -0.269    0.207   -1.296      0.195
##
## Variances:
##                       Estimate  Std.Err  z-value  P(>|z|)
##      .MOOD1            14.613    3.271    4.468     0.000
##      .MOOD4            18.948    2.535    7.474     0.000
##      .MOOD8             6.804    3.716    1.831     0.067
##      .SOCADJ1          11.862    7.545    1.572     0.116
##      .SOCADJ4          40.825    7.226    5.650     0.000
##      .SOCADJ8           2.890   10.956    0.264     0.792
##      iMOOD             25.436    3.646    6.975     0.000
##      sMOOD              2.234    1.230    1.817     0.069
##      iSOCADJ           83.591   12.869    6.496     0.000
##      sSOCADJ           12.081    3.401    3.552     0.000
```

We can see now that the estimated residual covariance between $MOOD_4$ and $SOCADJ_4$ is -10.371, which is statistically significant with p-value < 0.01.

We can see the improvement of these two model fittings between *fitHK2* and *fitHK1* using the likelihood ratio test:

```
# Perform the likelihood ratio test
anova(fitHK1, fitHK2)
```

```
## Scaled Chi-Squared Difference Test
##    (method = "satorra.bentler.2001")
##
## lavaan NOTE:
##     The "Chisq" column contains standard test statistics,
##     not the robust test that should be reported per model.
##     A robust difference test is a function of two standard
##     (not robust) statistics.
##
##            Df    AIC    BIC  Chisq  Chisq diff  Df diff
## fitHK2     6  13740  13823   7.15
## fitHK1     7  13771  13850  40.28       25.2         1
##            Pr(>Chisq)
## fitHK2
## fitHK1  0.00000052 ***
## ---
## Signif. codes:
## 0 '***' 0.001 '**' 0.01 '*' 0.05 '.' 0.1 ' ' 1
```

As seen from the *anova*, the modified model *fitHK2* has smaller *AIC* and *Chisq* statistics. The likelihood ratio test yielded an extremely low p-value, which indicates that the modified model is significantly better fitted to the data than the original model. The chi-square difference is 25.179, which is not too different from the modification index of 29.565 predicted by the modification index in the fitting the original model.

We can further examine other fit measures. There are obvious improvements for the Comparative Fit Index (CFI) (0.999) and Tucker-Lewis Index (TLI) (0.998), Root Mean Square Error of Approximation (RMSEA) (0.016), and Standardized Root Mean Square Residual (SRMR) (0.023). Therefore, the modified joint LGC model fits quite well.

9.3.3.4 Final Model Based on Model Parsimony

In the modified model, there are several estimates that are clearly not significant statistically. They are:

1. Three covariances. One of them is within-outcome, i.e., iMOOD ∼ ∼ sMOOD and the other two are between-outcome, i.e., iMOOD ∼ ∼ sSOCADJ and sMOOD ∼ ∼ iSOCADJ.

2. One intercept for sSOCADJ.

3. One error variance for SOCADJ8.

Note that the covariance estimate between sMOOD and sSOCADJ (sMOOD ∼ ∼ sSOCADJ) has an associated *p*-value of 0.057, which is judged as marginally significant and therefore is not included in the first (covariance) list above. It is also interesting to note that iMOOD ∼ ∼ sMOOD in the covariance list was also non-significant in the single outcome model for MOOD and the two estimates in the second and third lists were also non-significant in the single outcome model for SOCADJ.

For the sake of model parsimony, one can fix these parameters to zero and refit the joint LGC model as follows:

```
# Joint LGC Model for Both MOOD and SOCADJ
mod4HK3 <- '
  # Intercept and Slope with MOOD
  iMOOD =~ 1*MOOD1 + 1*MOOD4 + 1*MOOD8
  sMOOD =~ 0*MOOD1 + 1*MOOD4 + 2.33*MOOD8
  # Intercept and Slope with SOCADJ
  iSOCADJ =~ 1*SOCADJ1 + 1*SOCADJ4 + 1*SOCADJ8
  sSOCADJ =~ 0*SOCADJ1 + 1*SOCADJ4 + 2.33*SOCADJ8
  # Add the error covariance based on modification indices
  MOOD4 ~~ SOCADJ4
  # Constrain the insignificant paths
  # 1. covariances
  iMOOD ~~ 0*sMOOD
  iMOOD ~~ 0*sSOCADJ
  sMOOD ~~ 0*iSOCADJ
  # 2. Intercept
  sSOCADJ ~ 0
```

```
  # 3. Error variance
  SOCADJ8 ~~ 0*SOCADJ8

# Call growth to fit the model to the data
fitHK3 <- growth(mod4HK3, data = dCancer, estimator ="MLR",
              missing="ml" )


# Print the model fitting
summary(fitHK3, fit.measures = TRUE)


## lavaan 0.6-12 ended normally after 181 iterations
##
##   Estimator                                       ML
##   Optimization method                        NLMINB
##   Number of model parameters                      16
##
##                                            Used       Total
##   Number of observations                     386         405
##   Number of missing patterns                  15
##
## Model Test User Model:
##                                        Standard      Robust
##   Test Statistic                         13.756      10.488
##   Degrees of freedom                         11          11
##   P-value (Chi-square)                    0.247       0.487
##   Scaling correction factor                           1.312
##     Yuan-Bentler correction (Mplus variant)
##
## Model Test Baseline Model:
##
##   Test statistic                        801.198     529.602
##   Degrees of freedom                         15          15
##   P-value                                 0.000       0.000
##   Scaling correction factor                           1.513
##
## User Model versus Baseline Model:
##
##   Comparative Fit Index (CFI)             0.996       1.000
##   Tucker-Lewis Index (TLI)                0.995       1.001
##
##   Robust Comparative Fit Index (CFI)                  1.000
##   Robust Tucker-Lewis Index (TLI)                     1.001
##
## Loglikelihood and Information Criteria:
##
```

```
## Loglikelihood user model (H0)            -6852.169   -6852.169
## Scaling correction factor                              1.695
##     for the MLR correction
## Loglikelihood unrestricted model (H1)    -6845.291   -6845.291
## Scaling correction factor                              1.539
##     for the MLR correction
##
## Akaike (AIC)                            13736.337   13736.337
## Bayesian (BIC)                          13799.631   13799.631
## Sample-size adjusted Bayesian (BIC)     13748.865   13748.865
##
## Root Mean Square Error of Approximation:
##
## RMSEA                                        0.025       0.000
## 90 Percent confidence interval - lower       0.000       0.000
## 90 Percent confidence interval - upper       0.062       0.047
## P-value RMSEA <= 0.05                        0.840       0.966
##
## Robust RMSEA                                             0.000
## 90 Percent confidence interval - lower                   0.000
## 90 Percent confidence interval - upper                   0.059
##
## Standardized Root Mean Square Residual:
##
## SRMR                                         0.035       0.035
##
## Parameter Estimates:
##
## Standard errors                           Sandwich
## Information bread                         Observed
## Observed information based on              Hessian
##
## Latent Variables:
##                     Estimate  Std.Err  z-value  P(>|z|)
## iMOOD =~
##   MOOD1               1.000
##   MOOD4               1.000
##   MOOD8               1.000
## sMOOD =~
##   MOOD1               0.000
##   MOOD4               1.000
##   MOOD8               2.330
## iSOCADJ =~
##   SOCADJ1             1.000
##   SOCADJ4             1.000
##   SOCADJ8             1.000
```

```
##   sSOCADJ =~
##      SOCADJ1           0.000
##      SOCADJ4           1.000
##      SOCADJ8           2.330
##
## Covariances:
##                     Estimate  Std.Err  z-value  P(>|z|)
##   .MOOD4 ~~
##     .SOCADJ4          -10.634    2.447   -4.346    0.000
##   iMOOD ~~
##      sMOOD             0.000
##      sSOCADJ           0.000
##   sMOOD ~~
##      iSOCADJ           0.000
##   iMOOD ~~
##      iSOCADJ          -18.775    2.912   -6.447    0.000
##   sMOOD ~~
##      sSOCADJ           -0.853    0.499   -1.710    0.087
##   iSOCADJ ~~
##      sSOCADJ          -14.849    4.444   -3.341    0.001
##
## Intercepts:
##                     Estimate  Std.Err  z-value  P(>|z|)
##      sSOCADJ           0.000
##     .MOOD1             0.000
##     .MOOD4             0.000
##     .MOOD8             0.000
##     .SOCADJ1           0.000
##     .SOCADJ4           0.000
##     .SOCADJ8           0.000
##      iMOOD            21.405    0.313   68.351    0.000
##      sMOOD            -0.571    0.130   -4.398    0.000
##      iSOCADJ         100.592    0.433  232.135    0.000
##
## Variances:
##                     Estimate  Std.Err  z-value  P(>|z|)
##     .SOCADJ8           0.000
##     .MOOD1            17.371    2.273    7.642    0.000
##     .MOOD4            18.635    2.514    7.411    0.000
##     .MOOD8             9.009    3.606    2.499    0.012
##     .SOCADJ1          11.270    6.001    1.878    0.060
##     .SOCADJ4          41.626    5.327    7.814    0.000
##      iMOOD            22.007    2.719    8.093    0.000
##      sMOOD             1.126    0.775    1.454    0.146
##      iSOCADJ          81.989   13.035    6.290    0.000
##      sSOCADJ          12.776    1.874    6.818    0.000
```

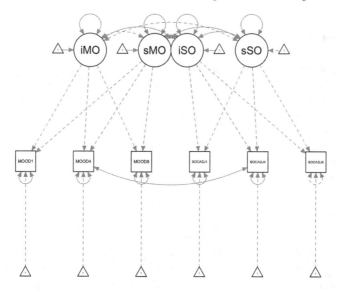

FIGURE 9.5
The Final Joint Latent Growth-Curve Model.

We can see that most parameter estimates are now significant at the 0.05 α-level. The rest are at least marginally significant by our judgment. In addition, the overall model fit measures appear to near-perfect, as indicated by the Comparative Fit Index (CFI) = 1.00, the Tucker-Lewis Index (TLI) = 1.00, the Root Mean Square Error of Approximation (RMSEA) = 0.00 and the p-value for testing the RMSEA $\leq$ 0.05 (test of close-fit) is 0.966, and the Standardized Root Mean Square Residual (SRMR) = 0.035. One might argue that this model might have been over-fitted. As our primary purpose here is just to demonstrate how LGC models can be fitted and improved, we leave this issue of "overfitting" as an open question to the readers.

Assume that we are now satisfied with this final joint LGC model. We can graphically depict the model in Figure 9.5 using R package *semPlot* as follows:

```
# Load the library
library(semPlot)
# Call "semPlots" to plot the estimated model
semPaths(fitHK3)
```

Based on this final joint LGC model, we can make some observations and conclusions:

- Estimated Covariances

  - Significant $MOOD_4 \sim \sim SOCADJ_4 = -10.634$ (p-value < 0.01). This indicates that the residuals of MOOD and SOCADJ in the 4-th month correlates significantly. Hence, the correlation between MOOD and SOCADJ in the 4-th month is not completely accounted for by the correlations between the random slopes and intercepts in the model.

- Significant iMOOD $\sim$ $\sim$ iSOCADJ $= -18.775$ (p-value $= 0.000$). This means that the *Intercept* of MOOD and *Intercept* of SOCADJ are negatively correlated, indicating bad initial MOOD (higher mood values represent worse mood) in the 1-st month was associated with lower level of initial social adjustment in the 1-st month. Alternatively, the better the MOOD at 1-month, the better the SOCADJ at the beginning.

- Significant iSOCADJ $\sim$ $\sim$ sSOCADJ $= -14.849$ (p-value $= 0.001$). This means that the *Intercept* of SOCADJ and *Slope* of SOCADJ are negatively correlated, indicating the better (or higher) the SOCADJ at 1-month, the less the improvement. Alternatively, the worse (or lower) the SOCADJ at 1 month, the more improvement of their SOCADJ during the 8-month post-surgery period.

- Estimated Intercepts or Means

  - Significant iMOOD and iSOCADJ. This means that the means of MOOD and SOCADJ are significantly different from zero at the beginning of 1-month.

  - Significant sMOOD $= -0.571$ (p-value $= 0.000$). This significant negative mean slope for MOOD means that on average the MOOD of women post-surgery is improving during the 8-month post-surgery period.

  - Fixed zero mean for sSOCADJ. A zero mean slope for SOCADJ means that on average the social adjustment of women are *not* improving during the 8-month post-surgery period. Note that this fixed zero was suggested by a model modification index.

- Estimated *Variances*

  - Significant error variances for $MOOD_1$, $MOOD_4$, $MOOD_8$, $SOCADJ_1$, $SOCADJ_4$. This indicates that there are significant *residual* variations among the women on these measurements.

  - Significant variances for iMOOD, iSOCADJ, sSOCADJ. This means that there are significant variations of initial MOOD and SOCADJ between women (i.e., as indicated by the variances of intercepts iMOOD and iSOCADJ). Also, the rate of improvement for SOCADJ between women are variable (i.e., as indicated by the variance of slope sSOCADJ).

  - Marginally significant variance for sMOOD. This means that rate of improvement for MOOD might not vary among women. More evidence is needed to substantiate such an inter-individual variation in the improvement of mood.

Combining the interpretations of the mean and variance estimates of sMOOD and sSOCADJ in this model, it suggests that while the mood of women post surgery on average improves over the 8-month period, the individual improvements does not seem to vary much. In contrast, the result suggests that while the social adjustment of women post surgery on average does not improve over the 8-month period, the individual improvements can vary quite a lot.

9.3.4   Joint LGC Model with Covariates

In the previous analysis with two outcomes, the variance estimates for the random intercept and slope for SOCADJ appear to be quite large as compared to their mean values. This shows that the initial social adjustment and the rate of improvement could be quite different between women. To explain such variations, one might attribute to different ages and treatment types among women. To implement this idea into the latent growth-curve model, one can include age groups and treatment types as covariates that predict the random intercepts and slopes for MOOD and SOCADJ. You can use the following code to specify a basic LGC model for the two outcomes with covariates:

```
# Joint LGC Model with Covariates
mod4HK4a <- '
  # Intercept and Slope with MOOD
  iMOOD =~ 1*MOOD1 + 1*MOOD4 + 1*MOOD8
  sMOOD =~ 0*MOOD1 + 1*MOOD4 + 2.33*MOOD8
  # Intercept and Slope with SOCADJ
  iSOCADJ =~ 1*SOCADJ1 + 1*SOCADJ4 + 1*SOCADJ8
  sSOCADJ =~ 0*SOCADJ1 + 1*SOCADJ4 + 2.33*SOCADJ8
  # Add regression paths to covariates
  iMOOD ~ AgeGrp + SurgTx
  sMOOD ~ AgeGrp + SurgTx
  iSOCADJ ~ AgeGrp + SurgTx
  sSOCADJ ~ AgeGrp + SurgTx
'

# Call growth to fit the model to the data
fitHK4a <- growth(mod4HK4a, data = dCancer, estimator ="MLR",
                  missing="ml" )
# Print the model fitting
summary(fitHK4a, fit.measures = TRUE, estimates = FALSE)


## lavaan 0.6-12 ended normally after 246 iterations
##
##   Estimator                                         ML
##   Optimization method                           NLMINB
##   Number of model parameters                        28
```

```
##
##    Number of observations                    405
##    Number of missing patterns                 16
##
## Model Test User Model:
##                                          Standard        Robust
##    Test Statistic                          43.106        41.402
##    Degrees of freedom                          11            11
##    P-value (Chi-square)                     0.000         0.000
##    Scaling correction factor                              1.041
##      Yuan-Bentler correction (Mplus variant)
##
## Model Test Baseline Model:
##
##    Test statistic                         835.261       665.566
##    Degrees of freedom                          27            27
##    P-value                                  0.000         0.000
##    Scaling correction factor                              1.255
##
## User Model versus Baseline Model:
##
##    Comparative Fit Index (CFI)              0.960         0.952
##    Tucker-Lewis Index (TLI)                 0.903         0.883
##
##    Robust Comparative Fit Index (CFI)                     0.961
##    Robust Tucker-Lewis Index (TLI)                        0.903
##
## Loglikelihood and Information Criteria:
##
##    Loglikelihood user model (H0)        -6849.812     -6849.812
##    Scaling correction factor                              1.475
##        for the MLR correction
##    Loglikelihood unrestricted model (H1)       NA            NA
##    Scaling correction factor                              1.352
##        for the MLR correction
##
##    Akaike (AIC)                         13755.624     13755.624
##    Bayesian (BIC)                       13867.733     13867.733
##    Sample-size adjusted Bayesian (BIC)  13778.886     13778.886
##
## Root Mean Square Error of Approximation:
##
##    RMSEA                                    0.085         0.083
##    90 Percent confidence interval - lower   0.059         0.057
##    90 Percent confidence interval - upper   0.112         0.110
##    P-value RMSEA <= 0.05                    0.015         0.019
```

```
##
##    Robust RMSEA                                      0.084
##    90 Percent confidence interval - lower            0.058
##    90 Percent confidence interval - upper            0.112
##
## Standardized Root Mean Square Residual:
##
##    SRMR                                    0.033      0.033
```

Note that we included the covariates in the basic model without adding those additional correlated error that has been suggested by previous modeling (e.g., covariance between $MOOD_4$ and $SOCADJ_4$). The main reason is that SEM modeling is a multivariate analysis that could be very sensitive to adding or removing variables from modeling. It, therefore, would be safe for us to start anew without adding extra modeling parameters when we work on a new set of variables in SEM modeling. Modifications can still be done after the important elements of the new model are first settled (such as the inclusion of covariates in the current analysis).

Back to the current fit of the basic model, the fit measures are somewhat unacceptable (e.g., TLI = 0.883, RMSEA = 0.083). We then examined the modification indices and found that adding the correlated error between $MOOD_4$ and $SOCADJ_4$ might improve the model fit. Notice that this is the same correlated error that has also been suggested for the modeling without covariates in previous sections. So now we are confident to revise the model by adding this correlated error, as implemented by the following code:

```
# Joint LGC Model with Covariates
mod4HK4b <- '
  # Intercept and Slope with MOOD
  iMOOD =~ 1*MOOD1 + 1*MOOD4 + 1*MOOD8
  sMOOD =~ 0*MOOD1 + 1*MOOD4 + 2.33*MOOD8
  # Intercept and Slope with SOCADJ
  iSOCADJ =~ 1*SOCADJ1 + 1*SOCADJ4 + 1*SOCADJ8
  sSOCADJ =~ 0*SOCADJ1 + 1*SOCADJ4 + 2.33*SOCADJ8
  # Add the error covariance based on modification indices
  MOOD4 ~~ SOCADJ4
  # Add regression paths to covariates
  iMOOD ~ AgeGrp + SurgTx
  sMOOD ~ AgeGrp + SurgTx
  iSOCADJ ~ AgeGrp + SurgTx
  sSOCADJ ~ AgeGrp + SurgTx
'

# Call growth to fit the model to the data
fitHK4b <- growth(mod4HK4b, data = dCancer, estimator ="MLR",
                  missing="ml" )
```

```
# Print the model fitting
summary(fitHK4b,fit.measures = TRUE)
```

```
## lavaan 0.6-12 ended normally after 258 iterations
##
##   Estimator                                         ML
##   Optimization method                           NLMINB
##   Number of model parameters                        29
##
##   Number of observations                           405
##   Number of missing patterns                        16
##
## Model Test User Model:
##                                             Standard      Robust
##   Test Statistic                               9.877       9.698
##   Degrees of freedom                              10          10
##   P-value (Chi-square)                         0.451       0.467
##   Scaling correction factor                                1.018
##     Yuan-Bentler correction (Mplus variant)
##
## Model Test Baseline Model:
##
##   Test statistic                             835.261     665.566
##   Degrees of freedom                              27          27
##   P-value                                      0.000       0.000
##   Scaling correction factor                                1.255
##
## User Model versus Baseline Model:
##
##   Comparative Fit Index (CFI)                  1.000       1.000
##   Tucker-Lewis Index (TLI)                     1.000       1.001
##
##   Robust Comparative Fit Index (CFI)                       1.000
##   Robust Tucker-Lewis Index (TLI)                          1.001
##
## Loglikelihood and Information Criteria:
##
##   Loglikelihood user model (H0)            -6833.198   -6833.198
##   Scaling correction factor                                1.468
##       for the MLR correction
##   Loglikelihood unrestricted model (H1)           NA          NA
##   Scaling correction factor                                1.352
##       for the MLR correction
##
##   Akaike (AIC)                             13724.396   13724.396
##   Bayesian (BIC)                           13840.508   13840.508
```

```
## Sample-size adjusted Bayesian (BIC)        13748.488   13748.488
##
## Root Mean Square Error of Approximation:
##
## RMSEA                                            0.000       0.000
## 90 Percent confidence interval - lower           0.000       0.000
## 90 Percent confidence interval - upper           0.053       0.052
## P-value RMSEA <= 0.05                            0.930       0.936
##
## Robust RMSEA                                                 0.000
## 90 Percent confidence interval - lower                       0.000
## 90 Percent confidence interval - upper                       0.053
##
## Standardized Root Mean Square Residual:
##
## SRMR                                             0.019       0.019
##
## Parameter Estimates:
##
## Standard errors                            Sandwich
## Information bread                          Observed
## Observed information based on              Hessian
##
## Latent Variables:
##                   Estimate  Std.Err  z-value  P(>|z|)
## iMOOD =~
##   MOOD1             1.000
##   MOOD4             1.000
##   MOOD8             1.000
## sMOOD =~
##   MOOD1             0.000
##   MOOD4             1.000
##   MOOD8             2.330
## iSOCADJ =~
##   SOCADJ1           1.000
##   SOCADJ4           1.000
##   SOCADJ8           1.000
## sSOCADJ =~
##   SOCADJ1           0.000
##   SOCADJ4           1.000
##   SOCADJ8           2.330
##
## Regressions:
##                   Estimate  Std.Err  z-value  P(>|z|)
## iMOOD ~
##   AgeGrp            1.625    0.620    2.623    0.009
```

```
##     SurgTx          -0.061    0.751   -0.082     0.935
## sMOOD ~
##     AgeGrp           0.201    0.259    0.773     0.439
##     SurgTx          -0.340    0.337   -1.007     0.314
## iSOCADJ ~
##     AgeGrp           0.969    0.991    0.977     0.328
##     SurgTx          -3.312    1.055   -3.139     0.002
## sSOCADJ ~
##     AgeGrp          -0.381    0.406   -0.939     0.348
##     SurgTx           1.261    0.487    2.591     0.010
##
## Covariances:
##                   Estimate  Std.Err  z-value   P(>|z|)
## .MOOD4 ~~
##     .SOCADJ4       -10.392    2.490   -4.173     0.000
## .iMOOD ~~
##     .sMOOD          -1.954    1.606   -1.217     0.224
##     .iSOCADJ       -22.766    3.890   -5.852     0.000
##     .sSOCADJ         2.173    1.533    1.417     0.156
## .sMOOD ~~
##     .iSOCADJ         1.626    1.748    0.930     0.352
##     .sSOCADJ        -1.324    0.744   -1.779     0.075
## .iSOCADJ ~~
##     .sSOCADJ       -14.826    4.445   -3.335     0.001
##
## Intercepts:
##                   Estimate  Std.Err  z-value   P(>|z|)
##     .MOOD1           0.000
##     .MOOD4           0.000
##     .MOOD8           0.000
##     .SOCADJ1         0.000
##     .SOCADJ4         0.000
##     .SOCADJ8         0.000
##     .iMOOD          19.031    1.668   11.407     0.000
##     .sMOOD          -0.258    0.780   -0.331     0.741
##     .iSOCADJ       105.393    2.285   46.121     0.000
##     .sSOCADJ        -1.953    1.134   -1.722     0.085
##
## Variances:
##                   Estimate  Std.Err  z-value   P(>|z|)
##     .MOOD1          14.757    3.220    4.583     0.000
##     .MOOD4          18.862    2.532    7.450     0.000
##     .MOOD8           7.158    3.693    1.938     0.053
##     .SOCADJ1        11.458    7.500    1.528     0.127
##     .SOCADJ4        40.854    7.113    5.744     0.000
##     .SOCADJ8         3.168   10.844    0.292     0.770
```

##	.iMOOD	24.708	3.540	6.979	0.000
##	.sMOOD	2.089	1.205	1.733	0.083
##	.iSOCADJ	81.976	12.515	6.550	0.000
##	.sSOCADJ	11.781	3.355	3.511	0.000

The fit of the revised model looks perfect: CFI = 1, TLI = 1, RMSEA = 0, SRMR = 0.019, and the chi-square for model fit = 9.698 (df = 10, p = 0.467). We, therefore, can proceed to interpret the estimates of this model. Or, if model parsimony is deemed important, we can fix some clearly non-significant estimates to zeros and then fit such a parsimonious model for interpretations, much like what we have done for previous analyses in this chapter. A parsimonious model with covariates is thus specified by the following code:

```
# Joint LGC Model with Covariates
mod4HK4 <- '
  # Intercept and Slope with MOOD
  iMOOD =~ 1*MOOD1 + 1*MOOD4 + 1*MOOD8
  sMOOD =~ 0*MOOD1 + 1*MOOD4 + 2.33*MOOD8
  # Intercept and Slope with SOCADJ
  iSOCADJ =~ 1*SOCADJ1 + 1*SOCADJ4 + 1*SOCADJ8
  sSOCADJ =~ 0*SOCADJ1 + 1*SOCADJ4 + 2.33*SOCADJ8
  # Add the error covariance based on modification indices
  MOOD4 ~~ SOCADJ4
  # Constrain the insignificant paths
  # 1. covariances
  iMOOD ~~ 0*sMOOD
  iMOOD ~~ 0*sSOCADJ
  sMOOD ~~ 0*iSOCADJ
  # 2. Intercept
  sMOOD ~ 0
  # 3. Error variance
  SOCADJ8 ~~ 0*SOCADJ8
  # Add regression paths to covariates
  iMOOD ~ AgeGrp + SurgTx
  sMOOD ~ AgeGrp + SurgTx
  iSOCADJ ~ AgeGrp + SurgTx
  sSOCADJ ~ AgeGrp + SurgTx
'
# Call growth to fit the model to the data
fitHK4 <- growth(mod4HK4, data = dCancer,estimator ="MLR",
                 missing="ml" )
# Print the model fitting
summary(fitHK4,fit.measures = TRUE)

## lavaan 0.6-12 ended normally after 249 iterations
##
```

```
## Estimator                                    ML
## Optimization method                      NLMINB
## Number of model parameters                    24
##
## Number of observations                       405
## Number of missing patterns                    16
##
## Model Test User Model:
##                                       Standard      Robust
## Test Statistic                          15.147      12.532
## Degrees of freedom                          15          15
## P-value (Chi-square)                     0.441       0.638
## Scaling correction factor                            1.209
##    Yuan-Bentler correction (Mplus variant)
##
## Model Test Baseline Model:
##
## Test statistic                         835.261     665.566
## Degrees of freedom                          27          27
## P-value                                  0.000       0.000
## Scaling correction factor                            1.255
##
## User Model versus Baseline Model:
##
## Comparative Fit Index (CFI)              1.000       1.000
## Tucker-Lewis Index (TLI)                 1.000       1.007
##
## Robust Comparative Fit Index (CFI)                   1.000
## Robust Tucker-Lewis Index (TLI)                      1.007
##
## Loglikelihood and Information Criteria:
##
## Loglikelihood user model (H0)        -6835.833   -6835.833
## Scaling correction factor                            1.442
##      for the MLR correction
## Loglikelihood unrestricted model (H1)       NA          NA
## Scaling correction factor                            1.352
##      for the MLR correction
##
## Akaike (AIC)                         13719.666   13719.666
## Bayesian (BIC)                       13815.759   13815.759
## Sample-size adjusted Bayesian (BIC)  13739.604   13739.604
##
## Root Mean Square Error of Approximation:
##
## RMSEA                                    0.005       0.000
## 90 Percent confidence interval - lower   0.000       0.000
```

```
## 90 Percent confidence interval - upper          0.047      0.036
## P-value RMSEA <= 0.05                            0.965      0.994
##
## Robust RMSEA                                                0.000
## 90 Percent confidence interval - lower                      0.000
## 90 Percent confidence interval - upper                      0.043
##
## Standardized Root Mean Square Residual:
##
## SRMR                                             0.027      0.027
##
## Parameter Estimates:
##
## Standard errors                          Sandwich
## Information bread                        Observed
## Observed information based on            Hessian
##
## Latent Variables:
##                Estimate   Std.Err   z-value   P(>|z|)
## iMOOD =~
##   MOOD1         1.000
##   MOOD4         1.000
##   MOOD8         1.000
## sMOOD =~
##   MOOD1         0.000
##   MOOD4         1.000
##   MOOD8         2.330
## iSOCADJ =~
##   SOCADJ1       1.000
##   SOCADJ4       1.000
##   SOCADJ8       1.000
## sSOCADJ =~
##   SOCADJ1       0.000
##   SOCADJ4       1.000
##   SOCADJ8       2.330
##
## Regressions:
##                Estimate   Std.Err   z-value   P(>|z|)
## iMOOD ~
##   AgeGrp        1.696      0.582     2.915     0.004
##   SurgTx        0.073      0.683     0.107     0.915
## sMOOD ~
##   AgeGrp        0.152      0.207     0.736     0.462
##   SurgTx       -0.434      0.173    -2.502     0.012
## iSOCADJ ~
##   AgeGrp        0.960      0.987     0.972     0.331
```

```
##      SurgTx           -3.337   1.057   -3.158    0.002
## sSOCADJ ~
##      AgeGrp           -0.390   0.404   -0.966    0.334
##      SurgTx            1.262   0.484    2.610    0.009
##
## Covariances:
##                     Estimate  Std.Err  z-value  P(>|z|)
## .MOOD4 ~~
##    .SOCADJ4         -10.734    2.445   -4.391    0.000
## .iMOOD ~~
##    .sMOOD             0.000
##    .sSOCADJ           0.000
## .sMOOD ~~
##    .iSOCADJ           0.000
## .iMOOD ~~
##    .iSOCADJ         -19.282    2.896   -6.658    0.000
## .sMOOD ~~
##    .sSOCADJ          -0.774    0.490   -1.579    0.114
## .iSOCADJ ~~
##    .sSOCADJ         -14.192    4.361   -3.254    0.001
##
## Intercepts:
##                     Estimate  Std.Err  z-value  P(>|z|)
##    .sMOOD             0.000
##    .MOOD1             0.000
##    .MOOD4             0.000
##    .MOOD8             0.000
##    .SOCADJ1           0.000
##    .SOCADJ4           0.000
##    .SOCADJ8           0.000
##    .iMOOD            18.674    1.408   13.258    0.000
##    .iSOCADJ         105.461    2.272   46.424    0.000
##    .sSOCADJ          -1.944    1.123   -1.731    0.083
##
## Variances:
##                     Estimate  Std.Err  z-value  P(>|z|)
##    .SOCADJ8           0.000
##    .MOOD1            17.629    2.268    7.771    0.000
##    .MOOD4            18.583    2.524    7.363    0.000
##    .MOOD8             9.402    3.608    2.606    0.009
##    .SOCADJ1          10.060    6.027    1.669    0.095
##    .SOCADJ4          41.946    5.290    7.929    0.000
##    .iMOOD           21.284    2.599    8.190    0.000
##    .sMOOD             0.945    0.763    1.238    0.216
##    .iSOCADJ         80.703   12.667    6.371    0.000
##    .sSOCADJ         12.557    1.877    6.689    0.000
```

We can see that all overall model fit measures are nearly perfect: CFI and TLI are both about 1.00, RMSEA and SRMR are perfect, and the chi-square for model fit = 12.532 (df = 15, p-value = 0.6389). The test of close fit of RMSEA (smaller than 0.05) has a p-value of 0.994. This is a near-perfect model and so we proceed to interpret the estimation results.

Observing the output corresponding to the *Regressions* component, we can make some conclusions as follows:

- For MOOD:

  - *AgeGrp* has a significant effect on the initial MOOD (i.e, 1.696 with p-value of 0.004). This means that the initial MOOD at the first month was significantly lower (by 1.696 units) for women over the age of 50 (who had better initial mood) than that for younger women, keeping other factors constant.

  - *AgeGrp* does not have a significant effect on the *slope* for MOOD. This means that the rate of changes in MOOD is not different for the two age groups.

  - *SurgTX* does not have a significant effect on the *Intercept* for MOOD. This means that the initial MOOD are not different for the two types of surgery.

  - *SurgTX* has a significant effect on the *Slope* for MOOD. Women with mastectomy appeared to improve their mood faster than those with lumpectomy (by 0.434 unit, p-value = 0.012) for the 8-month period, keeping other factors constant.

- For SOCADJ:

  - *AgeGrp* does not have a significant effect on either the *Intercept* and *Slope* for SOCADJ. This means that the two types of surgery appeared to have no effect on either the initial social adjustment or the rate of change across the 8-month period.

  - *SurgTX* has significant effects on both the *intercept* and *slope* for SOCADJ. Women with mastectomy had significantly lower SOCADJ at the beginning than those women with lumpectomy (by −3.337 units, p-value = 0.002). However, the rate of improvement in SOCADJ for women with mastectomy was faster (by 1.262 units, p-value = 0.009) for the 8-month period, keeping all other factors constant.

We now investigate whether the large individual differences in random slopes and intercepts could be accounted for by the covariates *AgeGrp* and *SurgTx* in the model. The estimated residual variances for the *latent* factors of iMOOD (21.284), iSOCADJ (80.703) and sSOCADJ (12.557) in the current model are not much different from the estimated variances obtained from the model in Section 9.3.3.4, which is without the covariates *AgeGrp* and *SurgTx*.

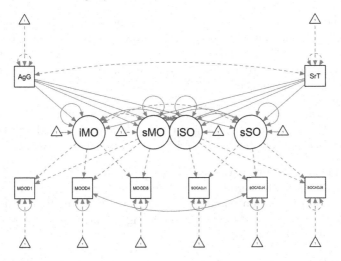

FIGURE 9.6
The Final Joint Latent Growth-Curve Model with Covariates of Age and Surgery Treatment.

This suggests that the age and the surgical type can only explain very small portions of variations of the random intercepts and slopes for MOOD and SOCADJ. Or, individual differences in the random intercepts and slopes are only minimally accounted for by their age and surgical type.

This final joint LGC model incorporating the two covariates of *AgeGrp* and *SurgTx* can be graphically plotted in Figure 9.6 using *R* package *semPlot* as follows:

```
# Load the library
library(semPlot)
# Call "semPlots" to plot the estimated model
semPaths(fitHK4)
```

9.4 Data Analysis Using SAS

As discussed in Section 9.2.4, a basic latent growth-curve analysis can be viewed as a confirmatory factor model and the estimation can be done by structural equation modeling techniques. With multiple outcomes and covariates for predicting the random slopes and intercepts, the modeling of latent growth-curve can be treated as a general structural equation model with exogenous observed variables that can have direct effects on the latent factors. Therefore, computationally, the latent curve analysis in this chapter can be handled by *PROC CALIS* as a general structural equation model.

However, because *PROC CALIS* itself is not tailored to latent growth-curve analysis, care must be taken when you specify a latent growth curve model in *PROC CALIS*. Here are some important considerations when fitting a latent growth curve model by a general structural equation modeling software (and in particular, by *PROC CALIS*).

- In general, random intercepts and slopes are treated as latent factors in SEM. This way, random intercepts and slopes are viewed as unobserved latent factor values (or scores) which vary among individuals.

- The mean structures (in addition to the covariance structures) must be modeled for a latent growth-curve analysis when using SEM techniques. If the full-information maximum likelihood (FIML) method is used (such as that in *PROC CALIS*), the modeling of mean structures would be included automatically.

- In most practical applications, the means of the random intercepts and random slopes are assumed to be nonzeros in latent growth-curve modeling. Therefore, the conventional default zero values for latent factor means (for representing latent intercepts and slopes) in some general SEM software (such as *PROC CALIS*) should be overridden by specifying them as free parameters.

- Because the mean structures of the observed measurement variables are already modeled by the (nonzero) means of random intercepts and slopes, the conventional default nonzero intercepts for observed variables in some general SEM software (such as *PROC CALIS*) should be overridden by fixing them to zeros.

As we discuss the *PROC CALIS* examples below, these important considerations will be considered and illustrated.

9.4.1 Latent Growth Curve for the Trajectory of Mood

A SAS data set *Dcancer* has been prepared for the analyses in this section. For women in the younger age group, AgeGrp = 1. Otherwise, AgeGrp = 0. Mastectomy surgery was coded with SurgTx = 1. Otherwise, SurgTx = 0. The following SAS code can be used to fit the latent growth-curve model for the trajectory of mood during the 8-month period:

```
/******* Basic Single outcome Latent Curve Model
        for Mood **********/
proc calis data="c:\Dcancer" method=FIML;
   path
      1        ==> MOOD1 MOOD4 MOOD8 = 3*0,
      iMOOD    ==> MOOD1 MOOD4 MOOD8 = 3*1,
      sMOOD    ==> MOOD1 MOOD4 MOOD8 = 0 1 2.33;
```

```
mean
   iMOOD sMOOD;
fitindex on(only)=[chisq df probchi RMSEA RMSEA_LL RMSEA_UL
       PROBCLFIT SRMR CFI BENTLERNNFI] noindextype;
run;
```

The procedure reads in a SAS data set *Dcancer* and uses the FIML estimation method. In the model, MOOD$_1$, MOOD$_4$, and MOOD$_8$ are observed measurements of mood at different time points. Two latent factors, iMOOD and sMOOD, are treated as latent factors representing random intercepts and slopes.

In the PATH statement, the 1-paths to the MOOD variables at different time points are set to fixed zeros, overriding the default free intercept parameters for these observed variables. Next, the path entry specifies random intercepts as paths from the latent factor iMOOD to the MOOD variables. All path coefficients are fixed to 1. The last path entry specifies random slopes as paths from the latent factor sMOOD to the MOOD variables. The path coefficients are fixed numbers 0, 1, and 2.33, respectively, which reflect the fixed time points of measurements.

In the MEAN statement, the latent factor means of iMOOD and sMOOD (representing the means of random intercept and random slope for MOOD) are set to be free parameters in the model, overridding the default fixed zero values.

In the FITINDEX statement, a selected set of fit indices are requested.

In the first output, some important modeling information is shown. Out of the 405 records (observations) read, 307 are complete cases without missing values in MOOD. There are 77 incomplete observations with at least one but less than 3 missing values. Because FIML estimation was requested and the mean structures were specified, both the means and covariances are modeled.

Modeling Information	
Full Information Maximum Likelihood Estimation	
Data Set	WC000001.DCANCER
N Records Read	405
N Complete Records	307
N Empty Records	21
N Incomplete Records	77
N Complete Obs	307
N Incomplete Obs	77
Model Type	PATH
Analysis	Means and Covariances

In the next output, the variables in the model are shown. It is important to confirm whether variables are classified as intended. In this case, it is confirmed that all MOOD variables are observed and iMOOD and sMOOD are latent factors representing the random intercept and random slope.

Variables in the Model		
Endogenous	Manifest	MOOD1 MOOD4 MOOD8
	Latent	
Exogenous	Manifest	
	Latent	iMOOD sMOOD
Number of Endogenous Variables = 3		
Number of Exogenous Variables = 2		

Because the FIML estimation is used, *PROC CALIS* analyzes the missing patterns and the mean profile profiles of the missing patterns, as shown in the following output:

		NVar			
	Pattern	Miss	Freq	Proportion	Cumulative
1	x..	2	25	0.0651	0.0651
2	x.x	1	23	0.0599	0.1250
3	xx.	1	15	0.0391	0.1641
4	.xx	1	7	0.0182	0.1823
5	..x	2	4	0.0104	0.1927

Rank Order of the 5 Most Frequent Missing Patterns
Total Number of Distinct Patterns with Missing Values = 6

NOTE: Nonmissing Pattern Proportion = 0.7995 (N=307)

Means of the Nonmissing and the Most Frequent Missing Patterns

		Missing Pattern				
Variable	Nonmissing (N=307)	1 (N=25)	2 (N=23)	3 (N=15)	4 (N=7)	5 (N=4)
MOOD1	21.11401	24.56000	22.91304	18.93333	.	.
MOOD4	20.96743	.	.	20.26667	24.85714	.
MOOD8	19.99023	.	20.43478	.	20.00000	18.25000

Five most frequent missing patterns are shown. The proportion of observations without any missing MOOD variable values is about 80%. The next table shows the mean profiles of the missing and non-missing patterns.

The fit summary is shown in the following output. These results are similar to that of the *lavaan*.

Fit Summary	
Chi-Square	1.7166
Chi-Square DF	1
Pr > Chi-Square	0.1901
Standardized RMR (SRMR)	0.0156
RMSEA Estimate	0.0432
RMSEA Lower 90% Confidence Limit	0.0000
RMSEA Upper 90% Confidence Limit	0.1508
Probability of Close Fit	0.3816
Bentler Comparative Fit Index	0.9977
Bentler-Bonett Non-normed Index	0.9930

The following table shows that all the path coefficients are fixed parameters (i.e., no standard errors and t statistics) in the model, as expected.

	Path		Estimate	Standard Error	t Value	Pr > |t|
1	===>	MOOD1	0			
1	===>	MOOD4	0			
1	===>	MOOD8	0			
iMOOD	===>	MOOD1	1.00000			
iMOOD	===>	MOOD4	1.00000			
iMOOD	===>	MOOD8	1.00000			
sMOOD	===>	MOOD1	0			
sMOOD	===>	MOOD4	1.00000			
sMOOD	===>	MOOD8	2.33000			

(Table title: PATH List)

Finally, the estimates of variances, covariances, and means are shown in the following output:

Variance Parameters

Variance Type	Variable	Parameter	Estimate	Standard Error	t Value	Pr > |t|
Exogenous	iMOOD	_Add1	25.83224	3.60420	7.1673	<.0001
	sMOOD	_Add2	2.26954	1.26279	1.7972	0.0723
Error	MOOD1	_Add3	14.30571	3.24017	4.4151	<.0001
	MOOD4	_Add4	18.61179	2.06786	9.0005	<.0001
	MOOD8	_Add5	6.82173	3.88471	1.7560	0.0791

Covariances Among Exogenous Variables

Var1	Var2	Parameter	Estimate	Standard Error	t Value	Pr > |t|
sMOOD	iMOOD	_Add6	-2.10908	1.49632	-1.4095	0.1587

Means and Intercepts

Type	Variable	Parameter	Estimate	Standard Error	t Value	Pr > |t|
Mean	iMOOD	_Parm1	21.47432	0.31707	67.7271	<.0001
	sMOOD	_Parm2	-0.59030	0.13215	-4.4669	<.0001

Previously with the *lavaan* fitting, the covariance estimate of iMOOD and sMOOD and the variance estimate of sMOOD were also not statistically significant and were fixed to zeros in a parsimonious model. The non-significance of these estimates are observed here in the *PROC CALIS* fitting. To fit the corresponding parsimonious model with *PROC CALIS*, you can use the following code:

```
/***** A Parsimonious Single outcome Latent Curve Model
       for Mood *****/
```

```
proc calis data="c:\Dcancer" method=FIML;
  path
    1          ==> MOOD1 MOOD4 MOOD8 = 3*0,
    iMOOD      ==> MOOD1 MOOD4 MOOD8 = 3*1,
    sMOOD      ==> MOOD1 MOOD4 MOOD8 = 0 1 2.33;
  mean
    iMOOD sMOOD;
  pvar
    sMOOD = 0;
  pcov
    iMOOD sMOOD = 0;
  fitindex on(only)=[chisq df probchi RMSEA RMSEA_LL RMSEA_UL
        PROBCLFIT SRMR CFI BENTLERNNFI] noindextype;
run;
```

Essentially, the PVAR statement fixes the variance of sMOOD to zero and the PCOV statement fixes the covariance between iMOOD and sMOOD to zero. This would result in the same fitted model as that of *lavaan*. Details are therefore omitted here.

Similarly, the fitting of the latent growth curve for SOCADJ with *PROC CALIS* can be done in the same manner. This would be left as an exercise for the readers.

9.4.2   Multiple-Outcome Latent Growth-Curve Modeling

To fit the MOOD and SOCADJ trajectories simultaneously, latent variables for the random intercepts and slopes are created for each set of the mood and social adjustment measurements at different time points. The following code for a two-outcome LGC model is a straightforward expansion of the code for the single-outcome case:

```
/******* Basic Joint Model for Two Response
        Trajectories ***********/
proc calis data="c:\Dcancer" method=FIML modification;
  path
    1          ==> MOOD1 MOOD4 MOOD8 = 3*0,
    1          ==> SOCADJ1 SOCADJ4 SOCADJ8 = 3*0,
    iMOOD      ==> MOOD1 MOOD4 MOOD8 = 3*1,
    sMOOD      ==> MOOD1 MOOD4 MOOD8 = 0 1 2.33,
    iSOCADJ    ==> SOCADJ1 SOCADJ4 SOCADJ8 = 3*1,
    sSOCADJ    ==> SOCADJ1 SOCADJ4 SOCADJ8 = 0 1 2.33;
  mean
    iMOOD sMOOD,
    iSOCADJ sSOCADJ;
  fitindex on(only)=[chisq df probchi RMSEA RMSEA_LL RMSEA_UL
        PROBCLFIT SRMR CFI BENTLERNNFI] noindextype;
run;
```

Again, all time points are coded as fixed path coefficients in the PATH statement. The regular intercepts are fixed to zeros in some 1-paths. The means of the random intercepts and slopes are set to be free parameters in the MEAN statement. The MODIFICATION option is used to output modification indices. The following fit summary confirms the same results as that of *lavaan*:

Fit Summary	
Chi-Square	40.2778
Chi-Square DF	7
Pr > Chi-Square	<.0001
Standardized RMR (SRMR)	0.0408
RMSEA Estimate	0.1110
RMSEA Lower 90% Confidence Limit	0.0792
RMSEA Upper 90% Confidence Limit	0.1454
Probability of Close Fit	0.0012
Bentler Comparative Fit Index	0.9577
Bentler-Bonett Non-normed Index	0.9093

The results of LM tests suggest that the residual covariance between $MOOD_4$ and $SOCADJ_4$ can be freely estimated for improving the model fit. The following code implements such an improved model:

```
/******* Modified Joint Model for Two Response
         Trajectories ***********/
proc calis data="c:\Dcancer" method=FIML;
   path
      1           ==> MOOD1 MOOD4 MOOD8 = 3*0,
      1           ==> SOCADJ1 SOCADJ4 SOCADJ8 = 3*0,
      iMOOD       ==> MOOD1 MOOD4 MOOD8 = 3*1,
      sMOOD       ==> MOOD1 MOOD4 MOOD8 = 0 1 2.33,
      iSOCADJ     ==> SOCADJ1 SOCADJ4 SOCADJ8 = 3*1,
      sSOCADJ     ==> SOCADJ1 SOCADJ4 SOCADJ8 = 0 1 2.33;
   pcov
      MOOD4 SOCADJ4;
   mean
      iMOOD sMOOD iSOCADJ sSOCADJ;
   fitindex on(only)=[chisq df probchi RMSEA RMSEA_LL RMSEA_UL
             PROBCLFIT SRMR CFI BENTLERNNFI] noindextype;
run;
```

Essentially, the PCOV statement now adds the residual covariance between $MOOD_4$ and $SOCADJ_4$ as a new parameter into the original model and obtains a better model fit. Estimation results similar to that of *lavaan* were obtained but are not shown here.

Finally, to improve model parsimony, some parameters were fixed in the previous analysis *lavaan*. You can implement the same final model for the joint LGC of MOOD and SOCADJ with the following *PROC CALIS* code:

```
/******* Final Parsimonious Model for Joint Modeling
         of Outcomes *********/
proc calis data="c:\Dcancer" method=FIML;
   path
      1           ==> MOOD1 MOOD4 MOOD8 = 3*0,
      1           ==> SOCADJ1 SOCADJ4 SOCADJ8 = 3*0,
      iMOOD       ==> MOOD1 MOOD4 MOOD8 = 3*1,
      sMOOD       ==> MOOD1 MOOD4 MOOD8 = 0 1 2.33,
      iSOCADJ     ==> SOCADJ1 SOCADJ4 SOCADJ8 = 3*1,
      sSOCADJ     ==> SOCADJ1 SOCADJ4 SOCADJ8 = 0 1 2.33;
   pcov
      MOOD4 SOCADJ4,
      iMOOD sMOOD   = 0,
      iMOOD sSOCADJ = 0,
      sMOOD iSOCADJ = 0;
   pvar
      SOCADJ8 = 0;
   Mean
      iMOOD sMOOD iSOCADJ,
      sSOCADJ = 0;
   fitindex on(only)=[chisq df probchi RMSEA RMSEA_LL RMSEA_UL
           PROBCLFIT SRMR CFI BENTLERNNFI] noindextype;
run;
```

As compared with the previous modified model, three exogenous covariances between the random slopes and intercepts are now fixed to zero in the PCOV statement. The residual variance of $SOCADJ_8$ is fixed to zero in the PVAR statement and the mean of sSOCADJ is also fixed to zero in the MEAN statement. The fit summary is shown in the following output and it confirms the same results obtained from *lavaan*.

Fit Summary	
Chi-Square	13.7556
Chi-Square DF	11
Pr > Chi-Square	0.2468
Standardized RMR (SRMR)	0.0348
RMSEA Estimate	0.0255
RMSEA Lower 90% Confidence Limit	0.0000
RMSEA Upper 90% Confidence Limit	0.0622
Probability of Close Fit	0.8396
Bentler Comparative Fit Index	0.9965
Bentler-Bonett Non-normed Index	0.9952

9.4.3 Joint LGC Model with Covariates

With *AgeGrp* and *SurgTX* serving as covariates of the random intercepts and slopes, additional paths must be added to the model specification, as shown in the following *PROC CALIS* code:

```
/*** Basic Joint Modeling of Response Trajectories
     With Covariates *****/
proc calis data="c:\Dcancer" method=FIML modification;
   path
       1            ==> MOOD1 MOOD4 MOOD8 = 3*0,
       1            ==> SOCADJ1 SOCADJ4 SOCADJ8 = 3*0,
       iMOOD        ==> MOOD1 MOOD4 MOOD8 = 3*1,
       sMOOD        ==> MOOD1 MOOD4 MOOD8 = 0 1 2.33,
       iSOCADJ      ==> SOCADJ1 SOCADJ4 SOCADJ8 = 3*1,
       sSOCADJ      ==> SOCADJ1 SOCADJ4 SOCADJ8 = 0 1 2.33,
       /* New paths from the covariates */
       iMOOD     <== AgeGrp SurgTx,
       sMOOD     <== AgeGrp SurgTx,
       iSOCADJ   <== AgeGrp SurgTx,
       sSOCADJ   <== AgeGrp SurgTx;
   pcov
       iMOOD sMOOD,
       iMOOD iSOCADJ,
       iMOOD sSOCADJ,
       sMOOD iSOCADJ,
       sMOOD sSOCADJ,
       iSOCADJ sSOCADJ;
   mean
       iMOOD sMOOD iSOCADJ sSOCADJ;
   fitindex on(only)=[chisq df probchi RMSEA RMSEA_LL RMSEA_UL
           PROBCLFIT SRMR CFI BENTLERNNFI] noindextype;
run;
```

In the PATH statement, four path-entries (representing eight paths) from AgeGrp and SurgTx to the random intercepts and slopes are now added. In addition, the residual correlations among the random intercepts and slopes are also specified in the PCOV statement. These residual correlations were assumed by default in *lavvan* but are default to zero in *PROC CALIS*. Therefore, we specify these residual correlations here for comparability.

Fit Summary	
Chi-Square	43.1056
Chi-Square DF	11
Pr > Chi-Square	<.0001
Standardized RMR (SRMR)	0.0326
RMSEA Estimate	0.0849
RMSEA Lower 90% Confidence Limit	0.0592
RMSEA Upper 90% Confidence Limit	0.1123
Probability of Close Fit	0.0146
Bentler Comparative Fit Index	0.9602
Bentler-Bonett Non-normed Index	0.8988

This model has a decent fit. To improve the model fit, the results of modification indices suggest adding the residual correlation between $SOCADJ_4$ and $MOOD_4$ to the model, which results in the following specification:

```
/*** Modified Joint Modeling of Response Trajectories
     With Covariates ****/
proc calis data="c:\Dcancer" method=FIML;
   path
      1          ==> MOOD1 MOOD4 MOOD8 = 3*0,
      1          ==> SOCADJ1 SOCADJ4 SOCADJ8 = 3*0,
      iMOOD      ==> MOOD1 MOOD4 MOOD8 = 3*1,
      sMOOD      ==> MOOD1 MOOD4 MOOD8 = 0 1 2.33,
      iSOCADJ    ==> SOCADJ1 SOCADJ4 SOCADJ8 = 3*1,
      sSOCADJ    ==> SOCADJ1 SOCADJ4 SOCADJ8 = 0 1 2.33,
      /* New paths from the covariates */
      iMOOD      <== AgeGrp SurgTx,
      sMOOD      <== AgeGrp SurgTx,
      iSOCADJ    <== AgeGrp SurgTx,
      sSOCADJ    <== AgeGrp SurgTx;
   pcov
      SOCADJ4 MOOD4, /* Added correlated error */
      iMOOD sMOOD,
      iMOOD iSOCADJ,
      iMOOD sSOCADJ,
      sMOOD iSOCADJ,
      sMOOD sSOCADJ,
      iSOCADJ sSOCADJ;
   mean
      iMOOD sMOOD iSOCADJ sSOCADJ;
   fitindex on(only)=[chisq df probchi RMSEA RMSEA_LL RMSEA_UL
           PROBCLFIT SRMR CFI BENTLERNNFI] noindextype;
run;
```

As can be seen, the first covariance entry in PCOV statement is the only new parameter added to the original model. The model fit is summarized in the following output:

Fit Summary	
Chi-Square	9.8768
Chi-Square DF	10
Pr > Chi-Square	0.4514
Standardized RMR (SRMR)	0.0191
RMSEA Estimate	0.0000
RMSEA Lower 90% Confidence Limit	0.0000
RMSEA Upper 90% Confidence Limit	0.0534
Probability of Close Fit	0.9298
Bentler Comparative Fit Index	1.0000
Bentler-Bonett Non-normed Index	1.0004

As mentioned previously, the fit of this model is nearly perfect.

If it is desirable to make the model more parsimonious, some clearly insignificant parameter estimates could be fixed to zero. The following *PROC CALIS* code specifies such a parsimonious model, similar to what we have done in the *lavaan* code:

```
/*** Parsimonious Joint Modeling of Response Trajectories
     With Covariates **/
proc calis data="c:\Dcancer" method=FIML;
   path
      1            ==> MOOD1 MOOD4 MOOD8 = 3*0,
      1            ==> SOCADJ1 SOCADJ4 SOCADJ8 = 3*0,
      iMOOD        ==> MOOD1 MOOD4 MOOD8 = 3*1,
      sMOOD        ==> MOOD1 MOOD4 MOOD8 = 0 1 2.33,
      iSOCADJ      ==> SOCADJ1 SOCADJ4 SOCADJ8 = 3*1,
      sSOCADJ      ==> SOCADJ1 SOCADJ4 SOCADJ8 = 0 1 2.33,
      iMOOD        <== AgeGrp SurgTx,
      sMOOD        <== AgeGrp SurgTx,
      iSOCADJ      <== AgeGrp SurgTx,
      sSOCADJ      <== AgeGrp SurgTx;
   pcov
      MOOD4 SOCADJ4,
      iMOOD sMOOD   = 0,
      iMOOD iSOCADJ,
      iMOOD sSOCADJ = 0,
      sMOOD iSOCADJ = 0,
      sMOOD sSOCADJ,
      iSOCADJ sSOCADJ;
   pvar
      SOCADJ8 = 0;
```

```
mean
   iMOOD iSOCADJ sSOCADJ,
   sMOOD = 0;
fitindex on(only)=[chisq df probchi RMSEA RMSEA_LL RMSEA_UL
        PROBCLFIT SRMR CFI BENTLERNNFI] noindextype;
run;
```

Three error covariances are fixed to zeros in the PCOV statement. the error variance of $SOCADJ_8$ is fixed to zero in the PVAR statement. Finally, the intercept for the random slope of MOOD (sMOOD) is also fixed to zero. As shown in the following fit summary table, the parsimonious model still has a near-perfect fit in all aspects.

Fit Summary	
Chi-Square	16.5921
Chi-Square DF	15
Pr > Chi-Square	0.3438
Standardized RMR (SRMR)	0.0276
RMSEA Estimate	0.0162
RMSEA Lower 90% Confidence Limit	0.0000
RMSEA Upper 90% Confidence Limit	0.0510
Probability of Close Fit	0.9433
Bentler Comparative Fit Index	0.9980
Bentler-Bonett Non-normed Index	0.9963

One motivation to include the covariates age (AgeGrp) and surgical type (SurgTx) into the model is to examine how much they can account for the variability of the random intercepts and slopes of the trajectories of MOOD and SOCADJ. The following table shows the R-squares (explained proportions of variances) of the endogenous variables in the model. While the R-squares are moderate to high for the MOOD and SOCADJ variables, the R-squares for the random intercepts and slopes are pretty small, indicating that the AgeGrp ad SurgTX do not explain much of the variability of random intercepts and slopes of the growth curves for MOOD and SOCADJ in women.

Squared Multiple Correlations			
Variable	Error Variance	Total Variance	R-Square
MOOD1	17.71760	39.76848	0.5545
MOOD4	18.61465	41.69494	0.5536
MOOD8	9.35405	37.01107	0.7473
SOCADJ1	10.02021	92.76830	0.8920
SOCADJ4	41.95109	107.59630	0.6101
SOCADJ8	0	82.76851	1.0000
iMOOD	21.23451	22.05088	0.0370
iSOCADJ	80.74535	82.74810	0.0242
sMOOD	0.96602	1.03508	0.0667
sSOCADJ	12.56361	12.86591	0.0235

9.5   Discussions

In this chapter, we explored the latent growth-curve modeling using the Hong Kong breast-cancer data from 405 women after their surgery to investigate the longitudinal changes over time. Two outcomes were available in this data set on *mood* and *social adjustment* at 1-month, 4-month and 8-month after surgery. The growth trajectories were modeled using a linear growth-curve model with the random *intercepts* and *slopes* treated as latent factors. We illustrated the latent growth-curve modeling for each outcome separately as well as jointly.

There are further extensions of the LGC modeling. Due to the longitudinal nature of the outcome, an auto-regressive latent growth-curve modeling can be explored on whether the $MOOD_1$, $MOOD_4$ and $MOOD_8$ (similarly for $SOCADJ_1$, $SOCADJ_4$ and $SOCADJ_8$) are auto-correlated with each other. For all the LGC analyses in this chapter, we assumed that these longitudinal measurements were independent of each other, which is an over-simplification.

We can use the auto-regressive LGC model to further investigate the auto-correlations among these three repeated measurements. As seen in Figure 9.7, we can investigate the first-order auto-regressive LGC model where we assume the measurements at time t only correlated with their previous measurements at $t-1$. Specifically, in this first-order auto-regressive LGC model, we assume the $MOOD_4$ is correlated with $MOOD_1$, $MOOD_8$ is correlated with $MOOD_4$, but not directly correlated with $MOOD_1$.

Since there are three repeated measurements, we can also investigate the second-order auto-regression LGC where the measurement at 8-month (i.e., $MOOD_8$) can be correlated with the measurements at 4-month (i.e., $MOOD_4$) and 1-month (i.e., $MOOD_1$). The implementations in *R*, *SAS* and *mplus* are very straightforward and we leave this auto-regressive LGC modeling as exercises.

There are also extensive discussions on the advantages and disadvantages between the LGC modeling and the statistical mixed-effects modeling with the within- and between-individual longitudinal trajectories can be captured by the random-intercept and random-slope modeling. There are two general *R* packages available. One is the package *nlme* (i.e., *nonlinear mixed-effects* modeling) developed and discussed by Pinheiro and Bates (2000) and the other one *lme4* (i.e, *linear mixed-effects* modeling) developed by Bates and his colleagues (https://cran.r-project.org/web/packages/lme4/index.html) which can provide functions for fitting and analyzing linear mixed-effects models using function *lmer*, generalized linear mixed-effects model using function *glmer* and nonlinear mixed-effects model using function *nlmer*. In Chapter 5 of Chen and Chen (2021a), this Hong Kong breast-cancer data was analyzed using these *R* packages.

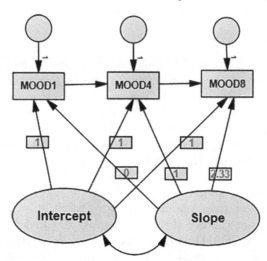

FIGURE 9.7
The Auto-Regressive Latent Growth-Curve Model for MOOD.

9.6   Exercises

1. Fit the first-order auto-regressive LGC model for the MOOD data as shown in Figure 9.4, and test whether the auto-correlation is significant.

2. Fit the second-order auto-regressive LGC model for the MOOD data, and test whether the auto-correlation is significant.

3. Fit the first-order auto-regressive LGC model for the SOCADJ data, and test whether the auto-correlation is significant.

4. Fit the second-order auto-regressive LGC model for the SOCADJ data, and test whether the auto-correlation is significant.

5. Fit the first-order auto-regressive joint LGC model for both the MOOD and SOCADJ data, and test whether the auto-correlations are significant.

6. Fit the second-order auto-regressive joint LGC model for both the MOOD and SOCADJ data, and test whether the auto-correlations are significant.

10

Full Longitudinal Mediation Modeling

In Chapter 3, we discussed the classical mediation analysis with implementation on *R* and *SAS*. Most of the classical mediation analyses are often based on cross-sectional data collected at a single time point to infer causal mediation effects. However, causal interpretations of effects presume some temporal order of the variables. That is, the causal effect of X on Y in the classical mediation analysis presumes that X occurs before Y and M, which is a mediator variable that occurs before Y. With cross-sectional data, there is no guarantee of such a temporal order of variables involved in modeling, thus jeopardizing the causal interpretation of the mediation effects. However, in a longitudinal study, data are collected in different time points and therefore the temporal order of some variables is known and affirmed. With an affirmed variable order in a mediation analysis, one can eliminate the temporal order as a roadblock to the causal interpretation of mediation effects. Therefore, Collecting longitudinal data is an important tool that supports a valid causal inference of mediation effects. This chapter illustrates an SEM application with longitudinal data.

Chen and Chen (2021b) presented a *full longitudinal mediation modeling* (FLMM) using three waves of longitudinal data from an NIH-funded project *Focus on Youth in the Caribbean*. This analysis was a theory-guided causal relationship from *HIV/AIDS knowledge* to *Condom self-efficacy*, further to *Condom-use intention* and implemented in *mplus* with *mplus* program given as an appendix. In this chapter, a full longitudinal mediation model was presented to analyze several potential longitudinal relationships, including (1) first-order auto-regressive longitudinal effects for the three variables from time 1 to time 2 and from time 2 to time 3; (2) cross-lagged effects among the three variables from time 1 to time 2 and from time 2 to time 3; and (3) second-order auto-regressive and second-order cross-lagged effects for the three variables from time 1 to time 3.

In this chapter, we will convert the analysis from *mplus* to *R/lavaan* and *SAS/CALIS* to present a step-by-step approach on how to implement the *FLMM* using the same data. With the *FLMM*, we can examine the auto-regressive mediation effects on self-enhancement effect over time, multi-path mediation process with cross-lagged effects and feedback effects simultaneously.

DOI: 10.1201/9781003365860-10

10.1   Data Descriptions

Data set is available in *dControlW123.dat* for 497 youth aged 9–14 at the beginning of the study.

This data set is a subset of the original study as described in Chen and Chen (2021b). Twelve variables are available as follows:

- *Gender* is the youth gender with 1 = Male and 2 = Female.

- *Age* is the age at the beginning of the study.

- *ID* is the identification of each youth.

- *HIVK1, HIVK2,* and *HIVK3* are the *HIV/AIDS knowledge* measured at 0, 6 and 12 months after study.

- *SELFC1, SELFC2* and *SELFC3* are the *Self-efficacy for condom use* measured at 0, 6 and 12 months after study.

- *INTENT1, INTENT2* and *INTENT3* are the *Intention to use condoms* measured at 0, 6 and 12 months after study.

As detailed in Chen and Chen (2021b), *HIV/AIDS knowledge* is the predictor variable in *FLMM* to measure youth's HIV/AIDS knowledge with higher scores for more knowledge. *Self-efficacy for condom use* is the mediator variable to measure youth's perceived ability to obtain, use, and/or convince their partner to use condoms, to ask for condoms in a store or clinic, and to refuse sex without a condom, with higher scores for stronger condom use self-efficacy. *Intention to use condoms* is the outcome with higher scores for a stronger intention to use a condom during sex. This causal chain from *HIV/AIDS knowledge* to *condom use self-efficacy* to *condom use intention* is supported by well-established *Protection Motivation Theory* in Rogers (1975) in promoting HIV protective behaviors.

We can load the data into *R* as follows:

```
# Read the data into R
dBahamas <- read.table("data/dControlW123.dat", quote='"',
                       na.strings="*")
# Name the variables
names(dBahamas) <- c("Gender","Age","HIVK1","HIVK2","HIVK3",
                     "SELFC1","SELFC2", "SELFC3",
                     "INTENT1","INTENT2","INTENT3","ID")
# Print the data dimension
dim(dBahamas)
```

```
## [1] 497  12
```

```
# Print the data summary
summary(dBahamas)
```

```
##      Gender            Age            HIVK1
##   Min.   :1.00    Min.   : 9.0    Min.   : 1.0
##   1st Qu.:1.00    1st Qu.:10.0    1st Qu.:10.0
##   Median :2.00    Median :10.0    Median :12.0
##   Mean   :1.54    Mean   :10.4    Mean   :12.2
##   3rd Qu.:2.00    3rd Qu.:11.0    3rd Qu.:14.0
##   Max.   :2.00    Max.   :14.0    Max.   :20.0
##                                   NA's   :4
##      HIVK2            HIVK3           SELFC1
##   Min.   : 1.0    Min.   : 1.0    Min.   :1.00
##   1st Qu.:11.0    1st Qu.:11.0    1st Qu.:1.33
##   Median :13.0    Median :14.0    Median :2.17
##   Mean   :12.8    Mean   :13.7    Mean   :2.37
##   3rd Qu.:16.0    3rd Qu.:16.0    3rd Qu.:3.17
##   Max.   :20.0    Max.   :20.0    Max.   :5.00
##   NA's   :33      NA's   :65      NA's   :5
##      SELFC2           SELFC3          INTENT1
##   Min.   :1.0     Min.   :1.0     Min.   :1.0
##   1st Qu.:2.0     1st Qu.:2.0     1st Qu.:1.0
##   Median :3.0     Median :3.0     Median :3.0
##   Mean   :2.8     Mean   :2.8     Mean   :2.8
##   3rd Qu.:3.7     3rd Qu.:3.8     3rd Qu.:5.0
##   Max.   :5.0     Max.   :5.0     Max.   :5.0
##   NA's   :37      NA's   :68      NA's   :43
##      INTENT2          INTENT3           ID
##   Min.   :1.0     Min.   :1.0     Min.   : 1
##   1st Qu.:2.0     1st Qu.:1.0     1st Qu.:125
##   Median :4.0     Median :4.0     Median :249
##   Mean   :3.4     Mean   :3.5     Mean   :249
##   3rd Qu.:5.0     3rd Qu.:5.0     3rd Qu.:373
##   Max.   :5.0     Max.   :5.0     Max.   :497
##   NA's   :53      NA's   :79
```

```
# Show the first 6 observations
head(dBahamas)
```

```
##   Gender Age HIVK1 HIVK2 HIVK3 SELFC1 SELFC2 SELFC3
## 1      1  12    13    15    15   3.00   2.50   1.00
## 2      1  11    10     1     7   1.33   3.00   3.00
## 3      1  10    13    10    14   1.00   2.00   2.33
```

```
## 4      1   10    11    16    15   1.67   3.00   3.67
## 5      1   10    16    17    18   4.17   3.83   3.50
## 6      1   10    12    10    10   3.67   3.50   3.00
##    INTENT1 INTENT2 INTENT3 ID
## 1       3       3       1  1
## 2       3       3       3  2
## 3       5       2       1  3
## 4       5       2       5  4
## 5       5       3       3  5
## 6       5       5       5  6
```

Notice that when the data file is saved into *dControlW123.dat*, there are missing values which are denoted by *. Due to these missing values, the data file *dControlW123.dat* contains symbol *"* to *quote* those variables with missing values. Therefore, to load this data into *R* for data analysis using *read.table*, we use option *quote='"'* with *na.strings="*"* to deal with these symbols and missing values.

10.2   The Full Longitudinal Mediation Model (FLMM) and the Decomposition of Total Effect

The basic paths (directional effects) of the FLMM for the current analysis includes:

(a) the first-order auto-regressive (AR) longitudinal effects for the three variables from time 1 to time 2 and from time 2 to time 3 (linked by the horizontal arrows for the three variables);

(b) cross-lagged effects among the three variables from time 1 to time 2 and from time 2 to time 3 (diagonal arrows); and

(c) second-order auto-regressive and second-order cross-lagged effects for the three variables from time 1 to time 3.

Some main components of the FLMM for the current analysis are depicted in Figure 10.1. All but the paths in (c) (i.e., second-order auto-regressive and second-order cross-lagged effects) are displayed to avoid clutters in the figure. However, all the paths in (a), (b), and (c) would be included in the modeling by using the *R* and *SAS* code in later sections.

In addition, the following variances and covariances are included in the basic FLMM:

(d) Variances of the exogenous time-1 variables *HIVK1*, *SELFC1*, and *INTENT1* and error variances for time-2 and time-3 variables;

(e) Covariances among the exogenous time-1 variables *HIVK1*, *SELFC1*, and *INTENT1* and covariances among the error terms for the time-3 variables *HIVK3*, *SELFC3*, and *INTENT3* (i.e., correlated errors).

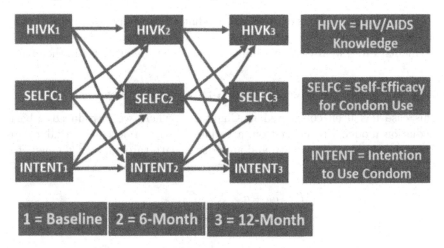

FIGURE 10.1
Full Longitudinal Mediation Model for Bahamas Study.

Although parameters in (d) and (e) can be represented as bi-directional paths (or arrows) in Figure 10.1, they are omitted to avoid clutters. Again, these variances and covariances are included by default or by explicit specifications in later modeling by using the *R* and *SAS* code.

In Chapter 3, we introduce a basic mediation model with a single mediator. The effect from a variable X to an outcome variable Y is characterized by a direct path $(X \to Y)$ and an indirect pathway through a single mediator M $(X \to M \to Y)$. In the current model, however, there are multiple mediators and multiple pathways for accounting the (total) effect of a time-1 variable on a time-3 variable. For example, *HIVK1* can "reach" to *INTENT3* via the following directional pathways:

- *HIVK1 → HIVK2 → INTENT3*
- *HIVK1 → SELFC2 → INTENT3*
- *HIVK1 → INTENT2 → INTENT3*
- *HIVK1 → INTENT3*

The first three pathways involve, respectively, three intermediate time-2 variables: *HIVK2*, *SELFC2*, and *INTENT2*. While each of these three pathways represents a distinct indirect (or mediated) effect of *HIVK1* on *INTENT3*, the last pathway represents the remaining effect or the direct effect of *HIVK1* on *INTENT3*. The three distinct indirect effects are also known as *specific indirect effects* because each of them represents a specific mediation process that is characterized by a distinct mediator. The combined mediation effect from the three specific indirect effects is the so-called *total indirect effect*. As a result, for models involving multiple mediation pathways, such as the current one, the total effect is the sum of its direct effect and its total indirect effects.

The next few sections explain how mediation effects are defined and computed when there are multiple mediators and multiple mediation pathways in the model. First, we review the simplest case of a single mediator in the model.

10.2.1 A Single Mediator in a Single Pathway

This case has been considered in Chapter 3, where we introduced a basic mediation model. The effect from a variable X on an outcome variable Y is characterized by a direct path and an indirect path through a single mediator M. Essentially, two model equations describe the process:

$$
\begin{aligned}
M &= a_0 + a \times X + e_M \\
Y &= b_0 + b \times M + c' \times X + e_Y
\end{aligned}
\tag{10.1}
$$

where a_0 and b_0 are intercepts, a, b, and c' are slopes or effects, and e_M and e_Y are error terms. Substituting the first equation into the second equation yields:

$$
Y = (b_0 + ba_0) + (ba + c') \times X + (b \times e_M + e_Y)
\tag{10.2}
$$

This equation suggests a decomposition of total effect of X on Y into two components:

$$
\begin{aligned}
\text{Indirect effect} &= b \times a \\
\text{Direct effect} &= c'
\end{aligned}
\tag{10.3}
$$

10.2.2 Multiple Mediators in a Single Mediation Pathway

Let us expand the basic mediation model to include M_1 and M_2 as two sequential mediators in a single pathway between X and Y. In addition to a direct path from X to Y, the mediation process (indirect pathway) is now represented schematically as the following pathway:

$$
X \to M_1 \to M_2 \to Y
\tag{10.4}
$$

Assuming that no other paths are specified in the model, the following set of equations represents such a mediation model with the two sequential mediators:

$$
\begin{aligned}
M_1 &= a_0 + a \times X + e_{M1} \\
M_2 &= d_0 + d \times M_1 + e_{M2} \\
Y &= b_0 + b \times M_2 + c' \times X + e_Y
\end{aligned}
\tag{10.5}
$$

where a_0, d_0, and b_0 are intercepts, a, d, b, and c' are slopes or effects, and e_{M1}, e_{M2} and e_Y are error terms. By substituting M_1 and M_2 in the first two equations successively into their next equations, the following equation is obtained:

$$Y = (b_0 + bd_0 + bda_0) + (bda + c') \times X + (bd \times e_{M1} + b \times e_{M2} + e_Y) \quad (10.6)$$

Now, the "combined" coefficient for the term with X suggests that the total effect of X on Y can be decomposed into the following two additive parts:

$$\text{Indirect effect} = b \times d \times a$$
$$\text{Direct effect} = c' \quad (10.7)$$

Because a, d, and b are the effects for the $X \to M_1$, $M_1 \to M_2$, and $M_2 \to Y$ paths, respectively, the product of these effects appears to be a quite intuitive measure of the overall mediation effect for the $X \to M_1 \to M_2 \to Y$ pathway. As compared with Equation (10.3) for the single mediator case, the indirect effect computation in Equation (10.7) includes an additional intermediate effect d into the product. In fact, such a product formula can be generalized to more than two sequential (continuous) mediators in the context of linear structural equation model. That is, it can be shown that the indirect effect of a specific mediation pathway is computed as the product of all intermediate effects along the mediation pathway.

10.2.3 Multiple Mediators in Multiple Mediation Pathways

We now consider the case where multiple mediators represent distinct mediation processes. For example, in addition to a direct path from X to Y in a mediation model, the following represents three specific mediation pathways from X to Y, that are characterized, respectively, by three distinct mediators M_1, M_2, and M_3:

$$X \to M_1 \to Y$$
$$X \to M_2 \to Y$$
$$X \to M_3 \to Y \quad (10.8)$$

Assuming that no other paths are specified in the model, the following set of equations represents such a mediation model with the three specific mediation pathways:

$$M_1 = a_{01} + a_1 \times X + e_{M1}$$
$$M_2 = a_{02} + a_2 \times X + e_{M2}$$
$$M_3 = a_{03} + a_3 \times X + e_{M3}$$
$$Y = b_0 + b_1 \times M_1 + b_2 \times M_2 + b_3 \times M_3 + c' \times X + e_Y \quad (10.9)$$

where a_{01}, a_{02}, a_{03}, and b_0 are intercepts, $a_1, a_2, a_3, b_1, b_2, b_3$, and c' are slopes or effects, and e_{M1}, e_{M2}, e_{M3}, and e_Y are error terms. By substituting M_1,

M_2, and M_3 in the first three equations into the last equation, the following equation is obtained:

$$Y = (b_0 + b_1 a_{01} + b_2 a_{02} + b_3 a_{03})) + (b_1 a_1 + b_2 a_2 + b_3 a_3 + c') \times X$$
$$+(b_1 \times e_{M1} + b_2 \times e_{M2} + b_3 \times e_{M3} + e_Y) \qquad (10.10)$$

Now, the "combined" coefficient for the term with X suggests that the total effect of X on Y can be decomposed into the following two additive parts:

$$\text{Total indirect effect} = b_1 \times a_1 + b_2 \times a_2 + b_3 \times a_3$$
$$\text{Direct effect} = c' \qquad (10.11)$$

Equation (10.11) shows that the total indirect effect is the sum of the following three specific indirect effects:

$$\text{Specific indirect effect via } M_1 = b_1 \times a_1$$
$$\text{Specific indirect effect via } M_2 = b_2 \times a_2$$
$$\text{Specific indirect effect via } M_3 = b_3 \times a_3 \qquad (10.12)$$

Recall that in the single mediator case, the indirect effect (which is also the total indirect effect), is computed as $b \times a$ (see Equation (10.3)). In the current case with three specific mediation pathways, each specific direct effect is of the same product form, as seen in Equation (10.12). Thus, the total indirect effect formula in Equation (10.11) for multiple mediation paths is an extension of the formula for the single mediator case by summing up the mediation effects in the specific mediation pathways. Again, it can be shown that this summation formula can be generalized to two or more mediation pathways in the linear structural equation model context.

10.2.4   Computing Specific Indirect Effects in Applications

In applications, the computation of the specific indirect effects might not be automatically available in the output of an SEM software. In this situation, you can compute these effects by applying the formulas that are based on the generalization of Equations (10.7) and (10.11). The R or SAS code in later sections demonstrates such a computational application. To summarize, you can use the following steps to compute the specific and total indirect effects:

(1) Identify all the distinct mediation pathways of an effect of interest, say X on Y.

(2) For each mediation pathway $X \to ... \to Y$, compute the specific indirect effect by multiplying all effect estimates along the mediation pathway (i.e., a generalization of the product formula in Equation (10.7)).

(3) Compute the total indirect effect by summing up all specific mediation effects obtained in Step 2 (i.e., a generalization of the summation formula in Equation (10.11)).

10.2.5   Limitations

Finally, we caution the readers about the limitations of the formulas presented here for computing mediation effects (including indirect and total effects). First, these formulas can only be applied to linear models with continuous variables. If any of the effect (X), outcome (Y), or mediator variables are categorical variables (and hence nonlinear models might be more appropriate) or interaction effects are present, the formulas presented in this chapter are not applicable.

Another limitation of these formulas is that they are not applicable to models with "feedback" (looped-back) effects. Graphically, a feedback effect occurs when a variable in the path diagram for a model has a unidirectional pathway that can lead back to itself. With a feedback effect in a model, computation of direct and indirect effects is possible only when the parameter values (or estimates) satisfy a certain complex mathematical condition, which is stated in terms of complex eigenvalues of a particular matrix and is beyond the scope of the current discussion. When such a condition is satisfied, then mediation effects are computable but the formulas presented in this chapter are not accurate.

10.3   Data Analysis Using *R*

In *mplus*, multi-path mediation analysis can be implemented by *Model indirect:* with the code of *X ind Z* which will find all the intermediate paths from X to Z (see Appendix for *Mplus*) internally and automatically. As far as we know, at present *R/lavaan* does not have this function to automatically find all the multiple mediation pathways. We thus would need to manually code the computation of mediation effects that correspond to specific indirect longitudinal effects.

In this chapter, we will first illustrate the model fitting for *FLMM* in Section 10.3.1, and then present the mediation analysis in Section 10.3.2.

10.3.1   Fit the FLMM

This section fits the FLMM model with *R/lavaan*. Modeling results are first interpreted, followed by a detailed investigation of mediation effects.

10.3.1.1   R/Lavaan Implementation

The FLMM can be readily implemented with *R/lavaan* as follows:

```
# The FLMM for Bahamas Youth
FLMMBahamas <- '
```

```
# Longitudinal effects for HIV: 1st-order auto-regressive(AR)
HIVK2 ~ HIVK1; HIVK3 ~ HIVK2;
# Longitudinal effects for HIV: 2nd-order auto-regressive
HIVK3 ~ HIVK1;

# Longitudinal effects for SELFC: 1st-order AR
SELFC2 ~ SELFC1; SELFC3 ~ SELFC2;
# Longitudinal effects for SELFC: 2nd-order AR
SELFC3 ~ SELFC1;

# Longitudinal effects for INTENT: 1st-order AR
INTENT2 ~ INTENT1; INTENT3 ~ INTENT2;
# Longitudinal effects for INTENT: 2nd-order AR
INTENT3 ~ INTENT1;

# 1st-order Cross-lags
HIVK3 ~ SELFC2; HIVK3 ~ INTENT2;
HIVK2 ~ SELFC1; HIVK2 ~ INTENT1;

SELFC3 ~ HIVK2; SELFC3 ~ INTENT2;
SELFC2 ~ HIVK1; SELFC2 ~ INTENT1;

INTENT3 ~ HIVK2; INTENT3 ~ SELFC2;
INTENT2 ~ HIVK1; INTENT2 ~ SELFC1;

# 2nd-order cross-lags
HIVK3    ~ SELFC1; HIVK3    ~ INTENT1;
INTENT3 ~ SELFC1; INTENT3 ~ HIVK1;
SELFC3   ~ HIVK1;  SELFC3   ~ INTENT1;

# Correlations for better model fitting
SELFC1   ~~ HIVK1; SELFC1   ~~ INTENT1;
INTENT1 ~~ HIVK1; SELFC2 ~~ INTENT2;
'
```

With this *ModBahamas*, we can then call *sem* with *full information maximum likelihood (fiml)* estimation to deal with missing values from the data as follows:

```
# Model fitting
fit.Bahamas <- sem(FLMMBahamas, data = dBahamas, missing="fiml")
# Print the model fitting
summary(fit.Bahamas, fit.measures=TRUE)
```

```
## lavaan 0.6-12 ended normally after 68 iterations
```

```
##
##   Estimator                                         ML
##   Optimization method                           NLMINB
##   Number of model parameters                        52
##
##   Number of observations                           497
##   Number of missing patterns                        21
##
## Model Test User Model:
##
##   Test statistic                                 3.926
##   Degrees of freedom                                 2
##   P-value (Chi-square)                           0.140
##
## Model Test Baseline Model:
##
##   Test statistic                               769.211
##   Degrees of freedom                                36
##   P-value                                        0.000
##
## User Model versus Baseline Model:
##
##   Comparative Fit Index (CFI)                    0.997
##   Tucker-Lewis Index (TLI)                       0.953
##
## Loglikelihood and Information Criteria:
##
##   Loglikelihood user model (H0)              -7954.164
##   Loglikelihood unrestricted model (H1)      -7952.201
##
##   Akaike (AIC)                               16012.328
##   Bayesian (BIC)                             16231.175
##   Sample-size adjusted Bayesian (BIC)        16066.125
##
## Root Mean Square Error of Approximation:
##
##   RMSEA                                          0.044
##   90 Percent confidence interval - lower         0.000
##   90 Percent confidence interval - upper         0.109
##   P-value RMSEA <= 0.05                          0.459
##
## Standardized Root Mean Square Residual:
##
##   SRMR                                           0.015
##
```

```
## Parameter Estimates:
##
##    Standard errors                              Standard
##    Information                                  Observed
##    Observed information based on                Hessian
##
## Regressions:
##                     Estimate  Std.Err  z-value  P(>|z|)
##    HIVK2 ~
##      HIVK1            0.399    0.047    8.449    0.000
##    HIVK3 ~
##      HIVK2            0.522    0.043   12.137    0.000
##      HIVK1            0.138    0.047    2.927    0.003
##    SELFC2 ~
##      SELFC1           0.384    0.039    9.766    0.000
##    SELFC3 ~
##      SELFC2           0.426    0.050    8.563    0.000
##      SELFC1           0.118    0.044    2.696    0.007
##    INTENT2 ~
##      INTENT1          0.086    0.047    1.813    0.070
##    INTENT3 ~
##      INTENT2          0.276    0.053    5.203    0.000
##      INTENT1          0.156    0.050    3.082    0.002
##    HIVK3 ~
##      SELFC2          -0.003    0.154   -0.017    0.987
##      INTENT2         -0.021    0.099   -0.209    0.834
##    HIVK2 ~
##      SELFC1           0.250    0.134    1.863    0.062
##      INTENT1          0.147    0.099    1.479    0.139
##    SELFC3 ~
##      HIVK2           -0.001    0.014   -0.092    0.927
##      INTENT2          0.125    0.032    3.955    0.000
##    SELFC2 ~
##      HIVK1            0.023    0.014    1.685    0.092
##      INTENT1          0.088    0.030    2.949    0.003
##    INTENT3 ~
##      HIVK2            0.027    0.023    1.143    0.253
##      SELFC2           0.109    0.083    1.317    0.188
##    INTENT2 ~
##      HIVK1            0.064    0.022    2.840    0.005
##      SELFC1           0.319    0.064    4.977    0.000
##    HIVK3 ~
##      SELFC1           0.160    0.136    1.175    0.240
##      INTENT1          0.077    0.096    0.808    0.419
##    INTENT3 ~
```

```
##    SELFC1           0.066    0.072    0.911    0.362
##    HIVK1            0.043    0.026    1.682    0.093
##  SELFC3 ~
##    HIVK1            0.047    0.015    3.030    0.002
##    INTENT1          0.057    0.031    1.844    0.065
##
## Covariances:
##                    Estimate  Std.Err  z-value  P(>|z|)
##  HIVK1 ~~
##    SELFC1           0.684    0.176    3.876    0.000
##  SELFC1 ~~
##    INTENT1          0.341    0.090    3.794    0.000
##  HIVK1 ~~
##    INTENT1         -0.067    0.247   -0.273    0.785
## .SELFC2 ~~
##    .INTENT2         0.368    0.071    5.202    0.000
## .HIVK3 ~~
##    .SELFC3          0.189    0.129    1.467    0.142
##    .INTENT3         0.051    0.220    0.233    0.816
## .SELFC3 ~~
##    .INTENT3         0.308    0.071    4.364    0.000
##
## Intercepts:
##                    Estimate  Std.Err  z-value  P(>|z|)
##    .HIVK2           6.992    0.660   10.589    0.000
##    .HIVK3           4.811    0.714    6.734    0.000
##    .SELFC2          1.400    0.195    7.192    0.000
##    .SELFC3          0.216    0.231    0.932    0.351
##    .INTENT2         1.638    0.313    5.229    0.000
##    .INTENT3         0.751    0.389    1.931    0.054
##    HIVK1           12.192    0.146   83.506    0.000
##    SELFC1           2.370    0.054   44.277    0.000
##    INTENT1          2.789    0.076   36.937    0.000
##
## Variances:
##                    Estimate  Std.Err  z-value  P(>|z|)
##    .HIVK2          10.630    0.699   15.198    0.000
##    .HIVK3           8.145    0.564   14.439    0.000
##    .SELFC2          0.911    0.060   15.149    0.000
##    .SELFC3          0.846    0.059   14.437    0.000
##    .INTENT2         2.310    0.155   14.899    0.000
##    .INTENT3         2.308    0.161   14.361    0.000
##    HIVK1           10.526    0.670   15.708    0.000
##    SELFC1           1.413    0.090   15.655    0.000
##    INTENT1          2.603    0.173   15.073    0.000
```

10.3.1.2   Model Fitting Measures

As seen from the model fitting, we found a satisfactory model fitting with chi-square model fit statistic at 3.926 (df = 2, p = .140). The RMSEA is 0.044 (90% CI = [0.000, 0.109]). The CFI is 0.997 and the TLI is 0.955 with the SRMR at 0.015. This replicates the results in Chen and Chen (2021b) obtained from using *mplus*.

10.3.1.3   Self-Enhancement Process Over Time with Auto-Regressive Effects

As seen from the output on longitudinal effects, the *HIVK1* (i.e., HIV/AIDS knowledge at baseline) had a statistically significant relation with *HIVK2* at the 6-month (0.399, $p = 0.000$) and 12-month (0.138, $p = 0.020$), respectively. *HIVK2* at 6-month was also significantly related to *HIVK3* at 12-month (0.522, $p = 0.000$). This significant auto-regressive effect suggests a self-enhancement process driving the accumulation of HIV/AIDS knowledge over time, as depicted in Figure 10.2.

The same idea about investigating specific longitudinal effects can be applied to assess *Condom use self-efficacy* and *Intention to use a condom*. Similar to our *HIVK* results, we found significant positive auto-regressive relationships between *SELFC1*, *SELFC2*, and *SELFC3* across the three time points as well as *INTENT1*, *INTENT2*, and *INTENT3* across the three time points, which suggests a self-enhancement mechanism for the development of condom use self-efficacy **and** intention to use a condom over time.

10.3.1.4   Cross-Lag Effects

There are multiple paths to examine the cross-lag effects in FLMM as seen in Figure 10.1. For example, as seen from the output, *HIVK1* (HIV/AIDS knowledge at baseline) was not statistically related to *SELFC2* (the self-efficacy for condom use at 6-month) (0.023, $p = 0.071$), but it was significantly related to *SELFC3* at 12-month (0.047, $p = 0.004$). This result suggests a delayed effect by which HIV/AIDS knowledge takes more than 6-month to show its effect on enhancing condom use self-efficacy, as suggested by the cognitive theory

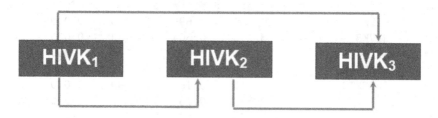

FIGURE 10.2
Testing Self-Enhancement Model of HIV/AIDS Knowledge Over Time.

(Rogers, 1975). Other cross-lag effects can be similarly considered based on the FLMM model.

10.3.1.5 Feedback Effects

With FLMM, we can test different feedback effects as discussed in Chen et al. (2010). As seen in Figure 10.3, there are the HIVK-SELFC loop, the SELFC-INTENT loop and the long HIVK-INTENT-SELFC loop. We can then examine the feedback relationships where three feedback loop relationships are presented. First, it depicts the feedback relationship between *HIV/AIDS knowledge* and *condom use self-efficacy*. The causal relationship from *HIV/AIDS knowledge* to *condom use self-efficacy* can be assessed with *HIVK1* and *SEFC2*; the reverse impact of *condom use self-efficacy* on *HIV/AIDS knowledge* can be assessed with *SELFC1* and *HIVK2*. Second, we can test the feedback loop between *condom use self-efficacy* and the *intention to use condoms*. Third, it depicts the long feedback loop between *HIVK, INTENT,* and *SELFC.*

For example, as seen from the output, the *INTENT1* (i.e., the intention to use condoms at baseline) was significantly related with *SELFC2* at 6-month (0.088, $p = 0.004$) and *INTENT2* at 6-month was significantly related to *SELFC3* at 12-month (0.125, $p = 0.000$). These two coefficients quantified the direction (positive) and strength of the two feedback loops. These results indicate the existence of a positive feedback impact by the intention to use condoms on condom use self-efficacy as shown in Figure 10.3.

Note that the feedback effects or looped relationships for the FLMM in this section are not the same as those mentioned in Section 10.2.5, where the limitations of the computational formulas for specific and total indirect effects are described. The feedback effects in this section refer to the autoregressive or lagged effects within a set of variables that are measured at different time points. Each of these feedback effects refers to a pathway from one variable measured at a time point to another variable measured at a later time point. Indeed, in the FLMM, there is no pathway for any variable that would lead back to itself (i.e., no looping in the path diagram). Therefore, the

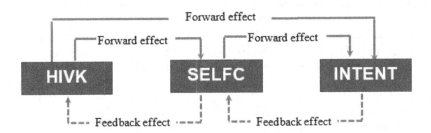

FIGURE 10.3
Potentials of Different Feedback Effects.

limitations about total and specific indirect effect computations described in 10.2.5 do not apply to the FLMM here.

10.3.2 Mediation Analysis

For mediation analysis in *R/lavaan*, we can *label* paths for studying various mediation processes in the current FLMM, much like what we did in Chapter 3. In the current FLMM, we can analyze the following mediation processes from a time-1 variable to a time-3 variable, such as:

- *HIVK1 → HIVK3*
- *HIVK1 → SELFC3*
- *HIVK1 → INTENT3*
- *SELFC1 → HIVK3*
- *SELFC1 → SELFC3*
- *SELFC1 → INTENT3*
- *INTENT1 → HIVK3*
- *INTENT1 → SELFC3*
- *INTENT1 → INTENT3*

For illustration purpose, we analyze only the mediation process from *HIV/AIDS knowledge (HIVK1)* to *Intentions to use a condom during sex (INTENT3)*. First, we need to identify all possible pathways from *HIVK1* to *INTENT3*. The list of the possible pathways is:

(1) *HIVK1 → HIVK2 → INTENT3*
(2) *HIVK1 → SELFC2 → INTENT3*
(3) *HIVK1 → INTENT2 → INTENT3*
(4) *HIVK1 → INTENT3*

Pathways (1)–(3) are specific mediated or longitudinal pathways that comprise the total indirect effect and (4) is a direct pathway that captures the remaining effect (or the direct effect) of *HIVK1* on *INTENT3*.

Next, for each of the first three specific longitudinal pathways, we compute the specific indirect effect as the product of the corresponding effects along the pathway. To facilitate the computation of specific indirect effects, we will label the effects of all directed paths in the following *R* code. The effect of *HIVK1* on *SELFC2* (for the *HIVK1 → SELFC2* path) is s2h1, the effect of *SELFC2* on *INTENT3* (for the *SELFC2 → INTENT3* path) is i3s2, and so on. With this labeling scheme, the computation of the specific indirect effects (see Equation (10.7)) and the direct effect is given by the following formulas:

(1) Specific indirect effect due to the *HIVK1 → HIVK2 → INTENT3* pathway = h2h1 × i3h2

(2) Specific indirect effect due to the *HIVK1 → SELFC2 → INTENT3* pathway = s2h1 × i3s2

(3) Specific indirect effect due to the *HIVK1* → *INTENT2* → *IN-TENT3* pathway = i2h1 × i3i2

(4) Direct effect due to the *HIVK1* → *INTENT3* pathway = i3h1

Finally, the total indirect effect is the sum of the three specific indirect effects (see Equation (10.11)), and the total effect is the sum of the direct and total indirect effect.

The FLMM and the corresponding mediation analyses can now be readily implemented with *R/lavaan* as follows:

```
# The FLMM for Bahamas Youth
medFLMMBahamas <- '
# Longitudinal effects for HIV: 1st-order auto-regressive(AR)
HIVK2 ~ h2h1*HIVK1; HIVK3 ~ h3h2*HIVK2;
# Longitudinal effects for HIV: 2nd-order auto-regressive
HIVK3 ~ h3h1*HIVK1;

# Longitudinal effects for SELFC: 1st-order AR
SELFC2 ~ s2s1*SELFC1; SELFC3 ~ s3s2*SELFC2;
# Longitudinal effects for SELFC: 2nd-order AR
SELFC3 ~ s3s1*SELFC1;

# Longitudinal effects for INTENT: 1st-order AR
INTENT2 ~ i2i1*INTENT1; INTENT3 ~ i3i2*INTENT2;
# Longitudinal effects for INTENT: 2nd-order AR
INTENT3 ~ i3i1*INTENT1;

# 1st-order Cross-lags
HIVK3 ~ h3s2*SELFC2; HIVK3 ~ h3i2*INTENT2;
HIVK2 ~ h2s1*SELFC1; HIVK2 ~ h2i1*INTENT1;

SELFC3 ~ s3h2*HIVK2; SELFC3 ~ s3i2*INTENT2;
SELFC2 ~ s2h1*HIVK1; SELFC2 ~ s2i1*INTENT1;

INTENT3 ~ i3h2*HIVK2; INTENT3 ~ i3s2*SELFC2;
INTENT2 ~ i2h1*HIVK1; INTENT2 ~ i2s1*SELFC1;

# 2nd-order cross-lags
HIVK3    ~ h3s1*SELFC1; HIVK3    ~ h3i1*INTENT1;
INTENT3 ~ i3s1*SELFC1; INTENT3 ~ i3h1*HIVK1;
SELFC3  ~ s3h1*HIVK1;  SELFC3  ~ s3i1*INTENT1;

# Correlations for better model fitting
SELFC1  ~~ HIVK1; SELFC1 ~~ INTENT1;
INTENT1 ~~ HIVK1; SELFC2 ~~ INTENT2;
```

```
#
# Now the mediation analysis
#
# Define Direct, Indirect, and Total Effects with Multiple
  Mediation Path
# (a) HIVK1-HIVK2-INTENT3
mM.indirect1 := h2h1*i3h2
# (b) HIVK1-SELFC2-INTENT3
mM.indirect2 := s2h1*i3s2
# (c) HIVK1-INTENT2-INTENT3
mM.indirect3 := i2h1*i3i2
# The indirect effect by summing up all specifica indirect effects
mM.t_indirect := h2h1*i3h2 + s2h1*i3s2 + i2h1*i3i2
mM.direct    := i3h1
mM.total     := i3h1 + h2h1*i3h2 + s2h1*i3s2 + i2h1*i3i2
```

With this *medFLMMBahamas*, we can then call *sem* with *bootstrap* to estimate the standard errors as follows:

```
# Set random seed for bootstrapping
set.seed(3388)
# Model fitting
fit.medBahamas <- sem(medFLMMBahamas, data = dBahamas,
                  missing="fiml", se = 'bootstrap',
                  bootstrap = 1000)
# Print the model fitting
summary(fit.medBahamas, fit.measures=FALSE)
```

```
## lavaan 0.6-12 ended normally after 68 iterations
##
##    Estimator                                        ML
##    Optimization method                          NLMINB
##    Number of model parameters                       52
##
##    Number of observations                          497
##    Number of missing patterns                       21
##
## Model Test User Model:
##
##    Test statistic                                3.926
##    Degrees of freedom                                2
##    P-value (Chi-square)                          0.140
##
```

```
## Parameter Estimates:
##
##    Standard errors                                  Bootstrap
##    Number of requested bootstrap draws                 1000
##    Number of successful bootstrap draws                1000
##
## Regressions:
##                      Estimate  Std.Err  z-value  P(>|z|)
##    HIVK2 ~
##      HIVK1   (h2h1)     0.399    0.054    7.420    0.000
##    HIVK3 ~
##      HIVK2   (h3h2)     0.522    0.052   10.059    0.000
##      HIVK1   (h3h1)     0.138    0.058    2.393    0.017
##    SELFC2 ~
##      SELFC1  (s2s1)     0.384    0.039    9.836    0.000
##    SELFC3 ~
##      SELFC2  (s3s2)     0.426    0.054    7.930    0.000
##      SELFC1  (s3s1)     0.118    0.049    2.393    0.017
##    INTENT2 ~
##      INTENT1 (i2i1)     0.086    0.049    1.732    0.083
##    INTENT3 ~
##      INTENT2 (i3i2)     0.276    0.057    4.809    0.000
##      INTENT1 (i3i1)     0.156    0.049    3.147    0.002
##    HIVK3 ~
##      SELFC2  (h3s2)    -0.003    0.175   -0.015    0.988
##      INTENT2 (h3i2)    -0.021    0.090   -0.231    0.817
##    HIVK2 ~
##      SELFC1  (h2s1)     0.250    0.141    1.775    0.076
##      INTENT1 (h2i1)     0.147    0.099    1.476    0.140
##    SELFC3 ~
##      HIVK2   (s3h2)    -0.001    0.015   -0.086    0.932
##      INTENT2 (s3i2)     0.125    0.031    3.984    0.000
##    SELFC2 ~
##      HIVK1   (s2h1)     0.023    0.013    1.770    0.077
##      INTENT1 (s2i1)     0.088    0.031    2.842    0.004
##    INTENT3 ~
##      HIVK2   (i3h2)     0.027    0.021    1.283    0.199
##      SELFC2  (i3s2)     0.109    0.090    1.210    0.226
##    INTENT2 ~
##      HIVK1   (i2h1)     0.064    0.022    2.844    0.004
##      SELFC1  (i2s1)     0.319    0.064    5.005    0.000
##    HIVK3 ~
##      SELFC1  (h3s1)     0.160    0.146    1.095    0.274
##      INTENT1 (h3i1)     0.077    0.089    0.868    0.385
```

```
##    INTENT3 ~
##      SELFC1  (i3s1)   0.066    0.080    0.823    0.410
##      HIVK1   (i3h1)   0.043    0.024    1.801    0.072
##    SELFC3 ~
##      HIVK1   (s3h1)   0.047    0.015    3.066    0.002
##      INTENT1 (s3i1)   0.057    0.032    1.791    0.073
##
## Covariances:
##                     Estimate  Std.Err  z-value  P(>|z|)
##    HIVK1 ~~
##      SELFC1           0.684    0.168    4.079    0.000
##    SELFC1 ~~
##      INTENT1          0.341    0.095    3.573    0.000
##    HIVK1 ~~
##      INTENT1         -0.067    0.231   -0.292    0.771
##   .SELFC2 ~~
##     .INTENT2          0.368    0.072    5.141    0.000
##   .HIVK3 ~~
##     .SELFC3           0.189    0.126    1.502    0.133
##     .INTENT3          0.051    0.201    0.254    0.799
##   .SELFC3 ~~
##     .INTENT3          0.308    0.070    4.403    0.000
##
## Intercepts:
##                     Estimate  Std.Err  z-value  P(>|z|)
##    .HIVK2            6.992    0.701    9.972    0.000
##    .HIVK3            4.811    0.837    5.751    0.000
##    .SELFC2           1.400    0.183    7.658    0.000
##    .SELFC3           0.216    0.226    0.952    0.341
##    .INTENT2          1.638    0.319    5.132    0.000
##    .INTENT3          0.751    0.368    2.039    0.041
##     HIVK1           12.192    0.144   84.448    0.000
##     SELFC1           2.370    0.053   44.777    0.000
##     INTENT1          2.789    0.078   35.732    0.000
##
## Variances:
##                     Estimate  Std.Err  z-value  P(>|z|)
##    .HIVK2           10.630    0.781   13.611    0.000
##    .HIVK3            8.145    0.753   10.818    0.000
##    .SELFC2           0.911    0.055   16.673    0.000
##    .SELFC3           0.846    0.061   13.858    0.000
##    .INTENT2          2.310    0.104   22.289    0.000
##    .INTENT3          2.308    0.125   18.500    0.000
##     HIVK1           10.526    0.702   14.991    0.000
##     SELFC1           1.413    0.071   20.031    0.000
```

```
##      INTENT1           2.603    0.084    30.830    0.000
##
## Defined Parameters:
##                      Estimate  Std.Err  z-value  P(>|z|)
##      mM.indirect1      0.011    0.008    1.256    0.209
##      mM.indirect2      0.003    0.003    0.879    0.380
##      mM.indirect3      0.018    0.007    2.383    0.017
##      mM.t_indirect     0.031    0.011    2.680    0.007
##      mM.direct         0.043    0.024    1.800    0.072
##      mM.total          0.074    0.024    3.095    0.002
```

As seen from the *Defined Parameters*, the estimated *total indirect* effect from *HIVK1* to *INTENT3* is 0.031 with a *p*-value of 0.007 and the estimated *total* effect from *HIVK1* to *INTENT3* is 0.074 with a *p*-value of 0.002. Together, these results suggest that *HIV/AIDS knowledge* promotes future *Condom use intention.*

Looking more closely at the decomposition of the total effect, we can see that the direct effect contributes more than half of the total effect (.043 out of .074), even though the direct effect estimate is marginally significant (p = .072). This might imply that early *HIV/AIDS knowledge* does benefit later *Condom use intention* directly (i.e., not through the specified indirect pathways). More statistical evidence (e.g., with a larger sample size) needs to be shown to support a reliable direct effect.

Among the specific indirect effects, the *HIVK1* → *INTENT2* → *INTENT3* pathway contributes the most part of the total indirect effect of *HIV/AIDS knowledge* on *Condom use intention* (0.018 out of 0.031). In fact, only this specific indirect effect is statistically significant. Together with the interpretation of the direct effect, we might conclude that early *HIV/AIDS knowledge* benefits later *Condom use intention*, either directly or indirectly through earlier influence on *Condom use intention* (i.e., at time 2).

Note that there are slightly different estimated values between *R/lavaan* and *mplus* reported in Chen and Chen (2021b), which could be from different random starting seeds in the bootstrapping.

10.4   Data Analysis Using SAS

The *SAS* data set *DControlW123*, which includes the same set of variables that were used in this chapter, was prepared. Missing values in this data set were coded as ".", which is a common practice in the SAS system.

In this illustration, the same FLMM as the one analyzed by *R/lavaan* is fitted and the total effect of *HIV/AIDS knowledge* at time 1 on *Condom use intention* at time 3 is decomposed into direct and indirect effects. The following

SAS code inputs the *SAS* data set and estimate the specified FLMM by the full information maximum likelihood method:

```
proc calis data="c:\DControlW123" meanstr method=FIML;
   path
      /* Longitudinal effects for HIV: 1st-order auto-regressive(AR) */
      HIVK1   ==> HIVK2  = h2h1,
      HIVK2   ==> HIVK3,

      /* Longitudinal effects for HIV: 2nd-order auto-regressive */
      HIVK1   ==> HIVK3,

      /* Longitudinal effects for SELFC: 1st-order AR */
      SELFC1  ==> SELFC2,
      SELFC2  ==> SELFC3,

      /* Longitudinal effects for SELFC: 2nd-order AR */
      SELFC1  ==> SELFC3,

      /* Longitudinal effects for INTENT: 1st-order AR */
      INTENT1 ==> INTENT2,
      INTENT2 ==> INTENT3 = i3i2,

      /* Longitudinal effects for INTENT: 2nd-order AR */
      INTENT1 ==> INTENT3,

      /* 1st-order Cross-lags */
      SELFC2  ==> HIVK3,
      INTENT2 ==> HIVK3,
      SELFC1  ==> HIVK2,
      INTENT1 ==> HIVK2,

      HIVK2   ==> SELFC3,
      INTENT2 ==> SELFC3,
      HIVK1   ==> SELFC2   = s2h1,
      INTENT1 ==> SELFC2,

      HIVK2   ==> INTENT3  = i3h2,
      SELFC2  ==> INTENT3  = i3s2,
      HIVK1   ==> INTENT2  = i2h1,
      SELFC1  ==> INTENT2,

      /* 2nd-order cross-lags */
      SELFC1  ==> HIVK3,
      INTENT1 ==> HIVK3,
      SELFC1  ==> INTENT3,
```

```
     HIVK1   ==> INTENT3  = i3h1,
     HIVK1   ==> SELFC3,
     INTENT1 ==> SELFC3;

pcov
    /* Exogenous covariances */
    SELFC1 HIVK1,   SELFC1 INTENT1, INTENT1 HIVK1,
    /* Error Correlations/covariances for better model fitting */
    SELFC2 INTENT2, HIVK3 SELFC3, HIVK3 INTENT3, SELFC3 INTENT3;

fitindex on(only)=[chisq df probchi RMSEA RMSEA_LL RMSEA_UL
        PROBCLFIT SRMR CFI BENTLERNNFI] noindextype;
effpart
    HIVK1 SELFC1 INTENT1  ==> HIVK3 SELFC3 INTENT3;

parameters
    indirect1_h2 indirect2_s2 indirect3_i2 sum_indirect direct total;
indirect1_h2    = h2h1 * i3h2;
indirect2_s2    = s2h1 * i3s2;
indirect3_i2    = i2h1 * i3i2;
sum_indirect    = indirect1_h2 + indirect2_s2 +indirect3_i2;
direct          = i3h1;
total           = direct + sum_indirect;
run;
```

For the ease of references and comparisons, comments and parameter names in the SAS code follow that of the *lavaan* example.

In the PATH statement, all the direct paths in the model are specified. For clarity of presentation, only those paths that are relevant for the current computation of direct and indirect effects are labeled with path coefficients or effect parameters (i.e., *h2h1*, *i3i2*, and so on).

In the PCOV statement, three covariances among the exogenous variables are first specified, followed by the covariances of error terms for the endogenous variables in the model. Note that unlike the *lavaan* specification of the FLMM, the error covariances (or correlations) are not default parameters in the CALIS procedure and must be specified explicitly.

In the FITINDEX statement, some popular fit indices are requested in the output.

In the EFFPART statement, you request the total effect decomposition of each variable at Time 1 on all variables at Time 3. All these total effects are decomposed into the corresponding direct and total indirect effects. The *CALIS* procedure does not compute the specific indirect effects automatically. Therefore, specifications of additional parameters using the PARAMETER and programming statements are necessary to study the specific indirect effects.

Finally, in the PARAMETER statement and the programming statements, parameters for specific indirect effects, total indirect effect, direct effect, and total effect of *HIVK1* on *INTENT3* are defined and computed. See Section 10.2 and Equations (10.7) and (10.11) for explanations of the computation of all these effects. The three specific indirect effects and their corresponding pathways are explained as follows:

- *indirect1_h2* (indirect effect via HIVK2): *HIVK1* → *HIVK2* → *INTENT3*
- *indirect2_s2* (indirect effect via SELFC2): *HIVK1* → *SELFC2* → *INTENT3*
- *indirect3_i2* (indirect effect via INTENT2): *HIVK1* → *INTENT2* → *INTENT3*

The *sum_indirect* parameter represents the total indirect effects, which is the sum of *indirect1_h2*, *indirect2_s2*, and *indirect3_i2*. The direct effect (the *direct* parameter) of *HIVK1* on *INTENT3* is simply represented by the *i3h1* parameter. Finally, the total effect (i.e., the *total* parameter) is the sum of the direct and the total indirect effect.

The *CALIS* procedure displays the following results for the model fit:

Fit Summary	
Chi-Square	3.9263
Chi-Square DF	2
Pr > Chi-Square	0.1404
Standardized RMR (SRMR)	0.0147
RMSEA Estimate	0.0440
RMSEA Lower 90% Confidence Limit	0.0000
RMSEA Upper 90% Confidence Limit	0.1088
Probability of Close Fit	0.4585
Bentler Comparative Fit Index	0.9974
Bentler-Bonett Non-normed Index	0.9527

As discussed previously, the model fit is good. Next, the estimates of path coefficients or all direct effects in the model are shown. Some parameter labels were customized in our specifications (e.g., *h2h1*) and their values were used to compute additional parameter estimates.

	PATH List						
Path			Parameter	Estimate	Standard Error	t Value	Pr > |t|
HIVK1	===>	HIVK2	h2h1	0.39934	0.04726	8.4494	<.0001
HIVK2	===>	HIVK3	_Parm01	0.52225	0.04303	12.1365	<.0001
HIVK1	===>	HIVK3	_Parm02	0.13824	0.04723	2.9270	0.0034
SELFC1	===>	SELFC2	_Parm03	0.38386	0.03931	9.7657	<.0001
SELFC2	===>	SELFC3	_Parm04	0.42562	0.04970	8.5634	<.0001
SELFC1	===>	SELFC3	_Parm05	0.11828	0.04387	2.6962	0.0070
INTENT1	===>	INTENT2	_Parm06	0.08570	0.04726	1.8135	0.0698
INTENT2	===>	INTENT3	i3i2	0.27627	0.05309	5.2034	<.0001
INTENT1	===>	INTENT3	_Parm07	0.15555	0.05047	3.0819	0.0021
SELFC2	===>	HIVK3	_Parm08	-0.00255	0.15392	-0.0166	0.9868
INTENT2	===>	HIVK3	_Parm09	-0.02079	0.09948	-0.2090	0.8345
SELFC1	===>	HIVK2	_Parm10	0.24984	0.13409	1.8632	0.0624
INTENT1	===>	HIVK2	_Parm11	0.14657	0.09908	1.4794	0.1390
HIVK2	===>	SELFC3	_Parm12	-0.00129	0.01400	-0.0919	0.9268
INTENT2	===>	SELFC3	_Parm13	0.12517	0.03164	3.9556	<.0001
HIVK1	===>	SELFC2	s2h1	0.02346	0.01392	1.6855	0.0919
INTENT1	===>	SELFC2	_Parm14	0.08813	0.02988	2.9492	0.0032
HIVK2	===>	INTENT3	i3h2	0.02660	0.02328	1.1428	0.2531
SELFC2	===>	INTENT3	i3s2	0.10912	0.08288	1.3167	0.1880
HIVK1	===>	INTENT2	i2h1	0.06370	0.02243	2.8402	0.0045
SELFC1	===>	INTENT2	_Parm15	0.31881	0.06405	4.9772	<.0001
SELFC1	===>	HIVK3	_Parm16	0.15957	0.13582	1.1749	0.2401
INTENT1	===>	HIVK3	_Parm17	0.07730	0.09570	0.8077	0.4192
SELFC1	===>	INTENT3	_Parm18	0.06585	0.07230	0.9108	0.3624
HIVK1	===>	INTENT3	i3h1	0.04309	0.02562	1.6818	0.0926
HIVK1	===>	SELFC3	_Parm19	0.04659	0.01538	3.0296	0.0024
INTENT1	===>	SELFC3	_Parm20	0.05742	0.03113	1.8441	0.0652

The "Additional Parameters" table shows the estimates of parameters that are defined in the PARAMETER statements. In the current analysis, the components of the total effect of *HIVK1* on *INTENT3* are of interest. The total effect estimate is 0.074 ($p < 0.05$) and it is the sum of the direct effect (Direct = 0.043, $p = 0.09$) and the total indirect effect (Sum_indirect = 0.031, $p < 0.05$). Hence, although the direct effect appears to contribute more to the total effect than the sum of indirect effects, it is only marginally significant. More evidence (perhaps with a larger sample size) is needed for supporting a significant direct effect.

Additional Parameters					
Type	Parameter	Estimate	Standard Error	t Value	Pr > |t|
Dependent	indirect1_h2	0.01062	0.00939	1.1316	0.2578
	indirect2_s2	0.00256	0.00248	1.0339	0.3012
	indirect3_i2	0.01760	0.00709	2.4835	0.0130
	sum_indirect	0.03078	0.01186	2.5955	0.0094
	direct	0.04309	0.02562	1.6818	0.0926
	total	0.07387	0.02455	3.0087	0.0026

The total indirect effect is further decomposed into three specific indirect effects. The results show that only the indirect effect via *INTENT2* is statistically significant (Indirect3_i2 = 0.018, $p < 0.05$) and it also contributes to most of the total indirect effects. Certainly, all these results are similar to that of the *R/lavaan* model fitting. Therefore, the same interpretation applies here—that is, early *HIV/AIDS knowledge* benefits later *Condom use intention*, either directly or indirectly through earlier influence on *Condom use intention* at time 2.

The EFFPART statement specification of *CALIS* displays the following results:

Effects of HIVK1			
Effect / Std Error / t Value / p Value			
	Total	Direct	Indirect
HIVK3	0.3454	0.1382	0.2072
	0.0495	0.0472	0.0304
	6.9832	2.9270	6.8106
	<.0001	0.003422	<.0001
SELFC3	0.0640	0.0466	0.0174
	0.0156	0.0154	0.009197
	4.0953	3.0296	1.8965
	<.0001	0.002449	0.0579
INTENT3	0.0739	0.0431	0.0308
	0.0246	0.0256	0.0119
	3.0087	1.6818	2.5955
	0.002624	0.0926	0.009444

Effects of SELFC1			
Effect / Std Error / t Value / p Value			
	Total	Direct	Indirect
HIVK3	0.2824	0.1596	0.1229
	0.1397	0.1358	0.0916
	2.0219	1.1749	1.3420
	0.0432	0.2401	0.1796
SELFC3	0.3212	0.1183	0.2030
	0.0440	0.0439	0.0279
	7.3027	2.6962	7.2685
	<.0001	0.007014	<.0001
INTENT3	0.2025	0.0658	0.1366
	0.0682	0.0723	0.0374
	2.9670	0.9108	3.6568
	0.003007	0.3624	0.000255

Effects of INTENT1			
Effect / Std Error / t Value / p Value			
	Total	Direct	Indirect
HIVK3	0.1518	0.0773	0.0745
	0.1050	0.0957	0.0539
	1.4464	0.8077	1.3827
	0.1481	0.4192	0.1668
SELFC3	0.1055	0.0574	0.0480
	0.0333	0.0311	0.0159
	3.1686	1.8441	3.0287
	0.001532	0.0652	0.002456
INTENT3	0.1927	0.1555	0.0372
	0.0512	0.0505	0.0165
	3.7638	3.0819	2.2555
	0.000167	0.002057	0.0241

Each table displays the effect decompositions of a Time-1 variable on three Time-3 variables. For example, the first table decomposes the *HIVK1* total effects on *HIVK3*, *SELFC3*, and *INTENT3* into the corresponding direct effects and total indirect effect. Certainly, the results for *INTENT3* match that of the analysis of additional parameter estimates described previously.

Unfortunately, these tables do not decompose the total indirect effect into specific indirect effects. Therefore, these tables are best used as an preliminary overall analysis of effect decomposition when mediation with multiple pathways is present. For example, in the last table, the total, direct, indirect effects of *INTENT1* on *HIVK3* are all not statistically significant, implying a minimal interest of studying further the specific indirect effects.

In contrast, the total and the indirect effects of *INTENT1* on *SELF3* are both statistically significant and the total indirect effect contributes to about half of the total effect (i.e., 0.048 out of 0.106). In this case, it is of interest to investigate how specific pathways or indirect effects from *INTENT1* to *SELFC3* contribute to the total indirect effect. This investigation is left as an exercise to the readers.

10.5 Discussions

In this chapter, we demonstrated the full longitudinal mediation model (FLMM) by analyzing three waves of data from an educational HIV prevention program. The FLMM is applied to longitudinal data for testing causal relationships in social behavior research. FLMM has the advantage of studying specific mediation pathways so that more refined conclusions about the mediation process can be drawn.

With the FLMM, we can additionally examine the longitudinal self-enhancement effects as well as the feedback effects among the same or different measurements at different time points. Therefore, the full longitudinal mediation analysis provides an advanced method for producing informative empirical evidence in theory-guided causal relationship research.

10.6   Exercises

1. Use *lavaan* or *CALIS* to study the total effect decompositions of *SELFC1* on *SELFC3*, *INTENT1* on *SELFC3*, *INTENT1* on *INTENT3*, respectively, for the model and data used in this chapter. Interpret the results.

Appendix: Mplus Program

Note that this *mplus* program (file named *Bahamas3W-JSSWR.inp*) is included at the *Data* subfolder along with the *mplus* output (file named as *bahamas3w-jsswr.out*).

```
!!!!!!!!!!!!!!!!!!!!!!!!!!!!!!!!!!!!!!!!!!!!!!!!!!!!!!!!!!!!!!!!!!!!!
! Mplus Program to Analyze Bahamas Longitudinal data
!!!!!!!!!!!!!!!!!!!!!!!!!!!!!!!!!!!!!!!!!!!!!!!!!!!!!!!!!!!!!!!!!!!!!

DATA:
FILE IS dControlW123.dat;
VARIABLE:
NAMES ARE
    Gender Age HIVK1 HIVK2 HIVK3 SELFC1 SELFC2 SELFC3 INTENT1
        INTENT2 INTENT3 ID;
MISSING IS *;
USEVARIABLES ARE
  HIVK1 HIVK2 HIVK3 SELFC1 SELFC2 SELFC3 INTENT1 INTENT2
        INTENT3;
ANALYSIS:
BOOTSTRAP = 1000;
MODEL:
HIVK3 on HIVK1; HIVK3 on HIVK2; HIVK3 on SELFC2;
    HIVK3 on INTENT2;
HIVK2 on HIVK1; HIVK2 on SELFC1; HIVK2 on INTENT1;

SELFC3 on SELFC1; SELFC3 on SELFC2; SELFC3 on HIVK2;
    SELFC3 on INTENT2;
SELFC2 on SELFC1; SELFC2 on HIVK1;  SELFC2 on INTENT1;
```

```
INTENT3 on INTENT1; INTENT3 on INTENT2; INTENT3 on HIVK2;
    INTENT3 on SELFC2;
INTENT2 on INTENT1; INTENT2 on HIVK1;   INTENT2 on SELFC1;

! 1 to 3
SELFC3 on HIVK1; Intent3 on HIVK1;
hivk3 on selfc1; Intent3 on selfc1;
SELFC3 on intent1; hivk3 on intent1;

! correlations
SELFC1 with HIVK1; SELFC1 with INTENT1; INTENT1 with HIVK1;
!SELFC2 with HIVK2;
SELFC2 with INTENT2;   ! important one for model fitting
!INTENT2 with HIVK2;
!SELFC3 with HIVK3;
!SELFC3 with INTENT3;
!INTENT2 with HIVK3;


!!! Indirect effect
Model indirect:
HIVK3 ind INTENT1;
HIVK3 ind SELFC1;
HIVK3 ind HIVK1;
INTENT3 ind INTENT1;
INTENT3 ind SELFC1;
INTENT3 ind HIVK1;
SELFC3 ind INTENT1;
SELFC3 ind SELFC1;
SELFC3 ind HIVK1;

INTENT3 ind SELFC2 HIVK1;
OUTPUT: stdyx; CINTERVAL (BCBOOTSTRAP);
```

11

Multi-Level Structural Equation Modeling

Data produced from most of the research in social sciences and public health are multi-level structures, which are clustered and the data within cluster are correlated. In most instances, if not all, the classical structural equation modeling is not suited for the analysis of multi-level or clustered data to produce valid and efficient estimates because of the correlations existed in the multi-level data. Therefore, the structural equation modeling should be extended to incorporate the multi-level data structure. Extensive and more detailed discussions on SEM for multi-level data can be found in Silva et al. (2019) (supplementary materials and data can be downloaded from: `http://levente.littvay.hu/msem/`) and Hox et al. (2018).

In this chapter, we introduce the implementation in *lavaan* for analyzing multi-level data. We make use of the *GALO* data used in Hox et al. (2018). As analyzed in Chapter 15 of this book, the data were collected from 1377 pupils within 58 schools from an educational study by Schijf and Dronkers (1991). Therefore, in this data set, the 1377 pupils were nested within the 58 schools, where the 58 schools acted as clusters. Due to clustering, pupils' outcomes within the schools are more dependent on each other due to school-level factors (such as school environment, teachers' education level, school management practice, etc.). This data was used to illustrate *lavaan* multi-level SEM (`https://users.ugent.be/~yrosseel/lavaan/zurich2020/lavaan_multilevel_zurich2020.pdf`) and we will further use this data set and the *R* code from this file in this chapter to promote the application of multi-level structural equation modeling using *lavaan*.

11.1 Data Descriptions

As discussed in Hox et al. (2018), the *GALO* data were collected from 1559 pupils, but with complete data from 1377 pupils, within 58 schools. The data is available in *galo.dat*, which has the following variables:

- At pupil-level:
  - father's occupational status *focc*;
  - father's education *feduc*;

- mother's education *meduc*;
- pupil's sex *sex*;
- result of GALO school achievement test *GALO*;
- teacher's advice about secondary education *advice*;

- At the school level:

  - only the school's denomination *denom*.

Denomination is coded *1 = Protestant, 2 = nondenominational, 3 = Catholic* (categories based on optimal scaling). The data file has some missing values coded as *999*. We can load the data into *R* as follows:

```
# Read the data into R
dGalo <- read.table("data/Galo.dat", na.strings="999")
# Name the variables
names(dGalo) <- c("school", "sex", "galo", "advice", "feduc",
                  "meduc", "focc", "denom")
# Make the denom dummy variables
dGalo$denom1 <- ifelse(dGalo$denom == 1, 1, 0)
dGalo$denom2 <- ifelse(dGalo$denom == 2, 1, 0)
# Print the data summary
summary(dGalo)
```

```
##      school          sex             galo
##   Min.   : 1.0   Min.   :1.00   Min.   : 53
##   1st Qu.:16.0   1st Qu.:1.00   1st Qu.: 94
##   Median :30.0   Median :2.00   Median :103
##   Mean   :29.9   Mean   :1.51   Mean   :102
##   3rd Qu.:43.0   3rd Qu.:2.00   3rd Qu.:111
##   Max.   :58.0   Max.   :2.00   Max.   :143
##
##      advice          feduc          meduc
##   Min.   :0.00   Min.   :1      Min.   :1
##   1st Qu.:2.00   1st Qu.:1      1st Qu.:1
##   Median :2.00   Median :4      Median :2
##   Mean   :3.12   Mean   :4      Mean   :3
##   3rd Qu.:4.00   3rd Qu.:6      3rd Qu.:5
##   Max.   :6.00   Max.   :9      Max.   :9
##   NA's   :7      NA's   :89     NA's   :61
##       focc           denom          denom1
##   Min.   :1.0    Min.   :1.00   Min.   :0.00
##   1st Qu.:2.0    1st Qu.:2.00   1st Qu.:0.00
##   Median :3.0    Median :2.00   Median :0.00
##   Mean   :3.3    Mean   :2.01   Mean   :0.15
##   3rd Qu.:5.0    3rd Qu.:2.00   3rd Qu.:0.00
##   Max.   :6.0    Max.   :3.00   Max.   :1.00
```

```
##   NA's    :117
##       denom2
##   Min.    :0.000
##   1st Qu.:0.000
##   Median :1.000
##   Mean    :0.693
##   3rd Qu.:1.000
##   Max.    :1.000
##
```

The general research question is how the school's denomination impacts the *GALO* score and the teacher's advice after adjusting all other variables.

To answer this research question, Hox et al. (2018) analyzed this data with a sequential multi-level regression models and concluded that a multi-level structural equation model is more advantageous to include all hypothesized relations between independent, intervening and dependent variables simultaneously as illustrated in Figure 11.1.

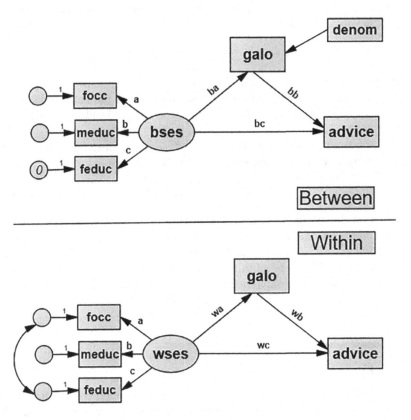

FIGURE 11.1
Multi-Level Mediation Model for GALO Data.

There are two components in this figure. The first component is the *Within* SEM where the *within social economic status* (denoted by *wses*) is a latent variable measured by the father's occupational status *focc*, father's education *feduc*, and mother's education *meduc*. A mediation model is then built on this latent variable with the result of GALO school achievement test *galo* as the mediator and the teacher's advice about secondary education *advice* as the outcome variable. Due to multi-level clustering structure, these pupils are nested within 58 schools and therefore, the second component is to model the *between*-school relationship with the school-level denomination *denom* as a covariate. The pupil's *sex* was not significant and not included in further analysis.

11.2 General Concept for Multi-Level Structural Equation Modeling

Typical implementation of multi-level SEM is to decompose the observed outcome into two components with one to describe the *within* and the other to describe *between* as follows:

$$y_{ij} - \bar{y}_{..} = (y_{ij} - \bar{y}_{.j}) + (\bar{y}_{.j} - \bar{y}_{..}) \qquad (11.1)$$

where j is the indicator for jth cluster (i.e., *school* in the above example) with $j = 1, \ldots, J$ ($J = 48$ in the above data set) and i is the indicator for ith pupil with $i = 1, \ldots, n_j$ within the cluster. $\bar{y}_{.j}$ is the cluster-level mean for cluster j and $\bar{y}_{..}$ is the overall mean.

The above equation would correspond to a decomposition of population covariance matrix into the within and between components:

$$\text{Cov}(y) = \Sigma_T = \Sigma_W + \Sigma_B \qquad (11.2)$$

Based on this decomposition, a corresponding likelihood can be constructed to estimate parameters associated with factor loadings, path coefficients, and residual variances in structural equation models. The standard errors and confidence intervals can be obtained from the theory of maximum likelihood estimation for statistical inference. This estimation is implemented in *lavaan* based on the methods in McDonald and Goldstein (1989).

11.3 Data Analysis Using *R*

The SEM illustrated in Figure 11.1 can be implemented it into *lavaan* (*R* code from https://users.ugent.be/~yrosseel/lavaan/zurich2020/lavaan_multilevel_zurich2020.pdf) as follows (with more explanations):

```
# DEfine the model described in the Figure
model <- '
 level: within
 # latent variable-wses
 wses    =~ a*focc + b*meduc + c*feduc
 # residual correlation
 focc    ~~ feduc
 # mediation model paths
 advice  ~  wc*wses + wb*galo
 galo    ~  wa*wses

 level: between
 # latent variable: bses
 bses    =~ a*focc + b*meduc + c*feduc
 # fixed residual variance
 feduc   ~~ 0*feduc
 # mediation model paths with denomination covariate
 advice  ~  bc*bses + bb*galo
 galo    ~  ba*bses + denom1 + denom2

# defined parameters for indirect effect
 wi := wa * wb
 bi := ba * bb
# define parameter for total effect
 wt := wc + wa * wb
 bt := bc + ba * bb
'
```

With this model construction, we can then call *lavaan* function *sem* and specify the multi-level structure with model option *cluster="school* to fit the data as follows:

```
# Call "sem* to fit the model
fit <- sem(model, data = dGalo, cluster = "school")

# Print the summary of the model fit
summary(fit,fit.measures=TRUE)

## lavaan 0.6-12 ended normally after 137 iterations
##
## Estimator                                       ML
## Optimization method                         NLMINB
## Number of model parameters                      29
## Number of equality constraints                   2
##
```

```
##                                              Used      Total
##    Number of observations                    1382       1559
##    Number of clusters [school]                 58
##
## Model Test User Model:
##
##    Test statistic                           21.763
##    Degrees of freedom                           18
##    P-value (Chi-square)                      0.243
##
## Model Test Baseline Model:
##
##    Test statistic                         2952.629
##    Degrees of freedom                           30
##    P-value                                   0.000
##
## User Model versus Baseline Model:
##
##    Comparative Fit Index (CFI)               0.999
##    Tucker-Lewis Index (TLI)                  0.998
##
## Loglikelihood and Information Criteria:
##
##    Loglikelihood user model (H0)         -14738.573
##    Loglikelihood unrestricted model (H1) -14727.692
##
##    Akaike (AIC)                          29531.146
##    Bayesian (BIC)                        29672.391
##    Sample-size adjusted Bayesian (BIC)   29586.623
##
## Root Mean Square Error of Approximation:
##
##    RMSEA                                     0.012
##    90 Percent confidence interval - lower    0.000
##    90 Percent confidence interval - upper    0.028
##    P-value RMSEA <= 0.05                     1.000
##
## Standardized Root Mean Square Residual (corr metric):
##
##    SRMR (within covariance matrix)           0.014
##    SRMR (between covariance matrix)          0.070
##
## Parameter Estimates:
##
##    Standard errors                        Standard
##    Information                            Observed
```

```
##   Observed information based on              Hessian
##
##
## Level 1 [within]:
##
## Latent Variables:
##                   Estimate  Std.Err  z-value  P(>|z|)
##   wses =~
##     focc      (a)    1.000
##     meduc     (b)    1.710    0.104   16.503    0.000
##     feduc     (c)    2.245    0.101   22.284    0.000
##
## Regressions:
##                   Estimate  Std.Err  z-value  P(>|z|)
##   advice ~
##     wses     (wc)    0.157    0.036    4.318    0.000
##     galo     (wb)    0.086    0.002   44.810    0.000
##   galo ~
##     wses     (wa)    5.554    0.571    9.732    0.000
##
## Covariances:
##                   Estimate  Std.Err  z-value  P(>|z|)
##  .focc ~~
##    .feduc           0.245    0.088    2.769    0.006
##
## Intercepts:
##                   Estimate  Std.Err  z-value  P(>|z|)
##    .focc            0.000
##    .meduc           0.000
##    .feduc           0.000
##    .advice          0.000
##    .galo            0.000
##     wses            0.000
##
## Variances:
##                   Estimate  Std.Err  z-value  P(>|z|)
##    .focc            1.182    0.066   17.868    0.000
##    .meduc           2.026    0.122   16.630    0.000
##    .feduc           1.547    0.173    8.933    0.000
##    .advice          0.574    0.022   25.520    0.000
##    .galo          125.172    5.124   24.429    0.000
##     wses            0.576    0.060    9.656    0.000
##
##
## Level 2 [school]:
##
```

```
## Latent Variables:
##                     Estimate  Std.Err  z-value  P(>|z|)
##   bses =~
##     focc       (a)    1.000
##     meduc      (b)    1.710    0.104    16.503   0.000
##     feduc      (c)    2.245    0.101    22.284   0.000
##
## Regressions:
##                     Estimate  Std.Err  z-value  P(>|z|)
##   advice ~
##     bses       (bc)   0.373    0.095    3.915    0.000
##     galo       (bb)   0.062    0.011    5.433    0.000
##   galo ~
##     bses       (ba)   6.935    0.850    8.162    0.000
##     denom1           -5.154    1.602   -3.216    0.001
##     denom2           -0.517    1.267   -0.408    0.683
##
## Intercepts:
##                     Estimate  Std.Err  z-value  P(>|z|)
##    .focc              3.254    0.090    36.000   0.000
##    .meduc             2.845    0.147    19.414   0.000
##    .feduc             3.869    0.187    20.696   0.000
##    .advice           -3.230    1.164   -2.774    0.006
##    .galo            103.352    1.277    80.944   0.000
##     bses              0.000
##
## Variances:
##                     Estimate  Std.Err  z-value  P(>|z|)
##    .feduc             0.000
##    .focc              0.033    0.016    2.030    0.042
##    .meduc             0.021    0.024    0.854    0.393
##    .advice            0.015    0.008    1.898    0.058
##    .galo              5.747    2.060    2.790    0.005
##     bses              0.360    0.080    4.488    0.000
##
## Defined Parameters:
##                     Estimate  Std.Err  z-value  P(>|z|)
##     wi                0.477    0.049    9.661    0.000
##     bi                0.429    0.093    4.601    0.000
##     wt                0.634    0.064    9.895    0.000
##     bt                0.802    0.083    9.624    0.000
```

As seen from the model fitting, this model fitted the data very satisfactory. The value of the Chi-square test statistic is 21.763 and the degrees of freedom is 18, which gives a p-value of 0.243. The other model-fitting measures are also very satisfactory as indicated by the Comparative Fit Index

(CFI) = 0.999, the Tucker-Lewis Index (TLI) = 0.998, the Root Mean Square Error of Approximation (RMSEA) = 0.012, and the Standardized Root Mean Square Residual (SRMR) = 0.014 (within covariance matrix) and 0.070 (between covariance matrix).

At the *within-school* level, all the factor loadings to *wses*, regression paths in the mediation model, and variances/covariances are highly statistically significant. Similarly, at the *between-school* level, the factor loadings to *bses*, regression paths in the mediation model, intercepts, and variances/covariances are highly statistically significant. For the covariate denomination *denom*, the parameter estimate associated with *denom1* is −5.154, which is statistically significant, indicating that the galo achievement test for pupils *Protestant* schools is significantly lower than that from non-Protestant schools. Since *denom2* is not significant, indicating that there is no statistically significant difference in the galo achievement test for pupils between *Nondenominational* and *denominational* schools.

Further from the mediation model, the estimated indirect effect from *bses* to *advice* at the school level is 0.429 with the total effect of 0.802, indicating that the galo achievement test score *galo* is a significant mediator. Similar conclusion can be made at the individual pupil level.

11.4 Discussions

In this chapter, we illustrated multi-level structural equation modeling using *lavaan*. As seen from the illustration, the implementation in *lavaan* is very straightforward with an inclusion of the option *cluster* in the *sem* function.

Note that at this point, *lavaan* can only perform two-level SEM and there is no SAS implementation on multi-level structural equation modeling and therefore there is no illustration of SAS in this chapter.

For readers who are interested in higher-level structural equation modeling, two commercial packages can be used. One is the *Mplus* and another is *Stata*. Both can fit very general multi-level structural equation models, with no formal limit to the number of levels.

11.5 Exercises

1. As presented in Chapter 3 (Silva et al., 2019), a multi-level confirmatory factor analysis model was used to analyze the 2015 wave of the *Programme for International Student Assessment (PISA)* for the Dominican Republic. Read this chapter and re-analyze this data

using *lavaan*. [Hint: data and R code are downloaded with data in *pisa_domrep_MPLUS_final.csv* and R code in *chapter3.R*]

2. As presented in Chapter 4 (Silva et al., 2019), a full multi-level structural equation model was used to analyze the 2004 Workplace Employment Relations Survey (WERS) teaching data set. Read this chapter and re-analyze this data using *lavaan*. [Hint: data and R code are downloaded with data in *WERS_MPLUS_final.csv* and R code in *chapter4.R*]

12

Sample Size Determination and Power Analysis

Studies should be well designed with adequate sample size and statistical power to address the research question or objective. The formal statistical basis for sample size determination and power calculation requires: (1) the research question or objective to be well defined; (2) the most relevant outcomes reflecting the research objective to be identified; (3) the specification of the effect size between treatment groups in terms of the outcome variables that is clinically important to detect; (4) specification of the magnitudes of the Type-I and Type-II decision errors; and (5) estimates of the effect size of the outcome variables.

There are numerous books and online resources on sample size determination and power calculations. In this chapter, we illustrate how to use R to determine sample size and calculate statistical power for structural equation models.

We start with simple case with two group comparison to illustrate the principles and then proceed to the general structural equation models. We will make use of the latent growth-curve (LGC) modeling in Chapter 9 to determine the sample size needed to detect the surgical effect between *Mastectomy* and *Lumpectomy* on *Mood* for women on breast-cancer post-surgery assessment. In Chapter 9, we analyzed data from 405 women on their mood (i.e., MOOD) and their social adjustment (i.e., SOCADJ) which were measured at three times longitudinally at 1, 4, and 8 months post-surgery. In the analysis, we found that there was a slight improvement on their *Mood* between the two types of surgeries, but the improvement was not statistically significant. In this chapter, we re-analyze this data and seek the required sample size to detect a significant effect on *Mood* with statistical power at 0.80.

12.1 Design Studies with Two-Treatment Comparison

12.1.1 A Brief Theory

In designing studies with two treatments, we are particularly interested in comparing the effectiveness of new intervention N to a conventional treatment C. This can be formulated in the hypothesis testing framework:

$$H_0 : \mu_2 - \mu_1 = 0 \text{ versus } H_a : \mu_2 - \mu_1 = \delta \neq 0. \quad (12.1)$$

where μ_2 and μ_1 represent the means of outcome for N and C, respectively, H_a reflects that the objective of the study is to demonstrate that the effectiveness of new intervention N exceeds that of the conventional treatment C by at least δ. The δ can be further interpreted as the *expected minimal comparative treatment effect size* at the design stage.

In this hypothesis testing, a Type-I error occurs if H_0 is rejected when it is true. A Type-II error occurs if H_0 is accepted when it is false. It is therefore desirable that the magnitudes of the Type-I (α) and Type-II (β) decision errors regarding H_0 are chosen to be small. Historically, α is chosen to be 0.05; whereas, β is usually chosen to be 0.20, which would give a power at 0.80 based on the definition of statistical power $= 1 - \beta$.

To determine sample size and calculate statistical power, the *estimates of the means and variances* are required to be known in prior and these estimates may be obtained from the literature or from previous studies. The well known per group sample size (n) formula is as follows:

$$n \geq 2(s^2/\delta^2)[z_{1-\alpha/2} + z_{1-\beta}]^2 \quad (12.2)$$

where s is an estimate of the standard deviation σ (assuming homogeneous variances), δ is the difference between groups which is clinically important to detect (i.e., the minimum expected treatment effect), and $z_{1-\alpha/2}$ and $z_{1-\beta}$ are the appropriate critical points of the standard normal distribution corresponding to the magnitudes of the Type-I and Type-II errors, respectively, explicitly reflects the prerequisites for sample size determination.

The above formula in Equation (12.2) typically reflects the interplay between the design characteristics: α, $1 - \beta$ (i.e., power), δ , s^2 and n. As δ decreases, n increases and vice versa; as power increases, n increases or vice versa; as s^2 increases, n increases; and as α decreases, n increases. The per group sample size n is interpreted as *the sample size required to detect δ with a power of $1 - \beta$ and a Type-I error of α is at least $2(s^2/\delta^2)[z_{1-\alpha} + z_{1-\beta}]^2$.* More discussion on sample size determination and power calculation can be found at Chen et al. (2017).

12.1.2 R Implementation

Suppose that we would like to design a study to detect the difference between a new behavioral intervention and a conventional practice-at-usual in increasing health wellness. Historically, we have known the mean measurement for the outcome of health measure is $\mu_1 = 1$ and standard deviation $s = 0.5$. The new behavioral intervention is designed to increase the health wellness to $\mu_2 = 1.2$. Using the above formula in Equation (12.2), we can design this new study and calculate the sample size required as follows:

```
# Data known
mu2 = 1.2; mu1= 1; sd=0.5; alpha=0.05; beta=0.2
# Mean difference
delta = mu2-mu1
# Required sample size
n = 2*sd^2/delta^2 *(qnorm(1-alpha/2)+qnorm(1-beta))^2
n
```

```
## [1] 98.1
```

This means that we would need a sample size of 99 per group to power the study at 0.80 to detect the new intervention effectiveness from $\mu_1 = 1$ to $\mu_2 = 1.2$.

There are many R functions to make this calculation, and we can make use of *power.t.test* for this purpose to get the sample size as follows:

```
power.t.test(sd=sd,delta=delta,power=.8)
```

```
##
##          Two-sample t-test power calculation
##
##                     n = 99.1
##                 delta = 0.2
##                    sd = 0.5
##             sig.level = 0.05
##                 power = 0.8
##           alternative = two.sided
##
## NOTE: n is number in *each* group
```

As seen again, we need sample size 99 for each group to power this study. We can of course check whether this determined sample size 99 could really produce the power of 0.80 by use the same function as follows:

```
power.t.test(sd=sd,delta=delta,n=99)
```

```
##
##          Two-sample t-test power calculation
##
##                     n = 99
##                 delta = 0.2
##                    sd = 0.5
##             sig.level = 0.05
##                 power = 0.8
##           alternative = two.sided
##
## NOTE: n is number in *each* group
```

As can be seen, the power is indeed 0.80 with the sample size of 99.

12.1.3   Linkage to Structural Equation Modeling

The two-group comparison is the simplest study design, which is very different from designing a study with structural equation modeling. Structural equation models can be very complex and there is usually no such elegant theoretical formula to calculate sample size and the associated statistical power. For complex models, Monte-Carlo simulation-based (MCSB) approach is usually used for sample size determination and power calculation. We illustrate this MCSB approach using the above two-treatment comparison. With the understanding of this approach, we can then proceed to general structural equation modeling.

For this simple example, we can easily write a *R* code to implement this MCSB approach, however, this MCSB approach can be complex for general SEMs. The simulated structural equation modeling has been created as an *R* package *simsem* (http://simsem.org/) for general MCSB framework and we will illustrate the sample size determination and power calculation using this package in this chapter.

It is known that t-test for two-treatment comparison is a special case of analysis of variance (ANOVA) with multiple treatments, where ANOVA is a special case of regression. More generally, structural equation models can include multiple interconnected regression components as well as latent variable measurement models.

To illustrate the MCSB approach, let us re-formulate the above two-treatment comparison into regression framework as follows:

$$y = \beta_0 + \beta_1 TRT + \epsilon \tag{12.3}$$

where y is the outcome variable. TRT is the treatment indicator with *1* to denote the new intervention N and *0* to denote the conventional treatment C. β_0 is the mean for C, i.e., $\beta_0 = \mu_1$, and β_1 is the mean difference, i.e., $\beta_1 = \delta = \mu_2 - \mu_1$. ϵ is the error term assumed to be normally distributed with mean 0 and standard deviation σ.

The linear regression model in Equation (12.3) can be standardized into correlation for better structural equation modeling implementation. The correlation r can be calculated as follows

$$r = \frac{d}{\sqrt{(d^2 + a)}} \tag{12.4}$$

$$\text{var}(\hat{r}) = \frac{a^2 \text{var}(d)}{(d^2 + a)^3} \tag{12.5}$$

where $d = \delta/\sigma$ is the standardized effect and a corrects for imbalance in sample sizes in both treatments of n_1 and n_2 and is defined as $a = \frac{(n_1+n_2)^2}{n_1 n_2}$. Since this is a balanced design, a equals to 4 in this case. With this r, the variance of the standardized y can be calculated by $1 - r^2$.

We can use *R* library *compute.es* to make this conversion from *d* to *r* as follows:

```
# Get the library to calculate ES
library(compute.es)
# ES in d
d = delta/sd; d
```

```
## [1] 0.4
```

```
# convert to all other ES
d2ES = des(d, n.1=n, n.2=n, verbose = FALSE)
# Extract the r
r = d2ES$r;r
```

```
## [1] 0.2
```

```
var.r = d2ES$var.r;var.r
```

```
## [1] 0
```

```
# Calculate the variance of y
var.y = 1-r**2;var.y
```

```
## [1] 0.96
```

To use the *simsem* package, two model components are needed as inputs to the *sim* function. The first model component is the data generation model, which specifies the known parameters to generate the data as follows for the above two-treatment comparison:

```
# Data generation model #
datMod <- "
# Regression with known correlation of 0.2
y ~ 0.2*TRT
# Error variance
y ~~ 0.96*y
"
```

The second model component is the *estimation model* which is to specifies how *lavaan* to estimate the SEM parameters. This can be simply specified as follows:

```
# Analysis Model
estMod <- "
# REgresison
y ~ TRT
"
```

With the above two model components, we can then call *sim* function to perform the MCSB implementation as follows:

```
# Load the library
library(simsem)
```

```
# Call sim to do simulation
#simOut <- sim(nRep = 1000, generate=datMod, model=estMod, n=198,
#              lavaanfun = "sem", seed=123, silent=TRUE)
simOut <- sim(nRep = 1000, generate=datMod, model=estMod, n=198,
              lavaanfun = "sem", seed=123, silent=TRUE)
# Simulation output
summaryParam(simOut)
```

```
##         Estimate Average Estimate SD Average SE
## y~TRT             0.202       0.0693     0.0698
## y~~y              0.956       0.0969     0.0960
##         Power (Not equal 0) Std Est Std Est SD
## y~TRT                 0.816   0.201     0.0675
## y~~y                  1.000   0.955     0.0271
##         Std Ave SE Average Param Average Bias Coverage
## y~TRT       0.0671          0.20      0.00219    0.953
## y~~y        0.0267          0.96     -0.00433    0.942
```

Note in the above *sim* implementation, *nRep* is the number of replications which should be a large number (say > 10,000. In this example, 1000 is used to save time in running the simulation) for better results. Option *generate* is to link to the first model component for data generation and *model* is to link to the second model component for model fitting using *lavaanfun*. The option *n* is the sample size for power calculation and *seed* is for setting the random seed so the simulation can be duplicated with the same results. The last option *silent=TRUE* is to suppress the output.

As seen from the output, we can see several columns with two rows. The first row is for the estimate of r in y ~ TRT and the second row is for the variance of the standardized y in y ~~y. The columns are:

- Estimate average: is the average of parameter estimates across all *nRep* replications. They are very close to the simulated values of $r = 0.2$ and $var.y = 0.96$.

- Estimate SD: is the Standard Deviation of parameter estimates across all *nRep* replications.

- Average SE: is the average of standard errors across all replications

- Power (not equal 0): is the proportion of significant replications when testing whether the parameters are different from zero. This is what we want and it can be seen that the statistical power is 0.816, which is close to the 0.80 specified. The slight difference is probably due to the small *nRep* and this difference will be smaller if we increase the *nRep* to 10,000 or even larger number.

- Average Param: is the parameter values or average values of parameters if random parameters are specified.

- Average bias: is the difference between parameter estimates and parameter underlying data.

- Coverage: is the percentage of $(1 - \alpha)\%$ confidence interval covers parameters underlying the data, which should 95%.

12.2 Design Studies with Structure Equation Models

With the understanding on MCSB approach and the associated implementation in *simsem*, we can now proceed to illustrate the power calculation for general structural equation modeling. We use the LGC modeling in Chapter 9 as the example here in this chapter. We re-analyze this data and seek the required sample size to detect a significant effect on *Mood* with statistical power at 0.80.

12.2.1 Fit the LGC Model for MOOD

To determine the sample size to detect significant effect of surgery on MOOD, we need the data generation model component, which involves all the parameters to be known in advance. In order to get those parameters, we need to re-fit the data.

For this purpose, we again read the data into *R* as follows:

```
# Read the data into R
dCancer <- read.csv("data/hkcancer.csv", na.strings="*",header=T)
# Check the data dimension
dim(dCancer)
```

```
## [1] 405   10
```

```
# Print the first 6 observations
head(dCancer)
```

```
##    ID MOOD1 MOOD4 MOOD8 SOCADJ1 SOCADJ4 SOCADJ8 Age
## 1  1    15    NA    NA    95.9      NA      NA  70
## 2  2    16    25    22   114.9   105.1    90.4  47
## 3  3    37    26    25    80.7    95.3    95.3  47
## 4  4    19    16    15   112.8   108.6    99.0  52
## 5  5    13    16    14   115.0   105.0   101.0  43
## 6  6    21    28    19   106.5   115.0   107.5  34
##        SurgTx  AgeGrp
## 1  Mastectomy   Older
## 2  Mastectomy Younger
## 3  Mastectomy Younger
## 4  Mastectomy   Older
## 5  Mastectomy Younger
## 6  Lumpectomy Younger
```

The LGC model is implemented with *lavaan* syntax as follows:

```
# Model for Mood
mod4MOOD <- "
  # Intercept and Slope with fixed-coefficients
  iMOOD =~ 1*MOOD1 + 1*MOOD4 + 1*MOOD8
  sMOOD =~ 0*MOOD1 + 1*MOOD4 + 2.33*MOOD8
  # Regression with a labeled for simulation
  iMOOD ~ SurgTx
  sMOOD ~ a*SurgTx
"
```

We can then fit this LGC model using *lavaan* as follows:

```
# Load lavaan package
library(lavaan)
# Call growth function to fit the LGC
fitMOOD <- growth(mod4MOOD, data = dCancer,estimator ="MLR",
                 missing="fiml")
# Print the summary
summary(fitMOOD)
```

```
## lavaan 0.6-12 ended normally after 83 iterations
##
##   Estimator                                         ML
##   Optimization method                           NLMINB
##   Number of model parameters                        10
```

```
##
##    Number of observations                            405
##    Number of missing patterns                          8
##
## Model Test User Model:
##                                              Standard      Robust
##    Test Statistic                              1.822       1.880
##    Degrees of freedom                              2           2
##    P-value (Chi-square)                        0.402       0.391
##    Scaling correction factor                               0.969
##      Yuan-Bentler correction (Mplus variant)
##
## Parameter Estimates:
##
##    Standard errors                            Sandwich
##    Information bread                          Observed
##    Observed information based on               Hessian
##
## Latent Variables:
##                       Estimate  Std.Err  z-value  P(>|z|)
##    iMOOD =~
##      MOOD1              1.000
##      MOOD4              1.000
##      MOOD8              1.000
##    sMOOD =~
##      MOOD1              0.000
##      MOOD4              1.000
##      MOOD8              2.330
##
## Regressions:
##                       Estimate  Std.Err  z-value  P(>|z|)
##    iMOOD ~
##      SurgTx            -0.116    0.766   -0.152    0.879
##    sMOOD ~
##      SurgTx     (a)    -0.332    0.337   -0.987    0.324
##
## Covariances:
##                       Estimate  Std.Err  z-value  P(>|z|)
##   .iMOOD ~~
##     .sMOOD            -2.135    1.647   -1.296    0.195
##
## Intercepts:
##                       Estimate  Std.Err  z-value  P(>|z|)
##     .MOOD1             0.000
##     .MOOD4             0.000
##     .MOOD8             0.000
```

```
##     .iMOOD          21.683   1.402   15.468   0.000
##     .sMOOD           0.004   0.624    0.007   0.994
##
## Variances:
##                    Estimate  Std.Err  z-value  P(>|z|)
##     .MOOD1          14.307   3.451    4.146   0.000
##     .MOOD4          18.637   2.501    7.453   0.000
##     .MOOD8           6.745   4.044    1.668   0.095
##     .iMOOD          25.826   3.609    7.156   0.000
##     .sMOOD           2.272   1.307    1.738   0.082
```

12.2.2   Monte-Carlo Simulation-Based Power Calculation

With the parameter estimates from the above model fitting, we can then copy those estimates and build the data generation model component as follows:

```
dat.mod4MOOD <- "
  # Intercept and Slope with MOOD
  iMOOD =~ 1*MOOD1 + 1*MOOD4 + 1*MOOD8
  sMOOD =~ 0*MOOD1 + 1*MOOD4 + 2.33*MOOD8

  # residual variances for observed
  MOOD1 ~~ 14.307*MOOD1
  MOOD4 ~~ 18.637*MOOD4
  MOOD8 ~~ 6.745*MOOD8

  # Regression paths to covariates
  iMOOD ~ (-0.116)*SurgTx
  sMOOD ~  a*SurgTx + (-0.332)*SurgTx

  # latent Intercepts
  iMOOD ~ 21.683*1
  sMOOD ~ 0.004*1

    # latent variances/coVariances
  iMOOD  ~~ 25.826*iMOOD
  sMOOD  ~~ 2.272*sMOOD
  iMOOD  ~~ (-2.135)*sMOOD

    # mean and variance for SurgTX
  SurgTx ~ 0.5*1
  SurgTx ~~ 0.25*SurgTx

"
```

Note that in the above implementation, we used **sMOOD ˜ SurgTx +** (−0.332) **SurgTx** where *a* is labeled as the parameter of interest for power calculation. This label *a* is done in both the data generation model component (i.e., in *dat.mod4MOOD*) and estimation model component (i.e., in *mod4MOOD*).

We can now call *sim* to

```
# Call simsem library
library(simsem)
# Call sim for simulation
simOut1 <- sim(nRep = 1000, generate=dat.mod4MOOD,
               model=mod4MOOD, n=405, lavaanfun = "growth",
               seed=123, silent=TRUE)
# simulation output
summary(simOut1)
```

```
## RESULT OBJECT
## Model Type
## [1] "lavaan"
## ========= Fit Indices Cutoffs ============
##             Alpha
## Fit Indices     0.1       0.05      0.01      0.001
##       chisq     4.455     5.665     8.871     11.378
##       aic    7571.694  7588.105  7624.726  7667.762
##       bic    7611.733  7628.144  7664.765  7707.801
##       rmsea     0.055     0.067     0.092     0.108
##       cfi       0.994     0.991     0.983     0.975
##       tli       0.982     0.972     0.948     0.926
##       srmr      0.018     0.020     0.026     0.028
##             Alpha
## Fit Indices   Mean      SD
##       chisq     1.980    1.888
##       aic    7508.592   49.361
##       bic    7548.631   49.361
##       rmsea     0.016    0.025
##       cfi       0.998    0.004
##       tli       1.000    0.014
##       srmr      0.011    0.005
## ========= Parameter Estimates and Standard Errors ===========
##                Estimate Average Estimate SD  Average SE
## iMOOD~SurgTx             -0.088          0.614       0.607
## a                       -0.337          0.244       0.242
## MOOD1~~MOOD1            14.312          2.882       2.816
## MOOD4~~MOOD4            18.563          1.831       1.846
## MOOD8~~MOOD8             6.843          3.477       3.401
```

```
## iMOOD~~iMOOD              25.691        3.168        3.189
## sMOOD~~sMOOD               2.211        1.124        1.100
## iMOOD~~sMOOD              -2.072        1.278        1.284
## iMOOD~1                   21.664        0.440        0.429
## sMOOD~1                    0.007        0.167        0.171
##              Power (Not equal 0) Std Est Std Est SD
## iMOOD~SurgTx              0.056  -0.009        0.061
## a                        0.287  -0.126        0.129
## MOOD1~~MOOD1             1.000   0.357        0.067
## MOOD4~~MOOD4             1.000   0.437        0.031
## MOOD8~~MOOD8             0.536   0.195        0.099
## iMOOD~~iMOOD            1.000   0.996        0.005
## sMOOD~~sMOOD            0.512   0.968        0.147
## iMOOD~~sMOOD            0.359  -0.272        0.459
## iMOOD~1                  1.000   4.291        0.282
## sMOOD~1                  0.043   0.005        0.146
##              Std Ave SE Average Param Average Bias
## iMOOD~SurgTx      0.060        -0.116        0.028
## a                 0.195        -0.332       -0.005
## MOOD1~~MOOD1      0.066        14.307        0.005
## MOOD4~~MOOD4      0.031        18.637       -0.074
## MOOD8~~MOOD8      0.096         6.745        0.098
## iMOOD~~iMOOD      0.006        25.826       -0.135
## sMOOD~~sMOOD      0.311         2.272       -0.061
## iMOOD~~sMOOD     13.990        -2.135        0.063
## iMOOD~1           0.282        21.683       -0.019
## sMOOD~1           0.194         0.004        0.003
##              Coverage
## iMOOD~SurgTx   0.949
## a              0.955
## MOOD1~~MOOD1   0.946
## MOOD4~~MOOD4   0.947
## MOOD8~~MOOD8   0.949
## iMOOD~~iMOOD   0.956
## sMOOD~~sMOOD   0.940
## iMOOD~~sMOOD   0.949
## iMOOD~1        0.938
## sMOOD~1        0.956
## ========= Correlation between Fit Indices ============
##        chisq    aic    bic  rmsea    cfi    tli   srmr
## chisq  1.000 -0.012 -0.012  0.955 -0.939 -0.995  0.957
## aic   -0.012  1.000  1.000 -0.002 -0.014  0.016 -0.040
## bic   -0.012  1.000  1.000 -0.002 -0.014  0.016 -0.040
## rmsea  0.955 -0.002 -0.002  1.000 -0.932 -0.949  0.890
## cfi   -0.939 -0.014 -0.014 -0.932  1.000  0.941 -0.818
```

```
## tli    -0.995  0.016  0.016 -0.949  0.941  1.000 -0.957
## srmr    0.957 -0.040 -0.040  0.890 -0.818 -0.957  1.000
## ================== Replications =====================
## Number of replications = 1000
## Number of converged replications = 947
## Number of nonconverged replications:
##    1. Nonconvergent Results = 0
##    2. Nonconvergent results from multiple imputations = 0
##    3. At least one SE was negative or NA = 0
##    4. Nonpositive-definite latent or observed (residual)
##       covariance matrix
##       (e.g., Heywood case or linear dependency) = 53
```

We can see that the associated power for a (i.e., the surgery effect) is 0.287, which is not surprising since we have known that this parameter is not statistically significant with the existing sample size of 405. So the next question is what sample size can yield a power of 0.80.

The MCSB approach is needed to run for a series of sample sizes to get their associated powers. Then a sample size to power curve can be obtained to determine the sample size for power of 0.80. For this purpose, we use $n = \text{rep}(\text{seq}(400, 2000, \text{and } 200), 500)$ which is to run the MCSB for a sequence of sample sizes from 400 to 2000 by 200 (i.e., 400, 600, 800, 1000, 1200, 1400, 1600, 1800, and 2000) with each of these sample size for 500 simulations.

```
# Simulation for sequential sample size
simAll <- sim(nRep = NULL, generate=dat.mod4MOOD,
              model=mod4MOOD, n=rep(seq(400, 2000, 200),500),
              lavaanfun = "growth", seed=123, silent=TRUE) #,
              multicore=TRUE)
# Print the simulations
summary(simAll)
```

```
## RESULT OBJECT
## Model Type
## [1] "lavaan"
## ========= Fit Indices Cutoffs =============
##      N chisq   aic   bic rmsea   cfi   tli  srmr
## 1  400  6.19  7503  7545 0.061 0.993 0.978 0.019
## 2  800  6.04 14933 14979 0.053 0.994 0.983 0.016
## 3 1200  5.88 22363 22413 0.044 0.996 0.987 0.013
## 4 1600  5.73 29793 29847 0.036 0.997 0.992 0.011
## 5 2000  5.57 37223 37281 0.027 0.999 0.996 0.008
## ========= Parameter Estimates and Standard Errors ===========
##              Estimate Average Estimate SD Average SE
## iMOOD~SurgTx           -0.122       0.400      0.387
## a                      -0.327       0.163      0.154
```

```
## MOOD1~~MOOD1            14.298      1.831      1.794
## MOOD4~~MOOD4            18.662      1.230      1.178
## MOOD8~~MOOD8             6.700      2.223      2.163
## iMOOD~~iMOOD            25.761      2.094      2.033
## sMOOD~~sMOOD             2.265      0.722      0.700
## iMOOD~~sMOOD            -2.130      0.855      0.818
## iMOOD~1                 21.687      0.284      0.273
## sMOOD~1                  0.002      0.116      0.109
##               Power (Not equal 0) Std Est Std Est SD
## iMOOD~SurgTx             0.066     -0.012      0.039
## a                       0.607     -0.113      0.079
## MOOD1~~MOOD1            1.000      0.357      0.043
## MOOD4~~MOOD4            1.000      0.439      0.020
## MOOD8~~MOOD8            0.855      0.191      0.063
## iMOOD~~iMOOD           1.000      0.998      0.003
## sMOOD~~sMOOD           0.877      0.981      0.164
## iMOOD~~sMOOD           0.761     -0.272      0.097
## iMOOD~1                1.000      4.280      0.186
## sMOOD~1                0.058      0.000      0.086
##               Std Ave SE Average Param Average Bias
## iMOOD~SurgTx      0.038         -0.116        -0.006
## a                 0.329         -0.332         0.005
## MOOD1~~MOOD1      0.042         14.307        -0.009
## MOOD4~~MOOD4      0.020         18.637         0.025
## MOOD8~~MOOD8      0.061          6.745        -0.045
## iMOOD~~iMOOD     0.003         25.826        -0.065
## sMOOD~~sMOOD     1.757          2.272        -0.007
## iMOOD~~sMOOD     0.134         -2.135         0.005
## iMOOD~1          0.178         21.683         0.004
## sMOOD~1          0.169          0.004        -0.002
##               Coverage r_coef.n r_se.n
## iMOOD~SurgTx     0.949   -0.008 -0.933
## a                0.943   -0.022 -0.933
## MOOD1~~MOOD1     0.953   -0.007 -0.931
## MOOD4~~MOOD4     0.950   -0.022 -0.925
## MOOD8~~MOOD8     0.952    0.016 -0.933
## iMOOD~~iMOOD    0.951   -0.001 -0.929
## sMOOD~~sMOOD    0.955   -0.004 -0.935
## iMOOD~~sMOOD    0.948    0.012 -0.934
## iMOOD~1         0.948   -0.003 -0.935
## sMOOD~1         0.942    0.012 -0.935
## ========= Correlation between Fit Indices ============
##         chisq    aic    bic  rmsea    cfi    tli   srmr
## chisq  1.000 -0.007 -0.007  0.911 -0.780 -0.854  0.825
## aic   -0.007  1.000  1.000 -0.158  0.204  0.015 -0.427
```

```
## bic    -0.007  1.000  1.000 -0.158  0.204  0.015 -0.427
## rmsea   0.911 -0.158 -0.158  1.000 -0.875 -0.897  0.866
## cfi    -0.780  0.204  0.204 -0.875  1.000  0.930 -0.776
## tli    -0.854  0.015  0.015 -0.897  0.930  1.000 -0.818
## srmr    0.825 -0.427 -0.427  0.866 -0.776 -0.818  1.000
## n      -0.007  1.000  1.000 -0.158  0.204  0.015 -0.427
##                n
## chisq -0.007
## aic    1.000
## bic    1.000
## rmsea -0.158
## cfi    0.204
## tli    0.015
## srmr  -0.427
## n      1.000
## ================== Replications =====================
## Number of replications = 4500
## Number of converged replications = 4455
## Number of nonconverged replications:
##     1. Nonconvergent Results = 0
##     2. Nonconvergent results from multiple imputations = 0
##     3. At least one SE was negative or NA = 0
##     4. Nonpositive-definite latent or observed (residual)
##        covariance matrix
##        (e.g., Heywood case or linear dependency) = 45
## NOTE: The sample size is varying.
```

We can then obtain the power estimates for each of the sample size using function *getPower* as follows:

```
LGC.N = getPower(simAll)
#LGC.N
# Find the samplesize for 80% power
findPower(LGC.N,"N", 0.8)
```

```
## iMOOD~SurgTx              a MOOD1~~MOOD1 MOOD4~~MOOD4
##           NA           1710         Inf          Inf
## MOOD8~~MOOD8 iMOOD~~iMOOD sMOOD~~sMOOD iMOOD~~sMOOD
##          816          Inf          748         1161
##      iMOOD~1       sMOOD~1
##          Inf           NA
```

Therefore, we would need sample size of 1710 to obtain the power of 0.80. This can be graphically shown in Figure 12.1. In this figure, the dashed horizontal line indicates the power at 0.80 and the arrowed line from this

a

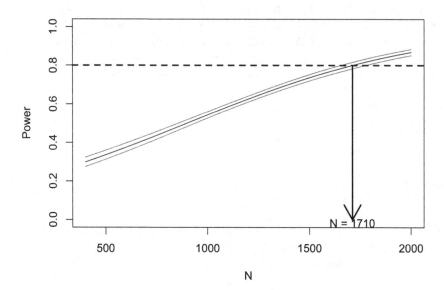

FIGURE 12.1
Power Curve for Different Sample Sizes.

horizontal line to $N = 1710$ indicates the sample size determined from the power curve.

```
# Call plotPower to plot the power to sample size
plotPower(simAll, powerParam="a")
# Add a horizontal line of 0.80
abline(h=0.8, lwd=2, lty=8)
arrows(1710, 0.8, 1710,0, lwd=2)
text(1710,-0.018, "N = 1710")
```

12.3   Discussions

In this chapter, we illustrated the sample size determination and power calculations using Monte-Carlo simulation-based approach. This approach is the most general to obtain the sample size and power for designing studies in any setting, including structural equation models.

There are many other online resources for this purpose. We list a few here for information purpose in this discussion and interested readers can further search online to find references and methods to better suit their own research questions.

- *Mplus* on *How to* with topic *Power Calculation*

Mplus is a power package for structural equation modeling and it can be used for sample size determination and power calculations. Examples and mplus code can be found at `https://www.statmodel.com/power.shtml`.

- *WebPower* Basic and Advanced Statistical Power Analysis

This is a collection of tools for conducting both basic and advanced statistical power analysis including correlation, proportion, t-test, one-way ANOVA, two-way ANOVA, linear regression, logistic regression, Poisson regression, mediation analysis, longitudinal data analysis, structural equation modeling and multi-level modeling. It also serves as the engine for conducting power analysis online at `https://webpower.psychstat.org`. *wp.sem.rmsea* for *Statistical Power Analysis for Structural Equation Modeling* at `https://rdrr.io/cran/WebPower/man/wp.sem.rmsea.html`.

As described, "Structural equation modeling (SEM) is a multivariate technique used to analyze relationships among observed and latent variables. It can be viewed as a combination of factor analysis and multivariate regression analysis. Two methods are widely used in power analysis for SEM. One is based on the likelihood ratio test proposed by Satorra and Saris (1985). The other is based on RMSEA proposed by MacCallum et al. (1996). This function is for SEM power analysis based on RMSEA". Example R code can be found on this webpage on how to use this function to calculate sample size and power.

- *Shiny App* for *analytical power calculations for structural equation modeling* without Monte-Carlo simulations

Jak et al. (2020) presented a tutorial and a Shiny app for conducting power analysis using structural equation models. In this tutorial, they explained how to do power calculations without Monte-Carlo methods for the Chi-square test and the RMSEA tests of (not-)close fit using the Shiny app *power4SEM*. This R Shiny function *power4SEM* can do power calculations for SEM using two methods that are not computationally intensive and that focus on model fit instead of the statistical significance of parameters. They illustrated the use of *power4SEM* with examples of power analyses for path models, factor models, and a latent growth model.

- Reliable and More Powerful Methods Yuan et al. (2017)

A reliable and more powerful method for power analysis in structural equation modeling was presented in Yuan et al. (2017), which is a Monte-Carlo simulation-based method. They provided a measure of effect size for characterizing the power property of different rescaled statistics with theoretical derivation. Robust methods were proposed to increase the power of the likelihood ratio statistic $T_{ml} = nF_{ml}$ and other statistics. Extensive simulation studies showed that their method could reliably control the Type-I errors and robust estimation methods effectively increase the power.

With All These Discussions and Illustrations, We Conclude This Book!

Thank You and We Hope That You Have Enjoyed Your Reading!

Bibliography

Akaike, H. (1987). Factor analysis and aic. *Psychometrika*, 52:317–332.

Arbuckle, J. L. (2007). *Amos™ 16 User's Guide*. SPSS, Chicago.

Baron, R. M. and Kenny, D. A. (1986). The moderator-mediator variable distinction in social psychological research: Conceptual, strategic, and statistical considerations. *Journal of Personality and Social Psychology*, 51(6):1173–1182.

Beauducel, A. and Herzberg, P. Y. (2006). On the performance of maximum likelihood versus means and variance adjusted weighted least squares estimation in CFA. *Structural Equation Modeling*, 13:186–203.

Bentler, P. M. (1995). *EQS: Structural Equations Program Manual*. Program Version 5.0. Multivariate Software, Encino, CA.

Bentler, P. M. and Bonett, D. G. (1980). Significance tests and goodness of fit in the analysis of covariance structures. *Psychological Bulletin*, 88:588–606.

Bentler, P. M. and Weeks, D. G. (1982). Linear structural equations with latent variables. *Psychometrika*, 45:289–308.

Bollen, K. A. (1989). *Structural Equations with Latent Variables*. John Wiley & Sons, Oxford.

Brown, T. A. (2015). *Confirmatory Factor Analysis for Applied Research*, Second Edition. Guilford Publications, New York.

Browne, M. W. (1984). Asymptotically distribution-free methods for the analysis of covariance structures. *British Journal of Mathematical and Statistical Psychology*, 37:62–83.

Browne, M. W. and Cudeck, R. (1993). Alternative ways of assessing model fit. In K. A. Bollen and J. S. Long, editor, *Testing Structural Equation Models*, pp. 136–162. Sage Publications, Newbury Park, CA.

Browne, M. W., MacCallum, R. C., Kim, C.-T., Andersen, B. L., and Glaser, R. (2002). When fit indices and residuals are incompatible. *Psychological Methods*, 7(4):403–421. https://doi.org/10.1037/1082-989X.7.4.403.

Byrne, B. M. (1994). Testing for the factorial validity replication, and invariance of a measuring instrument: A paradigmatic application based on Maslach Burnout Inventory. *Multivariate Behavioral Research*, 29:289–311.

Byrne, B. M. (2012). *Structural Equation Modeling with Mplus*. Routledge, New York.

Chen, D.-G. and Chen, J. K. (2021a). *Statistical Regression Modeling with R: Longitudinal and Multi-level Modeling*. Springer, Berlin/Heidelberg, Germany.

Chen, D.-G. and Chen, X. (2021b). Full longitudinal mediation modeling analysis of HIV/AIDS knowledge, self-efficacy and condom-use intention in youth. *Journal of the Society for Social Work and Research*, 13(1):161–177. https://doi.org/10.1086/718477

Chen, D.-G., Peace, K. E., and Zhang, P. (2017). *Clinical Trial Data Analysis Using R and SAS*, Second Edition. Chapman & Hall/CRC Press, Boca Raton, FL.

Chen, X., Stanton, B., Gomez, P., Lunn, S., Deveaux, L., Brathwaite, N., Li, X., Marshall, S., Cottrell, L., and Harris, C. (2010). Effects on condom use of an hiv prevention programme 36 months postintervention: A cluster randomized controlled trial among Bahamian youth. *International Journal of STD and AIDS*, 21(9):622–630.

Fillingim, R. B., Ohrbach, R., Greenspan, J. D., Knott, C., Dubner, R., Bair, E., Baraian, C., Slade, G. D., and Maixner, W. (2011). Potential psychosocial risk factors for chronic tmd: Descriptive data and empirically identified domains from the oppera case-control study. *The Journal of Pain*, 12(11):T46–T60.

Flora, D. B. and Curran, P. J. (2004). An empirical evalution of alternative methods of estimation for confirmatory factor analysius with ordinal data. *Psychological Methods*, 9:466–491.

Fox, J. and Weisberg, S. (2016). *Applied Regression Analysis and Generalized Linear Models*, Third Edition. Sage Publications, Newbury Park, CA.

Fox, J. and Weisberg, S. (2019). *An R Companion to Applied Regression*, Third Edition. Sage Publications, Newbury Park, CA.

Gana, K. and Broc, G. (2019). *Structural Equation Modeling with lavaan*. Wiley, Hoboken, NJ.

Gelman, A. and Hill, J. (2007). *Data Analysis Using Regression and Multi-level/Hierarchical Models*. Cambridge University Press, Cambridge, United Kingdom.

Greenspan, J. D., Slade, G. D., Bair, E., Dubner, R., Fillingim, R. B., Ohrbach, R., Knott, C., Mulkey, F., Rothwell, R., and Maixner, W. (2011). Pain sensitivity risk factors for chronic tmd: Descriptive data and empirically identified domains from the oppera case control study. *The Journal of Pain*, 12:T61–74.

Hayes, A. F. (2018). *Introduction to Mediation, Moderation, and Conditional Process Analysis- A regression-Based Approach*, Second Edition. The Guilford Press, New York.

Horn, J. L. and McArdle, J. J. (1992). A practical and theoretical guide to measurement invariance in aging research. *Experimental Aging Research*, 18(3):117–144.

Hox, J. J., Moerbeek, M., and van de Schoot, R. (2018). *Multilevel Analysis-Techniques and Applications*, Third Edition. Routledge: Milton Park, Abingdon.

Hu, L.T. and Bentler, P.M. (1995). Evaluating model fit. In R. H. Hoyle, editor, *Structural Equation Modeling: Concepts, Issues, and Applications*, pp. 76–99. Sage, London.

Hu, L. T. and Bentler, P. M. (1999). Cutoff criteria for ft indexes in covariance structure analysis: Conventional criteria versus new alternatives. *Structural Equation Modeling*, 6:1–55. http://dx.doi.org/10.1080/10705519909540118.

Hu, L., Bentler, P. M., and Kano, Y. (1992). Can test statistics in covariance structure analysis be trusted? *Psychological Bulletin*, 112:351–362.

Jak, S., Jorgensen, T. D., Verdam, M. G. E., Oort, F. J., and Elffers, L. (2020). Analytical power calculations for structural equation modeling: A tutorial and Shiny app. *Behavior Research Methods*, 53(4):1385–1406.

Jennrich, R. I. (1995). *An Introduction to Computational Statistics*. Prentice-Hall, Englewood Cliffs, NJ.

Jose, P. E. (2013). *Doing Statistical Mediation and Moderation*. Guilford Press, New York.

Jöreskog, K. G. (1970). A general model for analysis of covariance structures. *Biometrika*, 57:239–251.

Jöreskog, K. G. and Sörbom, D. (1989). *LISREL 7: A Guide to the Program and Applications*. ESPSS, Chicago, IL.

Jöreskog, K. G. and Sörbom, D. (1993). *LISREL 8: Structural Equation Modeling with the SIMPLIS Command Language*. Earlbaum, Hillsdale, NJ.

Li, Y., Huang, H., and Chen, Y.-Y. (2020). Organizational climate, job satisfaction, and turnover in voluntary childwelfare workers. *Children and Youth Services Review*, 119:105640.

MacCallum, R. C., Browne, M. W., and Sugawara, H. M. (1996). Power analysis and determination of sample size for covariance structure modeling. *Psychological Methods*, 1(2):130.

MacKinnon, D. P. (2008). *Introduction to Statistical Mediation Analysis*. Erlbaum, Mahwah, NJ.

Maslach, C. and Jackson, S. E. (1981). The measurement of experienced burnout. *Jounral of Organizational Behavior*, 2(2):99–113.

Maslach, C. and Jackson, S. E. (1986). *Maslach Burnout Inventory Manual, 2nd Edition*. Consulting Psychologists Press, Palo Alto, CA.

McArdle, J. J. and McDonald, R. P. (1984). Some algebraic properties of the reticular action model for moment structures. *British Journal of Mathematical and Statistical Psychology*, 37:234–251.

McDonald, R. P. (1978). A simple comprehensive model for the analysis of covariance structures. *British Journal of Mathematical and Statistical Psychology*, 31:59–72.

McDonald, R. E. and Goldstein, H. (1989). Balanced versus unbalanced designs for linear structural relations in two-level data. *British Journal of Mathematical and Statistical Psychology*, 42:214–232.

Miller, V. E., Chen, D.-G., Barrett, D., Poole, C., Golightly, Y. M., Sanders, A. E., Ohrbach, R., Greenspan, J. D., Fillingim, R. B., and Slade, G. D. (2020). Understanding the relationship between features associated with pain-related disability in people with painful temporomandibular disorder: An exploratory structural equation modeling approach. *Pain*, 161(12):2710–2719.

Muthen, B., du Toit, S. H. C., and Spisic, D. (1997). Robust inference using weighted least squares and quadratic estimating equations in latent variable modeling with categorical and continuous outcomes. Technical report, University of California, Los Angeles.

O'Brien, R. M. (1985). The relationship between ordinal measures and their underlying values: Why all the disagreement? *Quality and Quantity*, 19:265–177.

Ohrbach, R., Fillingim, R. B., Mulkey, F., Gonzalez, Y., Gordon, S., Gremillion, H., Lim, P.-F., Ribeiro-Dasilva, M., Greenspan, J. D., Knott, C., et al. (2011). Clinical findings and pain symptoms as potential risk factors for chronic tmd: Descriptive data and empirically identified domains from the oppera case-control study. *The Journal of Pain*, 12(11):T27–T45.

Olsson, U. (1979). Maximum likelihood estimation of the polychoric correlation coefficient. *Psychometrika*, 12:443–460.

Olsson, U., Drasgow, F., and Dorans, N. J. (1982). The polyserial correlation coefficient. *Psychometrika*, 47:337–347.

Pinheiro, J. C. and Bates, D. M. (2000). *Mixed-Effect Models in S and SPLUS*. Springer, New York.

Raftery, A. E. (1993). Bayesian model selection in structural equation models. In K. A. Bollen and J. S. Long, editor, *Testing Structural Equation Models*, pp. 163–180. Sage Publications, Newbury Park, CA.

Ren, Q., Li, Y., and Chen, D.-G. (2021). Measurement invariance of the Kessler psychological distress scale (K10) among children of Chinese rural-to-urban migrant workers. *Brain and Behavior*, 11(2):E2417.

Rogers, R. W. (1975). A protection motivation theory of fear appeals and attitude changes. *Journal of Psychology*, 91(1):93–114.

Rosseel, Y. (2012). Lavaan: An R package for structural equation modeling. *Journal of Statistical Software*, 48(2):1–36.

Satorra, A. and Saris, W. E. (1985). Power of the likelihood ratio test in covariance structure analysis. *Psychometrika*, 50(1):83–90.

Schijf, B. and Dronkers, J. (1991). De invloed van richting en wijk op de loop-banen in de lagere scholen van destad Groningen [The effect of denomination and neighborhood on education in basic schools in the city ofGroningen in 1971]. In I. B. H. van Abram, B. P. M. Creemersen, and A. van der Leij *Onderwijsresearchdagen: Curriculum*, pp. 3–14. University of Amsterdam, SCO, Amsterdam.

Schwarz, G. (1978). Estimating the dimension of a model. *Annals of Statistics*, 6:461–464.

Silva, B. C., Bosancianu, C. M., and Littvay, L. (2019). *Multilevel Structural Equation Modeling*. Sage Publications, Newbury Park, CA.

Slade, G. D., Bair, E., By, K., Mulkey, F., Baraian, C., Rothwell, R., Reynolds, M., Miller, V., Gonzalez, Y., Gordon, S., Ribeiro-Dasilva, M., Lim, P. F., Greenspan, J. D., Dubner, R., Fillingim, R. B., Diatchenko, L., Maixner, W., Dampier, D., Knott, C., and Ohrbach, R. (2011). Study methods, recruitment, sociodemographic findings, and demographic representativeness in the oppera study. *The Journal of Pain*, 12(11 Suppl):T12–26.

Sobel, M. E. (1982). Asymptotic confidence intervals for indirect effects in structural equation models. *Sociological Methodology*, 13:290–312.

Steiger, J. H. and Lind, J. C. (1980). Statistically based tests for the number of common factors. *Paper Presented at the Annual Meeting of the Psychometric Society*, Iowa City, IA.

Tucker, L. R. and Lewis, C. (1973). A reliability coefficient for maximum likelihood factor analysis. *Psychometrika*, 38:1–10.

Xie, Y. (2019). *Bookdown: Authoring Books and Technical Documents with R Markdown.* R package version 0.13.

Yuan, K.-H., Zhang, Z., and Zhao, Y. (2017). Reliable and more powerful methods for poweranalysis in structural equation modeling. *Structural EquationModeling: A Multidisciplinary Journal*, 24(3):315–330.

Index

Note: *Italic* page numbers refer to figures.

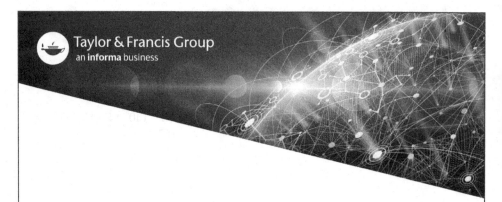

Printed in the United States
by Baker & Taylor Publisher Services